SMOKE AND TAILINGS

Mining and Society Series

Eric Nystrom, University of Nevada, Reno, *Series Editor*

Our world is a mined world, as the bumper sticker phrase "If it isn't grown, it has to be mined" reminds us. Attempting to understand the material basis of our modern culture requires an understanding of those materials in their raw state and the human effort needed to wrest them from the earth and transform them into goods. Mining thus stands at the center of important historical and contemporary questions about labor, environment, race, culture, and technology, which makes it a fruitful perspective from which to pursue meaningful inquiry at scales from local to global.

Books published in the series examine the effects of mining on society in the broadest sense. The series covers all forms of mining in all places and times, building from existing press strengths in mining in the American West to encompass comparative, transnational, and international topics. By not limiting its geographic scope to a single region or product, the series helps scholars forge connections between mining practices and individual sites, moving toward broader analyses of the global mining industry in its full historical and global context.

Seeing Underground: Maps, Models, and Mining Engineering in America
Eric C. Nystrom

Historical Archaeology in the Cortez Mining District: Under the Nevada Giant
Erich Obermayr and Robert W. McQueen

Mining the Borderlands: Industry, Capital, and the Emergence of Engineers in the Southwest Territories, 1855–1910
Sarah E.M. Grossman

The City That Ate Itself: Butte, Montana and Its Expanding Berkeley Pit
Brian James Leech

One Shot for Gold: Developing a Modern Mine in Northern California
Eleanor Herz Swent

The Weight of Gold: Mining and the Environment in Ontario, Canada, 1909–1929
Mica Jorgenson

Underground Leviathan: Corporate Sovereignty and Mining in the Americas
Israel G. Solares

Smoke and Tailings: An Environmental History of Butte & Anaconda, Montana, 1880-1930
Fredric L. Quivik

SMOKE AND TAILINGS

An Environmental History of Butte and Anaconda, Montana, 1880–1930

FREDRIC L. QUIVIK

UNIVERSITY OF NEVADA PRESS | *Reno & Las Vegas*

University of Nevada Press | Reno, Nevada 89557 USA
www.unpress.nevada.edu

Manufactured in the United States of America

First Printing

Cover design by Louise OFarrell
Cover photograph: National Archives at Denver, cat. # NAID 298153

Library of Congress Cataloging-in-Publication Data on file

ISBN 978-1-64779-199-5 (cloth)
ISBN 978-1-64779-200-8 (ebook)
LCCN: 2025005666

The paper used in this book meets the requirements of American National Standard for Information Sciences—Permanence of Paper for Printed Library Materials, ANSI/NISO Z39.48-1992 (R2002).

Contents

Illustrations

FIGURES

Tables

Preface

My introduction to Butte, Montana, came while I was a student in the late 1960s at St. Olaf College in Minnesota. I heard about Butte while working summers in Montana on a trail crew in the Lincoln Backcountry of the Helena National Forest, an area now designated as the Scapegoat Wilderness. Folks in Lincoln and working for the Forest Service were disparaging of Butte. They offered particulars, but I could tell that their negative attitudes toward the city and the people were largely because Butte was somehow "other." I got another insight into Butte one year when my college roommates, who played in the St. Olaf Band, came back from a tour performing in Montana cities. One tour stop was in Butte. They were told they should not leave the hotel at night, because Butte was a dangerous city. That somehow comported with inklings of Butte I had gathered in Lincoln.

That all changed in summer 1976, when I worked a summer internship for the Montana State Historic Preservation Office in Helena while studying as a graduate student in the Historic Preservation Program at Columbia University in New York City. The giant open-pit mine at Butte was expanding, planners were studying where to locate a new central business district for the community, and I was sent on several trips to Butte to photograph historic buildings that everyone believed were soon to be demolished. I found in Butte a magnificent historic district, with an impressive collection of historic buildings, and people who cherished their history and hoped to preserve their historic buildings.

As I was nearing completion of the graduate program at Columbia, I learned that a new federally funded organization was being established in Butte: the National Center for Appropriate Technology (NCAT). Its objectives would be to enable low-income communities around the United States to develop renewable-energy programs that could help them become more self-sufficient and less reliant on distant, large-scale technological systems. Having developed a strong environmental ethic growing up on a dairy farm in Minnesota and working for the Forest Service, I thought NCAT would be a job that would allow me to encourage reuse of existing buildings, thereby cutting construction costs and conserving resources. I got a job, and my wife, Melinda, and I moved to Butte.

Working at NCAT was energizing, but so was living in Butte, as I found myself at the confluence of several local and national developments. Melinda and I immediately joined the Butte Historical Society, whose members were advocating to preserve structures comprising Butte's built environment. The society also worked to preserve and document other aspects of the community's rich cultural heritage, its mining history, its many ethnicities, and its nationally significant labor history. Through the society, we also came to learn some of what sets Butte apart from the rest of Montana. Whereas most of the rest of the state is rural, agricultural, or forested, Butte has a long history of being both industrial and Montana's most urban place (it no longer is the state's largest city). While the rest of the state vacillated politically between Republicans and Democrats, Butte had become a Democratic stronghold, closely allied with Butte's strong labor unions. Butte residents were proud of their distinctiveness, with one wag calling Butte "the only island in the country completely surrounded by land."

When we moved to Butte in 1977, the city council had just voted down a formal mining-company-backed proposal to relocate the community's central business district, called Uptown Butte. By defeating the measure, the city council was taking a rare step in Butte history—resisting the will of the mining company that had long been the community's lifeblood. The decision meant that the local government had to turn its attention to revitalizing the declining Uptown.

After two years working at NCAT, I had an opportunity to take a part-time leave and work for a small agency within the National Park Service (NPS), the Historic American Engineering Record (HAER), to conduct a statewide survey of historic bridges in Montana. I worked alternating weeks, one week at NCAT and the next on the historic-bridge survey. The project was intended to serve as the prototype of an approach HAER could recommend to other states' highway departments so they could inventory their historic bridges and know which ones were eligible for listing in the National Register of Historic Places. Working with HAER introduced me to the society for Industrial Archeology (SIA), a scholarly organization whose members (historians, architects, engineers, archeologists, historic preservationists, museum people, and others) work to identify, document, preserve, and interpret historic resources that embody the industrial and engineering history of the nation.

I started attending the SIA's annual conferences, from which I gained inspiration for the kinds of heritage preservation that could be applicable for Butte. One SIA conference was held at Lowell, Massachusetts, where the NPS had just inaugurated the Lowell National Historical Park (NHP), a unit of the NPS that interpreted the role that the textile industry there, organized along the Falls of the Merrimac River, played in launching the Industrial Revolution in the U.S. By then I had also learned that the Butte mining district, from 1887 to the First World War, had been the largest producer of copper in the world. In addition to having an extensive historic district comprised of dwellings, commercial buildings, churches, and schools, Butte still possessed an impressive array of industrial structures from the period of underground mining, before the opening of the Berkeley Pit (open-pit mine) in 1955. For those reasons—the significance of Butte to the nation's mining history and impressive array of surviving historic buildings—Butte had been declared a National Historic Landmark (NHL) in 1961. I began to realize Butte had a kinship with Lowell.

When we moved to Butte, in 1977, Atlantic Richfield (ARCO), the giant oil company had just acquired the Anaconda Company, which despite having long been the dominant economic force in Butte, had recently come under severe financial strain because Chile had nationalized Anaconda's copper-mining properties there. Three years after buying the Anaconda Company, ARCO closed the smelter at nearby Anaconda, and two years later ceased all mining in Butte. Suddenly, Butte and Anaconda were without the long-time employment that had forged their identities. Both communities recognized the need to find new economic bases while preserving the physical vestiges of their heritage.

Then another major development cemented my historian's fascination with Butte. In late 1980, Congress passed and President Jimmy Carter signed the Comprehensive Environmental Remediation, Compensation, and Liability Act (CERCLA), commonly known as the Superfund law. CERCLA charged the U.S. Environmental Protection Agency (EPA) with identifying old industrial sites that harbored hazardous materials threatening to humans and the environment. The act also authorized the EPA to issue orders to the parties, responsible for those contaminations, to undertake cleanup efforts. Within a few years, the EPA had identified three Superfund sites at Butte,

Anaconda, and along the Clark Fork (a tributary of the Columbia River that has its headwaters at Butte). It soon became apparent that this collection of contaminated sites in the upper Clark Fork basin was the largest Superfund complex in the nation.

Inspired by the Lowell NHP, I began to envision a historical industrial park system for Butte and Anaconda with sites linked not by a river, as at Lowell, but by railroad lines that were the arteries of industrial operations at and between Butte and Anaconda. With others in the community, we obtained funding to develop a comprehensive master plan for such a concept. We identified interpretive themes that included the traditional focus on mining and labor history but also a theme of "Environmental Degradation and Remediation." I worked in Butte promoting development of the idea through the end of the 1980s. By that time, I had also left NCAT and started a historic preservation firm, Renewable Technologies Inc. (RTI). We specialized in projects involving historic engineering and industrial sites throughout the American West. Through those projects, I broadened my understanding of industrial history and deepened my appreciation for Butte's historical significance, not only for its mining history, as the town once called "The Richest Hill on Earth," but for its significance in the nation's environmental history.

In 1990, Melinda and I left Butte to return to school, she at the Lutheran Theological Seminary at Philadelphia, and I at the University of Pennsylvania, where I entered the PhD program in History and Sociology of Science to pursue a degree in history of technology. After three years in Philadelphia, we returned to Montana, where she began serving her first call as a Lutheran pastor, and I began research for my dissertation, which would be an environmental history of the copper-smelting technologies in Montana.

Shortly after I began that research, I received a phone call from an attorney with the U.S. Department of Justice (DOJ) to see if I would be interested in serving as an expert witness for the government in *United States v. ARCO*, the Clark Fork Superfund litigation. The historical research I would do in support of the litigation was almost identical to the research on which I had embarked for my dissertation. I conducted intensive historical research and report-writing for the DOJ on that case for the next four years and on and off for another four years. Along the way, I also wrote my dissertation, "Smoke and Tailings: An Environmental History of Copper Smelting Technologies in Montana, 1880–1930," and completed the PhD work at Penn in 1998.

I loved the research and writing on the Clark Fork Superfund case, so I let it be known to attorneys that I would welcome other projects like that. I was attracted to the work for two reasons. First, it is applied history. I had learned from years working in historic preservation that I found it deeply gratifying to apply my skills as a historian to actual physical projects. And the work was also rewarding because I felt I was contributing to the environmental remediation of contamination brought about by decades of large-scale industrial activity. Over the more than thirty years since I started working on the Clark Fork case, I have had the opportunity to work on cases involving sites from Long Island to Los Angeles and from Idaho to Oklahoma. Perhaps the most well-known case I worked on was the *Deepwater Horizon* spill in the Gulf of Mexico, in which I testified on behalf of the United States at trial in New Orleans. Through all those cases, I have come to learn and appreciate much more about the challenges involved in remediating the vast array of Superfund sites in the nation, about the legal complexities those cases reveal, and especially the magnitude of the Clark Fork Superfund site relative to other sites.

Across those decades, I also returned to Butte many times to visit friends; teach and make presentations; keep abreast of remediation efforts in Butte, Anaconda, and along the Clark Fork; and often to conduct my own research on Butte topics that continue to fascinate me. I also kept in touch with the preservation community in Butte and monitored developments in other parts of the country where folks are working to preserve and interpret cultural resources that embody our nation's industrial heritage. Then, in 2010, I had the opportunity to teach at Michigan Technological University in Houghton, located on the Keweenaw Peninsula. I was on the faculty of the graduate program in Industrial Heritage and Archaeology, and I worked with professionals in the NPS and others associated with the Keweenaw National Historical Park, which preserves and interprets cultural resources with the copper-mining industry on the Keweenaw.

This was a wonderful opportunity for me because of the ways the Keweenaw is related to Butte. It was the world's largest supplier of copper until 1887, when Butte claimed the title. Many immigrants from Europe worked in the mines on the Keweenaw before they and their families moved west to Butte. Most interesting to me, though, was that the mineral exploited there was native copper, meaning the copper came out of the ground as pure copper, not as a sulfide mineral as is the copper in Butte and most other large copper-mining districts. The Keweenaw is the only mining district in the world in which native copper was found in amounts sufficient to foster industrial-scale mining operations for decades. One consequence is that, although the Keweenaw's mining industry was on a scale comparable to Butte's mining industry, mining there did not have the same environmental impact as in Butte. Although mining on the Keweenaw had a profound physical impact, it did not leave chemical contamination on the scale of that in Butte. There is a small Superfund site associated with the Keweenaw's copper industry, but it is nowhere near the scale of the monumental complex of Superfund sites at Butte, Anaconda, and along the Clark Fork.

I have long believed that the Butte-Anaconda National Historic Landmark deserves its designation not only for the impressive historical architectural and industrial resources that survive in both communities but also for the environmental stories that Butte and Anaconda have to tell. The mining industry caused vast damages to urban, agricultural, forested, and riverine environments. Local populations began to resist that environmental harm as early as the 1890s, and the federal government joined the fray in 1910, but the corporate power of the mining industry was ascendent. Although there were important efforts to abate pollution, the industry largely prevailed in those contests, leading to contamination of such a vast area that, once CERCLA became law, as I have already mentioned, the complex of Superfund sites in southwest Montana became the nation's largest.

My dissertation described the history of mining, milling, and smelting at Butte and Anaconda, and the efforts of local citizens to resist the damage the industry was causing to their livelihoods and environment. I have worked since then to expand the scope of the dissertation for publication with the thought that the history of Butte and Anaconda is not just interesting on the page. People can travel to Butte, Anaconda, and the Clark Fork to see firsthand how copper was and still is mined, to see the environmental consequences of that mining activity, and especially to see the results of the huge undertaking that is the Clark Fork Superfund remediation. Some of that Superfund work is complete, some is still ongoing, and some will have to be maintained in perpetuity. Americans and others can visit Butte and Anaconda to experience—not just read about—the consequences that industrial activity has caused, the monumental work that the United States is now doing to rectify those harms through the Superfund, and the paths that we can choose to follow to build a sustainable future.

Acknowledgments

I owe a deep debt of gratitude to the members of the Butte Historical Society, who welcomed my wife and me into their midst when we moved to Butte in 1977, and who, through their passion, introduced me to the history of their community. I thank Alan Goddard, Ann Cote Smith, Al Hooper, Paul Bingaman, Ray Calkins, Alice Smith, Bonnie Stefanic, and Les and Margaret Cunningham. Alan was especially generous in introducing me to the back corners of Butte history. And I also thank new members of the society who arrived in Butte about the time we did: Janet and Jay Cornish, and Carolyn Smithson. New members arrived in subsequent years, including Bill Walker, Mary Murphy, and Brian Shovers. Through the Butte Historical Society, I met members of the Anaconda Historical Society, most notably Alice Finnegan. Together we worked on a rich array of projects—recording oral histories, producing publications, creating slideshows on Butte history, organizing ethnic dinners for the broader community, conducting architectural surveys of Butte's National Historic Landmark District, advocating for historic preservation, launching the Butte-Silver Bow Public Archives, and developing the initial master plan for preserving and interpreting the rich cultural resources in the industrial landscapes of Butte and Anaconda. Through those many projects, I caught the passion for Butte history myself, and as I delved deeper into scholarly pursuits, I came to appreciate the depth of opportunities that Butte history opened for scholarship.

Along the way, we in the Butte Historical Society received valuable assistance from the staffs of the Montana Historical Society and Montana's State Historic Preservation Office, including Marcella Sherfy, Bill Lang, Dave Walter, Pat Bick, Brian Cockhill, and others. They helped make the work of the Butte Historical Society better. We also had the opportunity to work with the Historic American Engineering Record (a small agency within the National Park Service), furthering my development as a scholar of Butte history with the help of T. Allan Comp, Isabel Hill, Jet Lowe, and

Gray Fitzsimons. While in Butte, I started a historic preservation consulting firm, Renewable Technologies Inc., that worked on projects in Butte and throughout the American West. I am grateful for my time working with RTI colleagues, Dennis Glick, Mark Fiege, Dale Martin, Bruce von Alten, Mary McCormick, Lon Johnson, Mitzi Rossillon, and Mark Hufstetler, as together we came to better understand the history of industrialization in the western regions of the United States.

Leaving Butte in 1990 to earn a PhD in History and Sociology of Science (HSS) at the University of Pennsylvania, I had the opportunity to work with an exceptional group of faculty and students in that program. My dissertation advisor, Thomas Parke Hughes, mentored me in several facets of becoming a historian of technology. Other especially supportive members of the faculty included Judy McGaw, Rob Kohler, Riki Kuklick, and Arnold Thackray. At Judy McGaw's suggestion, I invited Jeffrey Stine, of the Smithsonian Institution's National Museum of American History, to join my dissertation committee, and he was immensely helpful in my development as an environmental historian. Among the grad students at Penn with whom I had particularly fruitful exchanges of ideas were Jennifer Gunn, Nina Lerman, Mark Hamin, Eric Rau, and Atsushi Akera. I am also grateful for stimulating exchanges of research ideas with Timothy LeCain and Kent Curtis. While I was working on my dissertation at Penn, their dissertation work at the University of Delaware and the University of Kansas, respectively, was also on the environmental history of the copper industry at Butte and Anaconda. We shared insights then, and we have continued to share ideas about environmental history as the scopes of our scholarly interests have expanded. Through the HSS department, I also got to know Ruth Schwartz Cowan, who arrived to chair the department after my time as a student. Her encouragement and assistance were of great help to me.

As I was beginning focused research for my dissertation in late 1993, I received a telephone call from Michael Zevenbergen, an attorney with the Environmental and Natural Resources Division of the Department of Justice. He was working on the United States' suit against ARCO in the Clark Fork Superfund case, and he was looking for a historian who could research the milling technologies used at the silver mines in Butte in the early years. He had been referred to me by Brian Shovers, then a research librarian at the Montana Historical Society with whom I had collaborated on many history projects in Butte. There was a significant similarity in the research I was doing for my dissertation and the research he required, so I agreed to contract with the DOJ as an expert witness. The scope of my work for the DOJ quickly expanded to include copper and other concentrators in Butte and Anaconda and the smelters at those two cities. The opportunity to work not only with Zevenbergen but also with Sarah Himmelhoch, Rachel Jacobson, Matthew Morrison, and Alan Greenberg helped me to better understand the legal complexities of the smoke and tailings litigation in the early twentieth century and of the Comprehensive Environmental Remediation, Compensation, and Liability Act (CERCLA), under which the Clark Fork Superfund case was being tried. And through that engagement, I got to work with two expert metallurgists, Rex Bull and Terry McNulty (ARCO's expert metallurgist), who greatly expanded my technical education in the history of metallurgy.

Since that time, I have continued to serve as an expert witness in Superfund and other environmental litigation, and I have enjoyed the scholarly encouragement of others, including mining historians Eric Nystrom and Brian Leech, Pat Munday (on the Montana Tech faculty), and Joe Griffin (hydrologist living in Butte). When I taught at Michigan Technological University, colleagues—including Larry Lankton, Terry Reynolds, Patrick Martin, and Hugh Gorman—helped me to refine and reshape ideas that appear in this book. Sean Gohman and John Baeten are two of several grad students at Michigan Tech who contributed significantly to my thoughts on the environmental history of mining.

Libraries and archives hold treasures that provide the foundations for this book. Archivists at the National Archives, including the main archives when it was on the National Mall and I began this research, and then Archives II in the new facility in College Park, Maryland, as well as Seattle and Denver branches, helped me find valuable information, as did the Rare and Manuscript Collections at Cornell University. Dave Walter, Brian Cockhill, Ellie Arguimbau, and Brian Shovers helped me find my way through the massive Anaconda Copper Mining Company Records at the Montana Historical Society and other resources there. At the Butte-Silver Bow Public Archives, founding director Bill Walker and his successors Chris Daly, John Hughes, Mary Murphy, and especially Ellen Crain have been enthusiastic colleagues in my work, as have members of the staff, including Aubrey Jaap (current director), Lee Whitney, Kim Murphy Kohn, and Lindsay Mulcahy. Thanks to all archivists!

Last, but not least, I thank my wife, Melinda Quivik, for more than fifty years of support. She is a Lutheran pastor and liturgical scholar, and although our research and writing pursuits are quite different, we enjoy reading and critiquing each other's writing as together we have engaged in many other facets of our life together. I am especially grateful that, over the past two decades, since I finished the dissertation, she has never tired of asking, "How are the revisions going? When are you going to get your book published?" This work is for you, Pal.

SMOKE AND TAILINGS

Introduction

Job 28:1–12

28 "Surely there is a mine for silver,
and a place for gold to be refined.
2 Iron is taken out of the earth,
and copper is smelted from ore.
3 Miners put an end to darkness,
and search out to the farthest bound
the ore in gloom and deep darkness.
4 They open shafts in a valley away from human habitation;
they are forgotten by travelers,
they sway suspended, remote from people.
5 As for the earth, out of it comes bread;
but underneath it is turned up as by fire.
6 Its stones are the place of sapphires,
and its dust contains gold.
7 That path no bird of prey knows,
and the falcon's eye has not seen it.
8 The proud wild animals have not trodden it;
the lion has not passed over it.
9 They put their hand to the flinty rock,
and overturn mountains by the roots.
10 They cut out channels in the rocks,
and their eyes see every precious thing.
11 The sources of the rivers they probe;
hidden things they bring to light.

12 "But where shall wisdom be found?
And where is the place of understanding?"

> Butte—one of the world's richest copper producing centers, with around 22 billion pounds of extracted copper already removed and more on the way. Today, after 140 years of relentless industrialized mining, Butte is also the uppermost part of the largest Superfund complex in the United States, a place of staggering environmental ruin.
>
> Like Concord, Gettysburg, and Wounded Knee, Butte is one of the places America came from, and it is where we must return, in the manner of a pilgrimage, if we wish to grasp in full the implications of our appetite for metals.
>
> —Edwin C. Dobb, "Introduction: The Age of Sacrifice Zone"

The opening epigraph of this book is from the twenty-eighth chapter of the book of Job, which is in a part of the Old Testament (or Hebrew Bible) referred to as the wisdom literature. In this discourse, Job notes that mining occurs in places removed from human habitation and far from most people. In fact, mines are foreign environments to other animals as well for, unlike Earth's surface, mines produce no food. Mines are exploited by those seeking precious minerals, but the speaker questions where the wisdom is in that kind of striving for wealth.

Nevertheless, humans have a long history of mining: for nonmetallic minerals, like clay for pottery and sand for glass; for precious metals, like gold and silver; and for base metals like iron, copper, and lead. For millennia the impacts of mining were relatively small-scale and localized, especially because, as recounted in the book of Job, they were removed from human habitations. The fact that mining's impacts are out of sight from most people, however, has allowed the magnitude of those impacts to increase exponentially since the Industrial Revolution, making it possible for people throughout the industrialized world to enjoy the fruits of industrialization, often without being bothered by its negative consequences. Americans have been major beneficiaries of industrialization, yet Ed Dobb, in the second epigraph at the beginning of this book, wants us to appreciate the damage industrialization has wrought on a place like Butte, Montana.[1] Located in the Rocky Mountains near the Continental Divide, Butte, like so many other mining centers, is far from major human habitations and therefore beyond the perceptions of many.

By placing Butte in the company of Concord, Gettysburg, and Wounded Knee, Dobb suggests Butte should be one of those iconic places where we as Americans come to know ourselves better. Concord is celebrated in national lore, even though lives were lost on both sides, because it is where the first shots were fired as American patriots fought to free themselves from England, their colonial master. Gettysburg lives at a more complex reckoning in American history. The Battle of Gettysburg was the turning point in the United States' bloody Civil War, during which hundreds of thousands of Americans died, and many more were maimed. At the dedication ceremony at the cemetery for those killed in the Battle of Gettysburg, President Abraham Lincoln rearticulated the character of the nation as "dedicated to the proposition that all men are created equal," and he asserted that Union soldiers had died to preserve a "government of the people, by the people, for the people." But it is nearly impossible to celebrate what happened at Gettysburg without also realizing that a region of the United States, and a significant portion of its population, had fought the Civil War to defend slavery, an institution in which some Americans owned and abused other human beings as property. Wounded Knee embodies much of the history of the United States' treatment of the Indigenous people who lived in North America before European colonists arrived and occupied the land. The United States government signed treaties with American Indian peoples to force them onto reservations, often broke the treaties, and sometimes slaughtered people who protested the government's actions.

Concord, Gettysburg, and Wounded Knee are fraught with moral issues, and so, Dobb asserts, is Butte, as it embodies the ways Americans have abused a particular place on Earth, a place out of view from most of the population, in a two-centuries-long quest to derive the benefits of industrialization.[2] Jeremy Zallen has made the moral link between Butte and America's industrialization in a more recognizable way in his recent history of artificial lighting, *American Lucifers: The Dark History of Artificial Light, 1750–1865*.[3] Most of the book is about the strivings of Americans to produce reliable artificial light by means of whale oil, camphene (made from turpentine and alcohol), gas manufactured from coal, candles made from lard, kerosene distilled from petroleum, and phosphorous used to make matches. To achieve the moral link for each method of lighting, Zallen recounts the impacts on living beings at the far ends of the supply chains for those kinds of lighting: the whales and hogs, the enslaved people who harvested pine resins to make turpentine, workers in slaughterhouses, and children who dipped matchheads. His main story culminates with kerosene, the appearance of which coincided with the Civil War and the end of slavery, and which ended the use of enslaved labor to harvest pine resin

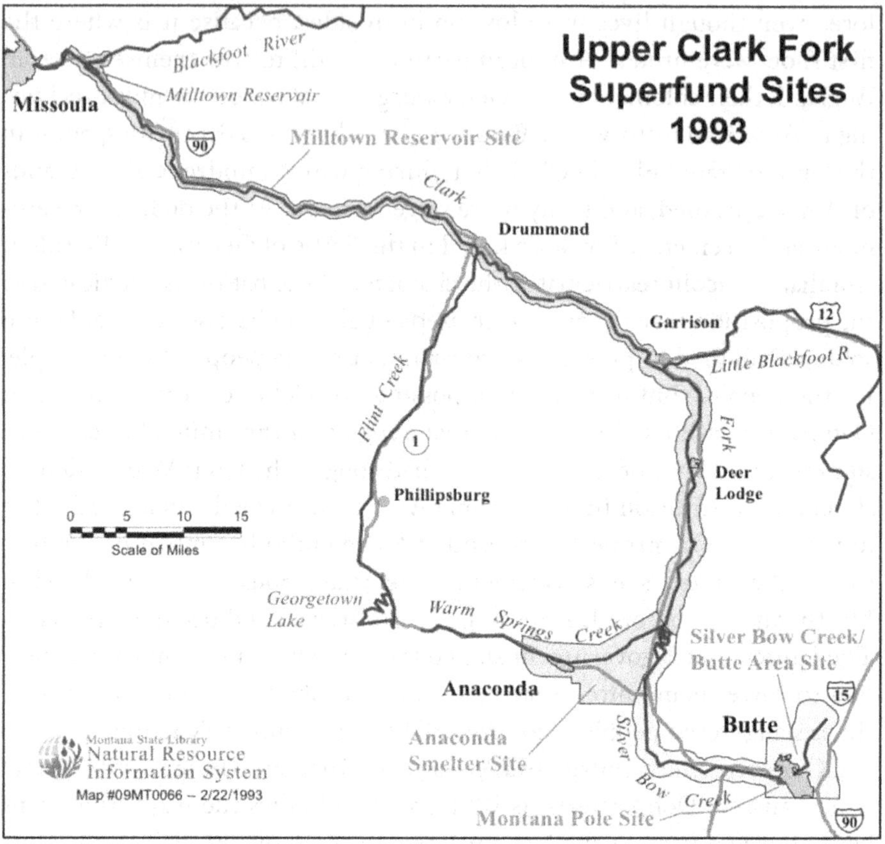

FIG. I.1. Map of the Upper Clark Fork Superfund Sites, 1993, showing the locations Butte, Anaconda, and Missoula, Montana, Silver Bow Creek, Warm Springs Creek, and the Clark Fork, and the four Superfund sites that comprise the Clark Fork Superfund complex. The Montana Pole Site is not directly related to copper mining, milling, and smelting at Butte and Anaconda. Image courtesy of the Montana State Library, Geographic Information.

and make turpentine and camphene. *American Lucifers* concludes, however, with an epilogue about electrification, and Zallen once again links the means of lighting at the consumer end, which seemed miraculous, clean, and safe, with the lives of the people who toiled to produce the copper for wiring the new electrical lighting systems. He specifically examines the dangerous and often deadly working conditions of the copper miners in Butte.[4]

More than twenty years before Zallen wrote his book, Ed Dobb published an important article in *Harper's Magazine*. In "Pennies from Hell,"

Dobb links the twenty million pounds of copper from the Butte hill, which American consumers had used, with the human and environmental costs that consumption has exacted on Butte, Anaconda, and the streams of the upper Clark Fork basin.[5] The human costs are enshrined in the 1917 Granite Mountain disaster, when 168 miners were killed by a fire burning in the Granite Mountain mine.[6] A memorial to the disaster stands atop the Butte hill, overlooking the Granite Mountain headframe.

Similar to the linkage Zallen has drawn between electric illumination and Butte miners, *Smoke and Tailings* aims to link the growth of copper consumption in the U.S. economy with the damage to the environment of southwest Montana. At the other end of America's 140-year electrification project is the nation's largest complex of Superfund sites. Located in Montana's mountains, those Superfund sites encompass the mining city of Butte, the smelter city of Anaconda, and the Clark Fork (a major tributary of the Columbia River) almost as far downstream as Missoula. This book examines the historical developments in Butte and Anaconda that exposed and concentrated such a vast array of hazardous materials—while feeding America's electrification, industrialization, and urbanization—that the two cities and their surroundings have required a monumental cleanup program, some components of which will have to continue in perpetuity.

The U.S. Department of the Interior has officially designated Butte and Anaconda together as a National Historic Landmark for their significance to the nation's mining history, including the communities' labor history. *Smoke and Tailings* closes with an argument to preserve and interpret these two national landmark communities and their landscapes also for their significance in illuminating the monumental environmental consequences of supplying the nation with copper. Americans ought to be given the opportunity to make pilgrimage to see and experience the consequences of industrialized mining.

* * *

"Butte is the black heart of Montana."[7] Like any shady character, Butte has been an attractive subject for writers, including historians, to investigate. Since Joseph Kinsey Howard coined that memorable phase in 1943, Butte has continued to attract scholarly inquiry. The catalog of noteworthy Butte histories is too long to list here, but some of the more important recent ones are worth naming as a backdrop to what this book intends to accomplish. Michael Malone's *The Battle for Butte* is still the best corporate history of Butte's mining industry to the point in 1906 when the Amalgamated Copper Company succeeded in consolidating nearly all the copper mines and mining companies into a single corporate enterprise.[8] Isaac Marcosson's commissioned history of *Anaconda* remains a valuable history of the Anaconda Copper Mining Company (ACM) to 1955, when it changed its name to the Anaconda Company.[9] Other important articles on the ACM's corporate history after the consolidation form a backdrop to histories on other facets of Butte history.

As a large industrial mining center in the American West at the turn of the twentieth century, Butte attracted a sizeable immigrant workforce, it was the scene of intense labor-management conflict, and it fostered a rich social and cultural history, all of which have inspired superb writings. Among them are Jerry Calvert on labor conflict and socialism in Butte during the 1910s, David Emmons on the Irish immigrants in Butte, Mary Murphy on the leisure lives of women and men in the mining city, Michael Punke on the 1917 Granite Mountain mine disaster, Dennis Swibold on the outsized influence the ACM had on Montana's newspapers, Janet Finn on childhood in Butte, Matthew Basso on race and masculinity in Montana's copper industry during World War II, and Brian Leech on community responses to the advent of open-pit mining in Butte in the mid-twentieth century.[10] Also important to note in this regard is Laurie Mercier's labor and social history of Anaconda, the smelter city west of Butte.[11]

My aim is to integrate the history of technology and environmental history of Montana's copper industry with the social, political, and corporate histories of Butte and Anaconda, in the vein of Duane Smith's pioneering *Mining American*, and which both Tim LeCain and Kent Curtis have achieved more recently.[12] While Curtis includes analyses of farmers' resistance to pollution from both smoke and tailings, LeCain looks only at smoke pollution. Their work follows in the wake of Donald MacMillan's *Smoke Wars*, about controversies over smelter smoke at Butte and Anaconda. Curtis and LeCain follow worthy environmental histories of smelter smoke at other locations, including John Wirth's history of battles over smelter smoke drifting across the United States' borders with both Canada and Mexico; Khaled Bloom's and Duncan Maysilles's histories of the conflicts over smoke from copper smelting in northern California and around Ducktown, Tennessee, respectively; Bode Morin's comparison of heritage and remediation responses around the copper-smelting industries on Michigan's Keweenaw Peninsula, Ducktown, and Butte and Anaconda; histories by Michael Mix and Bradley Snow about smoke from lead smelting at Kellogg, Idaho; and Marianne Sullivan's comparison of the public-health impacts of lead and copper smelting in Texas, Washington; and Idaho.[13] These fine histories focus mainly on the environmental and public-health impacts of smelter smoke and on the legal conflicts that ensued. I will, in addition, look more closely at how the mining industry at Butte and Anaconda also responded to challenges to its polluting methods by finding technical fixes that could reduce or mitigate the environmental impacts. I will also examine tailings more closely.

Smelter smoke has garnered much more interest among historians than has pollution from tailings, even though tailings have been the material requiring environmental remediation at many more Superfund sites than has been the outfall from smelter smoke. Curtis, Mix, and Snow provide some scholarly treatment of historical controversies over tailings, but an indication of the paucity of attention to tailings among environmental historians, so far, is evident in *Mining North America: An Environmental History since 1522*, a collection of essays edited by J. R. McNeill and George Vrtis. They mention mill tailings from mining for precious metals along Colorado's Front Range and give a more lengthy analysis of environmental problems with tailings on Minnesota's Mesabi Iron Range. Tailings from milling iron ore, however, is significantly different from the tailings produced by milling the ores of the nonferrous metals, both base and precious.[14] Smith's *Mining America* devotes considerable attention to the mining debris controversies in nineteenth-century California, when sands and gravels dislodged by hydraulic mining physically choked the Sacramento River and other streams (a topic well covered by Robert Kelley in 1959), but he says little about pollution or physical damage from mill tailings. Randall Rohe in 1986 called historians to pay more attention to the impacts of mining on the land, but the examples he gave were largely related to debris created by mining per se, making only brief mention of mill tailings.[15] Since that time, a few historians, for example, Gordon and Elwood Bakken, writing about tailings in the Missouri River from the giant concentrator and smelter at Great Falls, Montana, and John Sandlos and Arn Keeling, writing about tailings at the Pine Point mine in Canada's Northwest Territories, have provided historical analyses of the tailings' impacts on the environment.[16]

One group of scholars that has paid considerable attention to the history of the environmental impact of tailings has been historians and archeologists working in the field of cultural resource management (CRM). They regard altered environments, in part, as cultural landscapes that can inform the public about the history and values of the past and that therefore are worthy of identification, analysis, and, in some cases, preservation. Richard Francaviglia's *Hard Places* is the most comprehensive treatment of the subject, but because it is so broad, he does not delve into any one location with particular depth. Robert Spude and Bruce Noble, writing for the National Park Service, prepared a valuable guide to the evaluation of mining landscapes for people working in CRM. Scholarly articles on the environmental history of tailings, through the lens of cultural landscape studies, include Donald Hardesty's article on preserving toxic wastes as cultural sites, Paul White's writing on a mill and its tailings in Alaska, Sean Gohman's analysis of waste rock and tailings from copper mining on Michigan's Keweenaw Peninsula, and my own work on tailings in Montana and Idaho as cultural resources.[17] As the title of *Smoke and Tailings* suggests, the environmental history of tailings plays a significant role in this book.

Another purpose of this book is to get inside the "black box" of technology, to use Nathan Rosenberg's phrase. As he notes, historians can write quite satisfying histories on just about any topic—social, cultural, political, economic, military, environmental, and so on—by treating technology as a black box that plays a role in the dynamics and relationships of the story. Historians of technology open the black box to show how details of the technology within the box interact with the social, economic, or environmental details that historians are usually more comfortable addressing.[18] A wonderful model in this regard, in the realm of mining history, is *Cradle to Grave*, Larry Lankton's cultural history of the copper-mining district on the Keweenaw Peninsula, a history that also describes the technologies that were used in the mines, thus revealing what is often left invisible to the uninitiated.[19]

Likewise, Hugh Gorman's *Redefining Efficiency* shows the role of technology in resolving social conflicts about the environmental impacts of another extractive industry, petroleum. As the oil industry in the U.S. grew, and production expanded from Pennsylvania's oil fields to new fields in Ohio, West Virginia, New York, and then elsewhere in the U.S., so too did discharges of oil into local environments. Cross-country pipelines leaked. Refineries near population centers threatened neighbors with fire and explosions. Loading and unloading ships with crude oil and petroleum products polluted the nation's rivers and harbors. Environmental degradation led to calls to regulate the oil industry, but the industry was able to keep legislative bodies at bay until well into the twentieth century by arguing that the engineering ideals of efficiency offered a solution to the problem. Discharges of oil or finished product to the environment represented pollution to the public, but to engineers, discharges represented waste, defined as the irretrievable loss of resources, which in turn had an adverse effect on an enterprise's bottom line. Engineers were driven, so the argument went, to find ways to eliminate waste; it was in the economic interest of the oil companies to enable engineers to do so. Because the elimination of waste would have the additional benefit of reducing pollution, advocates for the industry urged legislatures to be patient. Economic forces would soon resolve the pollution problem without the nuisance of costly regulation.[20]

The oil industry did become more efficient through the nineteenth century and early years of the twentieth—as measured by the elimination of waste, an economic measure—but at the same time, the scale and scope of the industry grew, especially with the development of the internal combustion engine and advent of the automobile. The result of that tremendous growth was continuing and worsening pollution problems. Beginning in the 1920s and especially after World War II, legislatures realized that the ideal of efficiency would not abate the problem and that the oil industry had to be regulated. Responses by farmers, governments, mining companies, and metallurgical engineers to the problems caused by pollution from mining operations followed a similar pattern.

Katherine Aiken's 2021 article on the interactions among farmers, mine owners, governments, and engineers who devised new technologies in efforts to address the serious environmental problems caused by smelter smoke, suggests that those partial technological solutions in the Progressive Era paved the way for the smelting industry in the U.S. to continue spewing smelter smoke for another half century before the regulatory regimes inaugurated around 1970 began to place severe restrictions on smelter operations. That history left Superfund sites distributed throughout the mining regions of the American West.[21] To the story of failed environmental protection, *Smoke and Tailings* adds the problems created by mill tailings in understanding the complex dynamics that created the Clark Fork Superfund sites, the largest Superfund complex in the U.S.

Many of the contaminants being remediated today were placed into the environment after the Progressive Era reforms of the 1910s and after the Anaconda company had come to monopolize copper mining, milling, and smelting in Montana. The ACM had also come to own many of the sinks for waste disposal.[22] In this way, the mining industry internalized the costs of waste disposal, thus hiding it from the public.

The Butte-Anaconda National Historic Landmark District embodies all these themes: the advent and industrialization of hardrock mining in the American West, the social and labor histories of mining communities, the legal battles over pollution by the mining industry and the searches for technical fixes, the long-term environmental consequences of sustained industrialized mining, and the monumental effort to remediate that environmental damage. The Butte-Anaconda NHL is indeed a site to which Americans can make pilgrimage to understand the consequences of our appetite for metals.

CHAPTER 1

Historical Context

In *Mining Capitalism: The Relationship between Corporations and Their Critics*, Stuart Kirsch notes the significant power that the corporation, as an institution, exerts in our era: "Corporations organize much of the world's labor and capital, shape the material form of the modern world, and are a prime mover of globalization. But corporations are also responsible for a wide range of harmful effects, including the use of technologies with deleterious consequences for human health and the production of environmental hazards that threaten the planet."[1]

A fairly shallow understanding of U.S. history would assume that a century ago, when the overwhelming influence of corporations had become apparent, Americans were not much concerned with the environmental consequences of large industrial undertakings such as mining and smelting operations, but that would be an inaccurate reading of history. Around the turn of the twentieth century, the tremendous devastation caused by smoke and tailings from Montana's copper-mining industry, for example, was strenuously contested by local citizens who bore the brunt of industry's discharges of wastes into the environment. Moreover, even Theodore Roosevelt, the twenty-sixth president of the United States, whose presidency is perhaps best known for its advances of the conservation movement and for trust-busting (especially for initiating the eventual breakup of the Standard Oil trust), knew the damage a gigantic corporate enterprise was causing to national forests and agricultural lands in Montana. Yet Roosevelt and the Progressives proved almost helpless in staunching the environmental devastation of the upper Clark Fork by a giant corporation with close ties to the Standard Oil trust.

Before Theodore Roosevelt ascended to the presidency upon the assassination of President William McKinley in September 1901, he was already an active conservationist, having founded, for example, the Boone and Crockett Club in 1887 to promote conservation. When Roosevelt became president, Gifford Pinchot, another prominent conservationist, was already chief of the Interior Department's Forestry Division, and together Roosevelt and Pinchot would create the U.S. Forest Service and locate it in the U.S. Department of Agriculture.[2] Roosevelt and Pinchot were aware of the damage Montana's copper industry was doing to nearby agricultural lands and federal forest reserves, as we shall see. Despite being deeply concerned about the injury smoke and tailings were causing, Roosevelt ultimately was unwilling to bring the full legal authority of the U.S. government to bear against the industry and force it to abate the harmful waste products of its operations.

Similarly, Roosevelt was outspoken in his opposition to the political and economic power that giant corporate enterprises, known as trusts, were suddenly exerting on American life in the early twentieth century. He sought legislation that would rein in the power of trusts, using the Sherman Anti-Trust Act of 1890 to prosecute several trusts for restraint of fair trade. His administration's most notable suit in that vein was against the Standard Oil trust. When the U.S. Supreme Court upheld a lower court's ruling in 1911, leading to the dissolution of the Standard Oil behemoth, it cemented Roosevelt's reputation as a trustbuster.[3] Nevertheless, when the Roosevelt administration had the opportunity to take stern legal action against the Amalgamated Copper Company, a trust closely allied with Standard Oil, and that was causing so much damage to forests, crops, and livestock in southwest Montana, Roosevelt chose not to do so.

Amalgamated Copper and its major subsidiary, the Anaconda Copper Mining Company, did not, however, have free rein in polluting the environment of southwest Montana. In a complex set of social, political, legal, economic, technological, geological, and ecological environments, the corporation had its attorneys strenuously resist the legal challenges it faced, had its engineers develop technologies that reduced or mitigated the scope of devastation its operations were causing (even as those engineers were also reducing costs of operation so the company could remain profitable in a fiercely competitive global metals market), and had its public-relations arms reduce citizens' resentment of the trust for its monopolistic domination of Montana's government and to foster appreciation of the trust for the central contribution it was making to the young state's economic development. As a result, the history of Butte's mining industry is one that left the nation with its largest complex of Superfund sites. Yet that history could have been much worse for the environment had farmers in the Deer Lodge Valley and bureaucrats in Washington, D.C., not sought to blunt the copper industry's devastation, and had mining and metallurgical engineers not innovated impressive new means of reducing pollution *and* reducing operating costs to make the company more profitable.

In order to appreciate the cultural and economic climate in which these episodes unfolded in southwest Montana, it is important to understand the global juggernaut of which they were a part: industrialization. Americans and Europeans alike were in thrall to industrialization, which was dramatically changing the nature of Western civilization's material culture and making that material culture affordable to ever more people. Industrialization and its transportation systems were opening vast new markets to producers and allowing producers to access distant resources needed for production. And industrialization was creating vast numbers of new jobs for the huge pools of humanity that were migrating from peasant existences in rural areas to burgeoning cities.[4] Although many people recognized the dark sides of industrialization—brutal working conditions and terrible industrial pollution among them—most thought the benefits outweighed the costs and, indeed, thought the social costs and damage to resources could eventually be addressed through reform. That, as we shall see, was the calculation Theodore Roosevelt and his successors in the White House made when they decided not to halt the mining industry's devastation of farmers' and the public's property; the copper that Butte produced was too important to the ongoing industrialization of the U.S. economy.

Beginnings of the Industrial Revolution

Industrialization gave rise to a quantum leap in human economic activity (which is another way of saying a quantum leap in humans' interactions with and manipulations of natural environments) in a transition often called the Industrial Revolution. Although the beginnings of the Industrial Revolution were diverse and emerged in a variety of locales, for ease of thinking about it, we can say it began in England in the eighteenth century. It was the second of the two quantum leaps in humans' ability to manipulate and otherwise interact with their environments. The first was the Neolithic Revolution, beginning about eleven thousand years ago, when humans in disparate parts of the world made the transition from being hunter-gatherers—surviving on resources provided by the ecosystems in which they lived—to being agriculturalists, who manipulated environments in order to increase the availability of resources they needed to survive and for their material culture. They did so by encouraging certain plant and animal species to thrive and by discouraging other plants and animals from interfering with the resources the human groups were trying to exploit.[5] These ancient peoples manipulated what many scholars call the "organic" realm of Earth's environment. German environmental historian Rolf Peter Sieferle refers to it as the agrarian solar energy system, to emphasize the fact that all the energy driving the system ultimately came from the sun.[6]

One of the consequences of the agricultural revolution was that human societies and their material cultures grew more complex. Although the vast majority of people in agricultural societies exercised hard toil to raise crops and animals for survival, they did produce enough surplus that their societies could develop more complex hierarchical structures. Small, elite groups were in charge, exercising power while depending on the sweat and exertion of the vast majority to produce the resources the elites needed to survive. Importantly, the surpluses the peasants and slaves produced could also support other specialty niches within society, such as a warrior class and artisans, who did not have to toil all day in fields or pastures, and who therefore could develop skills to produce clothing, ornaments, buildings, and weapons. Most of the fine artifacts the specialized artisans made were for the ruling elites. In the centuries leading to the eighteenth, global markets for certain specialized goods, both farmed and fabricated, continued to grow, increasing the percentage of people who earned their livelihoods as artisans rather than toiling in fields, and merchant classes expanded as well. Merchants organized the flows of raw materials to artisans and of finished products to customers. It was in the realms of markets and the production of artifacts that the Industrial Revolution arose.[7]

The Industrial Revolution grew out of symbiotic relationships among three distinct paths of human development: 1) mechanization of production, 2) organization and management of production, and 3) harnessing alternative forms of energy to power mechanized production. Mechanization grew out of the division of labor. Turning wheat into bread, for example, requires the tasks of separating the chaff from the kernels of wheat, grinding the wheat into flour, mixing the flour into dough, and baking the dough to make bread. Sometime in the distant past, someone devised a hand quern, mechanizing and thereby improving productivity in grinding wheat into flour. In a similar vein, several tasks are necessary to convert wool fibers into cloth: the wool must be cleaned, and the fibers carded so they are all oriented in roughly the same direction. Then the fibers must be drawn and twisted to produce yarn, and the yarn must be woven into cloth. Long ago, someone devised a spinning wheel, a mechanical device that improved the productivity of persons drawing and twisting fibers into yarn.[8]

Prior to the Industrial Revolution, the household was the location where most production took place. Even if an artisan was making a product for a merchant in the context of the putting-out system (in which a merchant would put raw materials out into households, where people converted the raw materials into finished products and then returned the products to the merchant in exchange for payment and for more raw materials), the fabrication took place in a workshop that was part of the home, and the members of the household, including master, spouse, children, apprentices, enslaved workers, and/or paid workers, undertook the necessary tasks of production. Households worked according to their own schedules, attending to artisanal tasks and to the other daily and seasonal tasks necessary for survival, like cultivating a garden and cutting firewood. In the eighteenth century, some merchants in the textile trade built workhouses, or factories, for the carding and spinning of fibers into yarn. Instead of putting out raw materials to households, a merchant hired people to travel daily from their respective households to the factory to work using the merchant's machines, turning the merchant's raw materials into the merchant's finished products. People did that work in exchange for a wage. In factory settings, merchants and their surrogates enforced certain expectations, such as the hours during which work would be performed, and they could supervise the quality of work done. The factory system marked a radical change in the social organization of production.[9]

The third development of the Industrial Revolution, and the factor that made it truly revolutionary, as is well described by Sieferle in *The Subterranean Forest*, was the transition to using fossil fuels to power the mechanized production taking place in factories. Humans have a long history of harnessing forms of energy outside their own bodies to increase productivity. Across the millennia, humans have learned to use the muscle power of domesticated animals to do productive work. And as societies developed in particular ways, some people have learned to use coercion, through social institutions like slavery, to exploit the muscle power of other humans. People have also learned to harness inanimate sources of energy, like flowing water and the wind, to power mechanisms. A gristmill, with its waterwheel and grindstone, is basically a much larger quern, powered by water rather than muscles. The ultimate source of energy for all these activities is radiation from the sun, which allows plants to convert the solar radiation to chemical energy through photosynthesis, and which drives atmospheric currents and the hydrological cycle of evaporation, clouds, and precipitation feeding rivers with water that flows back to the sea to evaporate again. All the facets of this sun-powered system that humans have learned to exploit are encompassed in Sieferle's agrarian solar energy system.[10]

In England, as the pace of mechanization was accelerating and capitalists were moving more production to factories, innovators devised means for powering mechanized factories, first with waterpower and then, around the turn of the nineteenth century, with steam engines, the steam being generated by coal-fired boilers. It was the expansion of production by steam, generated by a seemingly infinite supply of coal, according to Sieferle, that made the industrialization taking place in England in the late eighteenth and early nineteenth centuries truly revolutionary. Andreas Malm, in *Fossil Capital*, makes the important point that the transition to coal in British industry was not made by the people who did the work in the textile mills but by the capitalists who owned the mills, because coal and steam power were easier to control than were workers or the streams that provided waterpower.[11]

English cities in the eighteenth century had only recently made the transition for household heating and cooking from using wood to using coal. This freed those cities from the limits that the availability of nearby firewood had imposed on them. Coal-fired steam did much the same for manufacturing, freeing it from the limits imposed on production by the availability of muscle power or wind and waterpower. Before the transition to coal, energy for productive activities had been dependent on Earth's current income of solar radiation and people's abilities to harness it. Coal, on the other hand, represented the sun's energy that had been converted by plants to chemical energy ages ago and then converted by geological processes into long-term storage in Earth's crust as a carbon-based mineral. Coal-fired steam made it possible for owners of textile mills and other factories to greatly increase the productivity of their workers and their machines while reducing costs of production, thereby increasing profits and expanding the economy.[12]

American Industrialization

When the United States began to industrialize, it followed the British pattern. American forests and streams offered an abundance of firewood and waterpower potential, so when the nation began to adopt the factory system and increase mechanization, early U.S. industrialization was driven largely by the waterwheel. The nation was also rich in coal resources, however, and it was not long until innovators found ways to take advantage of the mineral fuel. Developers of the anthracite beds in northeastern Pennsylvania built canals to reduce the cost of shipping their product to Philadelphia and New York, opening markets for anthracite as an attractive fuel for domestic heating. By the 1840s, entrepreneurs had developed blast furnaces for smelting iron along the canals, using anthracite as metallurgical fuel and beginning to free the American iron industry from the limits imposed by the need to harvest timber and sustainably make charcoal for metallurgical fuel. Meanwhile, west of the Allegheny Mountains, entrepreneurs in Pittsburgh learned to take advantage of the rich bituminous resources there to develop a manufacturing economy powered by coal-fired steam, and by the 1850s, Pittsburgh and surrounding areas were building an iron industry based on coke as the metallurgical fuel made from bituminous coal. By the outbreak of the Civil War, people put coal, iron, and steam power together to support an expanding network of railroads in the Union states, setting the stage for railroads and industrialization to accelerate into the former Confederate states and the American West after the war.[13] By the time mining capitalists found Butte's rich copper deposits in the 1870s, the process of industrialization had firmly taken hold of the U.S. economy, reaching into nearly all settled parts of the continent and shaping the ways new settlement would occur henceforth.

Industrialization created increased demand not only for iron but for other base metals as well, including copper, lead, and zinc, the production of which fostered continued acceleration in industrialization. There was a "great acceleration," to borrow a phrase from J. R. McNeill and Peter Engelke, in U.S. and global metals production beginning about 1870 or 1880.[14] McNeill and Engelke use the phrase in documenting the correspondence between accelerated burning of fossil fuels and the acceleration in production, consumption, and environmental impacts beginning in the years after the Second World War. A graph of the production of metals published in a 1913 issue of *Engineering & Mining Journal* shows the impact of industrialization in the previous century, a great acceleration (fig. 1.1).[15] To show the correspondence between production of metals and use of fossil fuels beginning about 1880, I have superimposed dots on the graph in figure 1.1 showing global coal production in those same years.[16]

The U.S. followed this pattern in metals production. Figure 1.2 shows

April 12, 1913 THE ENGINEERING & MINING JOURNAL 743

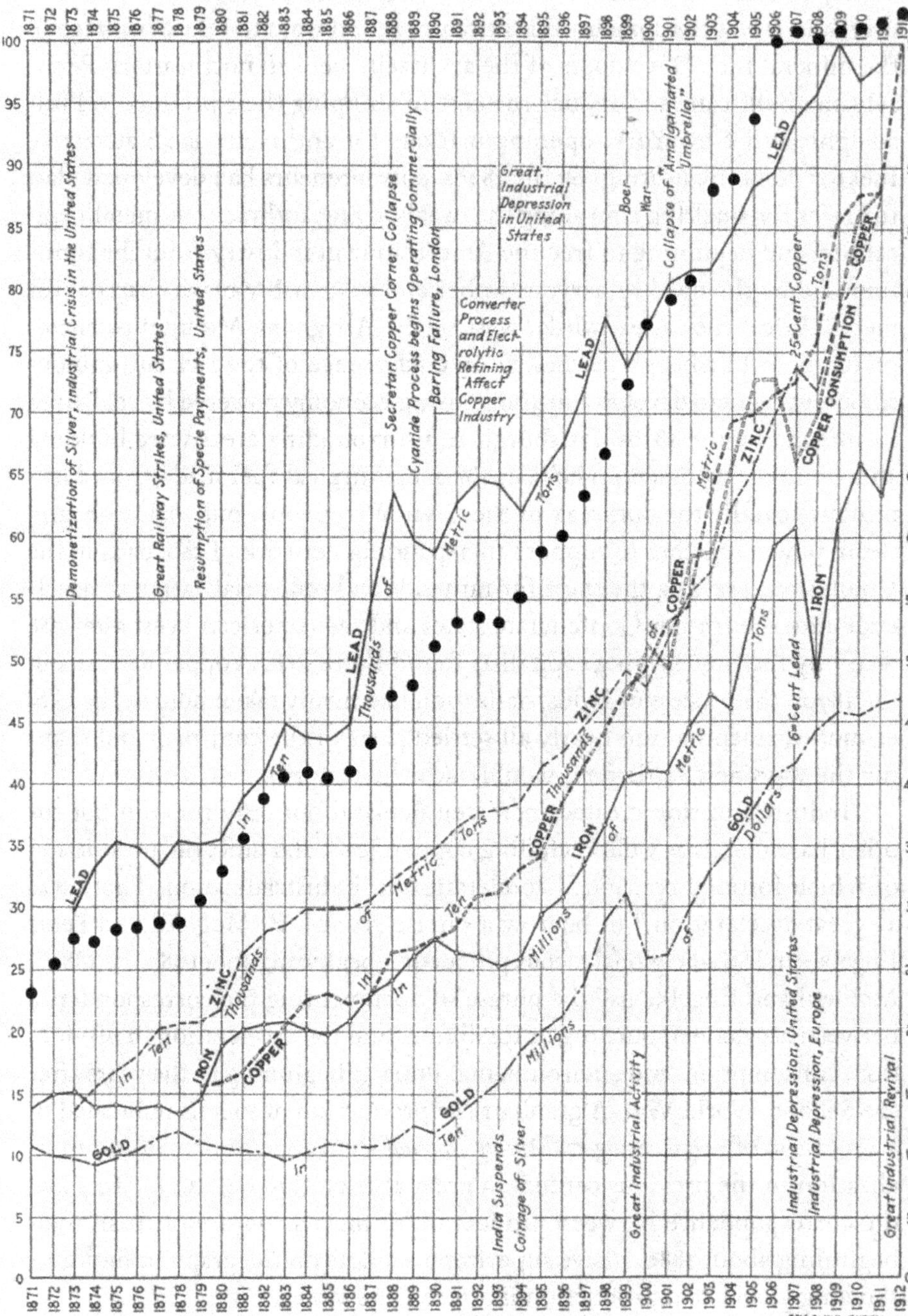

THE WORLD'S METAL PRODUCTION FOR 40 YEARS

U.S. copper production across the second half of the nineteenth century, with output growing gradually through 1870 and then increasing rapidly thereafter. The graph also includes the price of copper from 1860 onward, showing the curious fact that copper production grew despite the falling price for copper.[17] A graph of lead production in the U.S. would look similar. U.S. lead production was fairly steady across most of the nineteenth century, before it began to increase greatly in about 1870.[18]

Coupling price with production helps to illustrate the profound impact that industrialization had on copper production. Butte was just entering the global market for copper in about 1880 when the price of copper was in steep decline. The capitalists backing the Butte mines invested heavily in mechanization across the decade, however, and Butte's output increased despite falling prices. In 1887, Butte surpassed Michigan's Keweenaw Peninsula in production, becoming the world's leading supplier of copper. The companies in Butte had not taken copper's falling price as a signal to curtail the extraction of ore and the production of the metal; rather, they continued to invest heavily in capital improvements to reduce their costs, increase output, and grow more profitable despite falling prices.[19]

The mining companies were willing to invest in capital improvements, despite the disappointing copper price, because they understood the contingent nature in the definition of ore: "An ore is a metal-bearing mineral, or aggregate of such minerals, mixed with barren matter, called 'gangue,' and capable of being mined at a profit."[20] Three variables helped to determine whether a body of mineral qualified as ore: 1) its metals content and its other chemical and physical characteristics; 2) the costs of extracting, processing, and transporting the mineral in order to put the desired metal(s) on the market; and 3) the market price for the metal(s). Miners could only control the second of those variables by trying to install technologies that brought down costs and by trying to negotiate favorable labor, transportation, and other rates. Miners were therefore willing to risk their capital on

FIG. 1.1. (*Opposite*) The World's Metal Production for Forty Years. The graph, published in *The Engineering and Mining Journal* 95 (12 April 1913): 742, shows the global production of lead, zinc, copper, iron, and gold, which began to accelerate rapidly around 1875–1880 due to industrialization. Black dots applied to the graph by the author show global coal production for the same period, with data points derived from Parker, "Coal," in *Mineral Resources of the United States, 1910*, 62; Parker, "Coal," in *Mineral Resources of the United States, 1913*, 784.

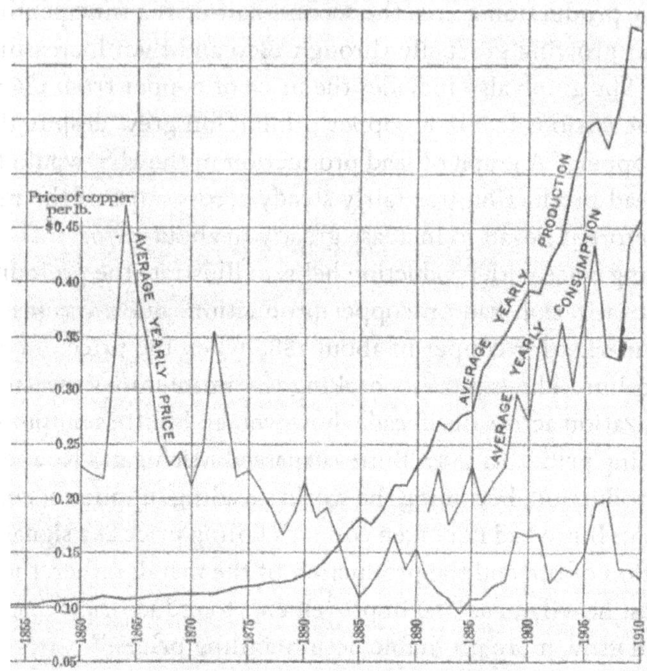

FIG. 1.2. Copper Production in the United States, 1855–1910. This graph shows copper production in the U.S. beginning in 1855, when it was quite small, to 1880, when it began to accelerate rapidly. The graph also shows the fluctuations in the price of copper from 1860 and the price spike during the Civil War. The price dropped below 10 cents per pound in the mid-1880s. *Mineral Resources of the United States, 1910* (Washington, D.C.: GPO, 1911), 165.

methods that might help lower costs in order that a metal-bearing mineral could indeed qualify as ore (i.e., be mined at a profit).

As copper production increased at Butte, there was an accompanying increase in the discharge of waste products (the gangue and the impurities like sulfur and arsenic in the metal-bearing mineral) from mining, milling, and smelting. It was from those wastes that severe environmental impacts derived. Metal-bearing mineral compounds exist in nature because of geological processes that have taken place over billions of years. To produce pure copper or other metals from ores, practitioners in the mineral industry have learned from centuries of trial and error (and more recently through scientific investigation) to subject ores to varieties of processes and series of treatments.[21] Thus, the processes with which Butte miners sought to ally themselves were in the chemical and geologic realms, not the organic, or agrarian solar realm.

As practiced by mining companies in the history narrated here, the first step in the reduction of ores was almost always to mechanically crush the ore to a size suitable for subsequent steps. The last steps almost always necessitated bringing the material to the melting point, at which elements separated into several assortments or agglomerations, one of which consisted of the desired metal in pure or near-pure form. The other elements—not sought but nevertheless present in the ore—separated into other agglomerations of waste, typically a molten mixture called slag and a gaseous mixture of smoke, including fumes and fine particulates. Between the first and last steps, miners and metallurgists usually employed intermediate steps intended to prepare the crushed ore, in stages, for the final steps. Intermediate steps could include grinding the ore to finer size; using mechanical means to segregate particles of greater specific gravity (generally bearing the desired metals) from particles of lesser specific gravity (the gangue); and roasting the material to effect chemical reactions at temperatures below the melting point. Whenever a step separated the material being treated into two or more streams, the stream(s) containing the desired metals were sent to subsequent steps in the process. The other streams were considered byproducts and usually discarded as waste. Most of the waste streams produced by the technologies of Butte's copper industry had impacts on organisms living in the Deer Lodge Valley.

None of the metallurgical activities involved in the reduction of ores were organic in character, other than the obvious fact that they were being accomplished by humans. The miners could and did remain focused on their inorganic enterprise. To be sure, as members of the organic realm, the miners had to satisfy certain organic needs: air to breathe, water to drink, food to eat, and clothing and shelter to protect their bodies from the elements. The miners happily imported into their workscapes the air, drinking water, food, clothing, and building materials they needed for survival.[22]

Miners mainly looked to their immediate environment not for food and fiber but as a resource to support metallurgical processes: air to supply oxygen for oxidation of minerals; water for cooling purposes and to facilitate conveying materials through the process; fuel to provide mechanical power and thermal energy to drive the processes. Regarding fuel, early miners relied on the organic realm in the form of nearby firewood, but shortly after mining was established at Butte they more often relied on the mineral realm in the form of coal.[23] Finally, the miners needed space to dump their wastes or byproducts. If too much waste accumulated at the "back door" of a mill or smelter, it would clog continuation of the processes, so miners and metallurgists needed ample room in which to dump their wastes far from their operations. Their immediate environment was valued in part, then, for the degree to which it could accommodate, absorb, or otherwise act as a sink to accept metallurgical wastes and remove them from interring with ongoing operations.[24]

Conflict arose in southwest Montana over discarding wastes because, even though the wastes were byproducts of inorganic processes, they had organic consequences. One of the first consequences concerned the air people breathed. Smoke given off by smelters is rich in sulfur dioxide, arsenic, and other substances unhealthy to humans. As we shall see, miners and smelterworkers were often willing to accommodate themselves to the fouled air in and around a smelter because their livelihoods depended on the smelter. As Butte grew, however, it came to be populated by people who did not recognize a direct link between polluted air and their own well-being. Such citizens sought to bring legal and political pressure to bear on the smelter operators during the smoke controversies of the 1890s, described in chapter 4, giving rise to the first serious environmental conflict involving the Montana copper industry.

The environmental conflict in southwest Montana took on national significance after the turn of the twentieth century as the polluting byproducts of the copper industry began to damage others' property on a monumental scale. The conflict stemmed from differing perceptions of how battling groups wished to interact with the environment. As just described, the miners, metallurgists, and their companies had come to use a vast territory as a dumping ground for the byproducts of their metallurgical endeavor, expecting Earth to absorb or otherwise harbor the wastes. By 1900, however, that waste-dumping had expanded to areas populated by people who made a living by interacting with the environment in the organic realm. Farmers and ranchers owned large tracts of land on which they planted crops and grazed livestock, following the pursuits of agricultural peoples before them. Most of the large blocks of forestland in southwest Montana were still part of the public domain, owned and administered by the U.S. government in order to conserve forest resources on behalf of the American people. By the early twentieth century, a strong conservation ethic had developed among some groups in the U.S., recognizing forests as a renewable resource important to the continuing vitality of American industry and the national economy and therefore as a resource that should be husbanded.

In 1900, the Montana copper industry had produced more than any of the world's other copper producers for more than a decade, continuing to discharge byproducts into the environment. Sulfur dioxide had an especially devastating effect on plant life, including both forests and agricultural crops. Arsenic killed livestock and wildlife by settling on forage that was then eaten. Tailings washed into the streams and rivers, killed aquatic life, and spread over riparian lands, destroying crops and pasture. In the early years of the twentieth century, the area adversely affected by inorganic byproducts of the Montana copper industry totaled several hundred thousand acres. Conflict arose between people working in the organic realm and those working in the inorganic realm. Associations of farmers and professionals employed by agencies of the U.S. government opposed the lawyers and engineers of the giant corporations who owned and operated the copper industry. Battles and negotiations between the groups often took place in the courts, rather than in armed conflict, as people sought relief in the law. But the conflict erupted in other venues as well, including the press, politics, and the professional societies of the experts who worked for both sides, as described in chapters 5, 6, and 7.

As Stuart Kirsch asserts at the head of this chapter, corporations play and have played the major role in marshalling the capital, skill, technology, and power necessary to cause the environmental harms that are at the center of this story, but this book is not intended to be a major investigation of how the industrial corporation came into being or how the American economy came to be so dominated by industrial corporations. I rely on a few works by business, economic, and legal historians to characterize the corporations that played important roles in this story, but I am largely content to let those corporations be characters in the story.

The environmental history of copper-smelting technologies in Montana, then, is basically a story of conflict between two groups over the land and resources in the upper Clark Fork basin. One group wanted to use the upper Clark Fork basin for exploiting mineral resources, for employing processes derived from the nonliving realm of nature to convert minerals into useful metals, and to dump byproducts with little regard for the impacts on the living parts of the upper Clark Fork ecosystem. Others wanted the upper Clark Fork basin as an ecosystem into which they could insert themselves as actors trying to optimize the agrarian solar energy system for their own purposes. The agricultural and silvicultural groups in the upper Clark Fork—the "organic" groups—found themselves in conflict with a group employing productive methods and working from a set of values that had little care for the biological realm of the upper Clark Fork. The conflicting groups had differing visions of what Butte, Anaconda, and the Deer Lodge Valley were. As folks in each group strived to make the area into what they imagined it to be, they found that their respective visions and methods were incompatible with each other.[25]

Summary of the Conflict in the Clark Fork Basin

Historically, productive enterprises typically relied on other producers for many of the supplies and services they needed to operate, like equipment, fuel, or transportation. In *Networks of Power*, on the development of technological systems, historian Thomas Hughes refers to this dynamic as enterprises relying on their "environment." He recounts how industrial corporations in the U.S., as they grew in scope and complexity, often sought to incorporate into their own systems of production many of the functions for which they had previously relied on their environment. Incorporating functions into their own systems gave corporate managers greater control over those functions.[26] The Anaconda Copper Mining Company (ACM) followed this trajectory as it acquired coal mines so that it would not have to rely on the environment (other coal companies) for fuel. When the ACM constructed its own Butte, Anaconda and Pacific Railway, it no longer had to rely on the environment (the independent Montana Union Railway) to provide a rail link between the Anaconda company's mines and smelters.[27]

The ACM followed a similar pattern in the ways it disposed of wastes. Initially waste disposal was externalized, as the ACM discharged wastes into environments others owned or on which others depended, leading to conflict with the people who had to bear those costs. Over time, however, the ACM came to incorporate more of its waste-disposal functions into its own industrial system, rather than relying on the environment for disposal, as described in chapters 5, 6, and 7. To some extent, the Anaconda company recovered byproducts for sale, making byproducts an obvious part of the company's cost/price accounting. In other instances, the company incorporated its waste-disposal functions into its own system simply by acquiring rights to property on which it dumped wastes, rather than depending on its neighbors to provide the dump. As long as the Anaconda company took advantage of its neighbors to provide the dumping ground, it could not control their responses to the company's byproduct discharges. Neighbors could and did take the ACM to court.

By 1905, Butte was producing nearly as much copper as the entire world had been producing in 1880, just twenty-five years earlier. In 1880, the world's mines produced about 160,000 metric tons of copper, and in 1905 Butte's mines produced 143,000 metric tons of copper.[28] Almost all of Butte's copper ore was milled and smelted twenty-six miles away at Anaconda, situated at the south end of the agriculturally productive Deer Lodge Valley. One way to conceptualize what was happening is to realize that the magnitude of all the combined environmental impacts in 1880 from milling and smelting copper ores throughout the entire world was suddenly, within a mere quarter century, concentrated in one valley in Montana. The devastation to farmers' crops and livestock was sudden as well. Following in the American tradition of self-government, as we will see, the farmers quickly called meetings, elected leaders and representatives, appealed to the Anaconda Copper Mining Company, and prevailed upon their state senator to introduce legislation that would give local officials the authority to stop the company from committing its destructive acts, as this book describes.

Farmers in Montana's Deer Lodge Valley at the turn of the twentieth century believed that they were entitled to protection, under the law, from the monumental damage that mining companies were causing to land, crops, and livestock, threatening their livelihoods. They depended on an environment relatively free of toxins for their families, crops, and animals to thrive. The farmers used the legal system, in good faith, to secure the protections they thought that the law should give them, but, as this book shows, they did not succeed, because the legal system now valued the broad economic well-being thought to be fostered by the mining industry more than it valued the rights of the farmers to conduct their lives free from the nuisances caused by industrial neighbors.[29]

The mining industry in Montana was able to press its case in the courts because of the tremendous wealth and power that corporate capitalism had concentrated in very few legal entities. The late nineteenth century was a period of such concentration, and some aspects of it concerned legislators and policymakers. On one hand, public officials believed that certain concentration was in the public interest, because it improved efficiency and reduced waste. On the other hand, officials were not blind to the abuse of power that monopolies could exercise. Striking a balance was the challenge Congress sought when it passed the Sherman Anti-Trust Act of 1890, and then the courts tried to refine that balance in the ensuing years when they heard cases filed under the act.[30] A decade later, President Theodore Roosevelt tried to improve that balance when he championed creation of a Bureau of Corporations, and when he and his administration decided to take certain giant trusts to court for violating the Sherman Anti-Trust Act. Montana was caught in this unfolding uncertainty when, a decade after the state was admitted to the Union in 1889, powerful capitalists formed a copper trust to consolidate Butte's copper industry, creating a corporation that could wield more political power in Montana than any collection of citizens, whether Republicans or Democrats, whether farmers or townsfolk.[31] Both sets of controversies—how to deal with unprecedented industrial pollution and how to deal with unprecedented corporate power—consumed the attentions of many people in Montana for a couple of decades, and the Roosevelt administration played a direct hand in several chapters of that story.

Along the way, Butte's mining companies engaged in what Barbara Freese calls corporate denial: "the lies, delusions, and rationalizations that emerge when people working in competitive, profit-driven group enterprises are faced with powerful evidence that they are causing harm."[32] She describes two kinds of denial: causality and responsibility. I narrate several episodes in which companies in Butte and Anaconda denied they were the cause of harm, and other times when they tried to minimize the harm they were causing. I also show the many instances in which the mining companies worked to convince courts, government entities, and the public that, even if the companies' operations *were* causing harm, they could neither do anything to decrease the damage nor should they be responsible to do so. When the farmers in the Deer Lodge Valley presented clear evidence of the harm tailings and smelter smoke were causing to their livestock and crops, for example, the companies offered alternative explanations, including that the farmers were engaging in poor agricultural practices. These are tactics of denial that Freese describes, as she presents eight case studies of corporations engaging in corporate denial. The history of corporate denials by the copper industry in southwest Montana can be added to the list that illustrates Freese's argument. She asks, "Are those offering such denials simply lying, or do they actually believe what they say? I will seldom try to parse outright lies from self-deceptions in this book."[33] *Smoke and Tailings* includes similar episodes, in which it is difficult to discern whether the historical actors were lying or actually believed what they were saying. We will also see instances in which the actors knew they were being deceitful.

By the end of the twentieth century's second decade, however, the conflicts over the industry's right to pollute the environment and over a single corporation's right to exercise almost unmitigated power were largely over. Despite Roosevelt's reputation as an avid and effective conservationist and as a trustbuster, the copper industry in southwest Montana continued discharging vast volumes of contaminants into the environment, and the Anaconda Copper Mining Company, the corporation that emerged from the various stages of consolidation in Butte to become one of the world's largest industrial corporations, was set to take a more prominent role in global mining affairs.

Yet despite those outcomes, the history documented here is not one of unambiguous victory by a corporate giant. At each turn in that history, the Anaconda Copper Mining Company was a determined manipulator of outcomes. But because the ACM's officials could not know in advance what an outcome would be, they undertook major initiatives on multiple fronts to try to improve the corporation's chances for success. One was on the pollution front. In order to improve the corporation's chances of long-term success, corporate officials authorized their engineers to investigate new means of reducing pollution, in the hopes of reducing the costs of operation-incurred legal challenges. As a result, the amounts of pollution the Anaconda company generated, going into the 1920s, were smaller, in comparison to production, than had been the case at the turn of the century. Nevertheless, the company's scale of operations was so huge that its pollution output was still tremendous, leading, after sixty more years of copper production, to the designation of the Clark Fork Superfund site, the largest complex of Superfund sites in the United States.[34]

Remediation of hazardous materials began in Butte and Anaconda under the Superfund program in the 1980s, and forty years later, after spending more than $1.6 billion, the work is not yet complete. The history of how these two communities in southwest Montana came to be so contaminated is emblematic of the histories that created Superfund sites throughout the U.S. and is a cautionary tale of the kinds of decisions Americans and citizens of developing nations throughout the world must confront in trying to balance environmental quality with industrial development that meets the demands of a market always hungry for more resources. Butte and Anaconda are therefore a National Historic Landmark to environmental degradation and renewal.

CHAPTER 2

Background to Mining at Butte

Humans have evolved to utilize not only the organic bounty the Earth offers but also the inorganic, nonliving resources and processes available in the geologic realm. Since the Stone Age, humans have gathered and shaped rocks, minerals, and gems of various sorts to fashion useful and ornamental objects. Since the Iron and Bronze Ages, people have gone beyond merely gathering and shaping inorganic resources, employing chemical and thermal processes to alter the character and properties of sand, rocks, and stones, yielding pottery, glass, and metals for new kinds of objects, opening entirely new realms of material culture. The new realms of material culture have in turn opened yet more possibilities for human interactions with environments, living and nonliving alike.

In his global history of technology, *Technics and Civilization*, Lewis Mumford distinguishes between the realms in which agriculturalists' (organic) and miners' (inorganic) endeavors operate. Both agriculture and mining must contend with the physical and chemical aspects of the physical world, but unlike agriculture, which must also contend with the living or biological aspects of the natural world, mining is not trying to foster life, other than, to a certain extent, the lives of miners while they are in the mine. Mumford calls mines "the first completely inorganic environment to be created and lived in by man."[1] He calls mines more inorganic than large cities. He also claims that "more closely than any other industry, mining was more bound up with the first development of modern capitalism. By the sixteenth century it had definitely set the pattern for capitalist exploitation." Mumford asserts that the mining industry in sixteenth-century Germany had already given rise to a number of improvements said to characterize the modern industrial organization: "the triple shift, the eight-hour day, the existence of guilds in the various metallurgical industries for social intercourse, charitable self-help and insurance," as well as other features of capitalist industry in the nineteenth century, such as "the division of classes, the use of the strike as a weapon of defense, the bitter class war, and finally the extinction of the guilds' power by a combination of the mine-owners."[2]

Unlike farming, mining had the potential to make the miner wealthy very quickly, sometimes even with little effort and without much skill.[3] Because of this potential, miners were and are willing to use methods Mumford compares with "the ruthlessness of modern warfare." To profit from mined material, the miner must either break it or melt it. After listing the sorts of implements used by the miner—the pick, the sledge, the crusher, the stamping machine, and the steam-hammer—he concludes, "The routine of the mine involves an unflinching assault upon the physical environment: every stage of it is a magnification of power."[4] These themes are at the heart of the history of Butte and Anaconda offered in this book, but first Butte should be placed in its physical setting.

Situated in southwest Montana, on the west side of the Continental Divide, Butte sits at the headwaters of the Clark Fork, a major tributary of the Columbia River draining Montana west of the Continental Divide. The Rocky Mountains in Montana are characterized by distinct ranges, some with peaks exceeding 10,000 feet, and broad mountain valleys. Between the various mountain ranges and valleys are expanses of lesser hills and drainages. The Clark Fork originates at the head of the broad and open Deer Lodge Valley at the confluence of two smaller streams, Silver Bow Creek and Warm Springs Creek. The former drains the hills and a valley north of the Highland Mountains, and the latter drains the hills on the north side of Pintler Range. Both creeks played major roles in Montana's mining industry.

In many ways, the history of mineral extraction from the Butte began seventy-eight million years ago, when a large mass of granite, otherwise known as quartz monzonite, formed as a molten mixture of minerals far below Earth's surface. The crystalline nature of the Butte granite indicates to geologists that the once molten mass cooled and solidified before it reached the surface. Changes in Earth's crust, brought about by the tectonic forces that build mountains and by the erosive forces of Earth's weather, eventually exposed the mass of granite to the light of day. In due time, Euro-American prospectors, looking for mineral wealth in the great western provinces of the United States, happened across the granite outcroppings on what is today the Butte hill, and they recognized signs that the hill held a promise of wealth.[5]

Before it reached Earth's surface, the mass of granite had already undergone significant changes. Known as the Boulder Batholith, the body is a large one, thirty miles wide and over sixty miles long, extending from south of Butte almost to Helena, with the town of Boulder near its center. The minerals that comprise the granite include quartz, feldspar, biotite, and hornblende.[6] As the minerals cooled, about sixty million years ago, the mass shrank, causing the first fissures in its structure. Southwestern Montana was underwater at the time, and the fissures provided loci for mineralizing solutions to concentrate copper and molybdenum, forming a mineral deposit geologically older than—and distinct from—the veins that made Butte famous throughout the mining world.[7] Several million years later, as the Earth's crust shifted, much larger fractures opened in the Boulder Batholith. This time, mineralizing solutions deposited a wider array of minerals containing copper, gold and silver, zinc, manganese, lead, arsenic, and other metals.[8]

Geologists offer several theories for the source or sources of the minerals, but the processes took place so long ago that the provenance of the hill's metal riches remains an open question. The results of that mineralizing process over fifty-five million years ago, however, remain one of the classic examples of mineral zonation in the world. The central zone of the Butte district contains copper minerals nearly free of zinc and manganese minerals; a surrounding intermediate zone contains copper minerals in conjunction with zinc; the peripheral zone contains minerals with little copper but relatively rich in manganese and silver. In a map view of the Butte mineral deposit, the zones form roughly concentric circles, while vertical sections through the Butte hill show that the zones increase in diameter with depth. Thus, the zones have assumed a conical shape within the Boulder Batholith.

In the millions of years since minerals were deposited in the vein structure, other transformations have occurred in the Boulder Batholith and in the surrounding areas of Earth's crust. The continuing exertion of forces against the mass of granite, still thousands of feet below surface, caused more fracturing, much of it running through, rather than along, the vein structure. Consequently, the ore deposits in the Butte hill had assumed greater complexity of shape and structure by the time they were discovered in the nineteenth century of the present era. That fracturing provided the physical basis for the litigation (aboveground) and hand-to-hand combat (belowground) over ownership of the ore deposits. But that is getting ahead of the story. It took tens of millions of years and the slow erosion of the overlying crust for the Butte granite to reach the Earth's surface.

In the meantime, at least two other significant geologic events shaped the Butte landscape. About fifty million years ago, a similar molten mass of rock formed beneath the crust, this time under such pressure that it erupted as a volcano rather than solidifying beneath the crust as the Boulder Batholith had done. That volcano, just northwest of the present townsite of Butte, forced molten material into fissures through the Butte granite, erupted to form surface rocks of a mixture of ash and lava, and left a volcanic cone, surrounding a rhyolite plug, still visible today as Big Butte, the small, inaccurately named hill from which Butte takes its name.

The other important geologic event that shaped the Butte landscape involved the plate tectonics of the North American continent. Perhaps twenty million years ago, Earth's crust fractured along a north-south line in what is now the western part of the continent, with the eastern side of the fracture ascending and the western side driving under it. In the Butte region, this fracture runs lengthwise through the Boulder Batholith. The Continental Fault is visible from Butte as the East Ridge, the local name for the section of the Continental Divide that forms the eastern edge of the Summit Valley, and Rampart Mountain, which forms the eastern edge of the headwaters of Silver Bow Creek. This shift in Earth's crust had important consequences for the local topography. East of the fault, the high mountain valley known as Elk Park, which once drained through Butte and eventually into the Pacific Ocean, tilted over time so that it now drains into the Boulder River and eventually the Gulf of Mexico. West of the fault, the relative dip of the crust created a depression that, over the millennia, has filled with alluvial material eroded from Rampart and the East Ridge. Butte sits on a hill that today is an outcrop of a portion of the Boulder Batholith that is west of the Continental Fault. This local topography is much as it was when the first humans arrived in the area, perhaps ten thousand years ago.

The earliest human residents of the Summit Valley had little interest in the mineralized vein structure lying beneath them, for they had no means to unlock the metals that the geological ages had placed there. But this is not to say they had no interest in the mineral resources of the region. Even though these early residents lived in hunter-gatherer societies—meaning they depended very directly on the plant and animal resources their environment provided them seasonally—they did use stone tools to enhance their ability to secure those resources and survive the elements. While much of their material culture was fashioned of wood, plant fiber, bone, and leather, they also relied on stone to perform certain functions for which substances from the organic realm were less suitable. Hunter-gatherers of southwest Montana knew which stones performed best and they knew where to find them. Not far west of Butte is a significant prehistoric quarry site. From those pits, ancient people excavated stone for tools, sending it through trade networks extending well beyond the region.[9]

Such was the use of mineral resources in southwestern Montana until the middle of the nineteenth century, when a series of events, triggered in 1492, swept through the mountains and valleys of the territory. A new nation, spawned in the long wake of Columbus's "discovery," acquired from France a vast parcel of the continent known as the Louisiana Purchase of 1802, paying little heed to any claims the current inhabitants had to the land. Almost immediately President Thomas Jefferson sent two army officers, Meriwether Lewis and William Clark, to investigate the United States' new resources. Lewis and Clark received broad scientific training to equip them to follow Jefferson's Enlightenment-inspired instructions: to observe and record an impressive array of data in the newly acquired lands, including the plant, animal, and mineral resources the territory offered. The resource that had first drawn American entrepreneurs into the region was fur, a commodity for which they had to compete with the British Hudson's Bay Company and North West Company, who had already established forts and relationships with Indigenous peoples.[10] Such international and corporate competition over control of resources in the American West continued even after the United States established firm political control over its western territory and attention shifted from furs, generated seasonally by the living ecosystem, to minerals, locked in Earth's crust by geologic processes eons before. That shift took place at the middle of the nineteenth century.

To be sure, population depletion of fur-bearing animals due to several decades of overexploitation played a role in discouraging settlement. But when gold was discovered in 1848 at Sutter's Mill in California, the ensuing gold rush brought an unprecedented population migration to California and the American West. Mineral exploitation became one of the chief motivating factors in development of Jefferson's new lands.[11]

Ever since Columbus's arrival, the promise of mineral wealth had been an important myth luring Europeans to American shores and drawing American settlers across the continent.[12] Base metals contributed early to some expanding settlement. Iron plantations helped populate Pennsylvania's forests during colonial times. Lead deposits lured miners to Missouri even during the Spanish colonial period. Native-copper deposits on

Michigan's Keweenaw Peninsula had supplied Indigenous peoples with that metal long before the arrival of Europeans. Then, beginning in the mid-nineteenth century, those native-copper deposits stimulated a significant rush of American mining enterprises to the Keweenaw Peninsula. But until the find at Sutter's Mill, no mineral discovery had provided the basis in reality for a major population migration.[13]

Before Sutter's discovery, land suitable for agricultural settlement had been the main resource drawing Americans westward. Territories and states in the upper Mississippi Valley had grown from populations in the tens of thousands in 1840 to populations in the hundreds of thousands in 1850.[14] Farther west, the fur trade had not led to significant permanent white settlement, but the 1840s did see some intrepid pioneers traverse the arid plains and the inhospitable Rocky Mountains to establish agricultural settlements west of the Continental Divide. During the first ten years of significant migration to Oregon's Willamette Valley, only about five thousand people made the trek, and in the years 1847–1848, five thousand Mormons settled around the Great Salt Lake. By 1850, the combined white population of Utah, Oregon, and California was only 117,271. Prior to the discovery of gold, California's estimated non-Indian population had been 15,000. In 1852, at which time California became a state, California had 225,000 non-Indians, of whom 100,000 lived in the mining regions. The gold discovery also stimulated a renewed effort to chart the mineral resources of California and the inland West scientifically.[15]

The mid-nineteenth-century agricultural settlements of the upper Midwest and far West were generally comprised of individuals and families who wished to start new lives and build new communities, and they worked hard with the soil or their nascent businesses to do so. Such settlers established communities—pockets of near self-sufficiency—from which they could try to make connections with the larger market economy. Mining communities, on the other hand, were highly transitory places, often located in settings where humans had little chance of scratching out a subsistence living. Populated first by men who hoped to find enough gold dust or nuggets to make them instantly rich, mining camps developed along somewhat different lines from farming communities. Often, people who made the most money in mining camps were those who provided goods and services to miners—at greatly inflated prices because of scarcity. Some gold -seekers struck it rich, while others quickly recognized that finding gold would take hard, concerted, and collaborative effort. Already in 1849, some groups of Californians had begun to work cooperatively, rather than individually, to divert water from streams to gravel beds where it could be used to separate the gold from the gravels. By 1850, actual companies had been formed to construct flumes and engage in hydraulic mining on a scale impossible for the single miner to accomplish.[16]

Some frustrated gold-seekers remained to work in such enterprises, while others, lured by the dream of striking it rich alone, fanned out, first through the California hills and then into Nevada and throughout the West in search of other easy pickings. Subsequent discoveries brought booms to other parts of the West: Nevada and Colorado in 1859, and Idaho and Montana in the 1860s. As Kent Curtis demonstrates, these discoveries of gold required preliminary development of infrastructure to link remote locations to the settled portions of the continent. In the case of Montana, that infrastructure was the Mullan Road, built by the United States government in 1859, under the supervision John Mullan, a lieutenant in the army, to provide an overland route between Fort Benton, the head of steamboat navigation on the Missouri River, and Walla Walla, an important navigation point along the Columbia River. The discovery of gold at the headwaters of the Clark Fork River along Silver Bow Creek in 1864 and the settlement of what would become Butte were parts of this larger continental pattern.[17]

The importance of mining to the establishment of Montana was not unlike that of neighboring states. According to an authoritative history of Montana by Michael Malone and Richard Roeder, "Montana's foundation, like that of several other western states, stands upon a golden cornerstone. The glitter of gold first attracted significant numbers of white men to this area, and their coming laid the basis of a community."[18] Montana's gold rush began in 1862 with the discovery of gold on Grasshopper Creek in the southwestern part of the territory. The boomtown of Bannack grew around the discovery, and from that base, prospectors explored the surrounding mountains and valleys hoping to discover more gold. Some did: in 1863, gold-seekers found placer deposits along Alder Gulch, seventy miles east of Bannack, and established Virginia City, Montana's richest gold-mining camp. In 1864, a party based in Virginia City found placer gold and staked claims along Silver Bow Creek, some sixty miles north of Bannack. When they returned to Virginia City for supplies, news of their discovery spread. Several other parties followed them back to Silver Bow Creek, and by autumn 1864, these few people had established a rude camp at the headwaters of Silver Bow Creek in what they called Town Gulch.[19] The settlement was located just north of what would become the Parrot mine.

During its first decade, as the next chapter describes, Butte's history could have been that of almost any small western mining camp: a few years of relatively booming prosperity, followed by a bust. Yet almost from the beginning, knowledgeable miners who picked at Butte's rock outcrops recognized an abundance of copper, manganese, and other base metals. But to recover wealth from such ores economically, the miners would have to bring to Butte a more complex technological dimension than they had been using. They had to bring the capability for smelting, which used fire to melt the ores, thus physically and chemically separating the metal from less valuable elements. Undaunted by the difficulty of the endeavor in a place as remote as Butte, a few miners built crude furnaces. They were not successful. But fifteen years after the discovery of gold in the Summit Valley, a new generation of entrepreneurs—better capitalized and more highly trained—did succeed at smelting copper. From that point onward, Butte departed from the path followed by other small mining camps and began its ascent to the position it held for about three decades as the world's most productive copper-mining district.

Sometime during that reign, Butte gained the moniker, "The Richest Hill on Earth," a name bearing an element of geological truth. The name stems from the boosterism of generations of miners whose hard work turned Butte's great mineral wealth into prosperity for the entire nation—a feat for which much respect and even awe is due them. As with so much boosterism, however, Butte's moniker ignores serious unintended consequences of the achievement, foremost among them the huge volumes of hazardous waste resulting from the extraction of metals. Those wastes will take center stage in this book.

CHAPTER 3

Beginnings of Butte's Mining Industry

> Butte City is beautifully located on an eminence near the junction of the left and right branches of Silver Bow Creek, and close to a stately grove of pine trees, beneath whose shelter has suddenly come into existence a town, comparatively small as yet, but destined ere long to be one of the most flourishing and prosperous in the territory of Montana.
>
> —Joe Bowers, January 7, 1865[1]

> The scenery changed rapidly, and we were soon in the heart of that celebrated formation of quartz-riven rock. Once upon a time, there must have been some hot work going on, for the country looks as if it had been melted, and set on end to cool.
>
> —*Montana Post*, December 9, 1865[2]

The two very different views of the Butte hill in 1865 suggest a tension that has been at work throughout the community's history. On one hand, people have brought visions of the pastoral to the place, hoping to find nurture in nature for their human activities. Certainly, trees and streams coupled with the bustle of hardworking people should be able to spawn a prospering community. On another hand, Butte has from its beginning as an American habitation engendered a sense of awe. Something not quite natural, certainly not organic, had been at work there. People who established Butte as a permanent city worked to realize their visions of what they imagined Butte was.[3] As Butte unfolded to become the largest mining center in the United States, that tension remained. The city was indeed a collection of hardworking people, but to many observers the place was utterly unnatural, perhaps even sinister. In time, people would stand in awe of the smoke that often enveloped the place. Today, that awe is often engendered in people who view the Berkeley Pit and the tailings pile that towers over it, or who see a space-station photo of Butte's landscape.

Southwest Montana has a semiarid climate, averaging less than twenty inches of precipitation per year. Not only are the broad mountain valleys open and treeless, but so are many of the lower hillsides. Higher hills, especially in the mountains, are more densely timbered. The dominant species of timber is lodgepole pine, with substantial stands of Douglas fir as well. The principal deciduous tree species in southwest Montana is the aspen, found in groves amid coniferous forests and in open meadows alike. Caleb Irvine, the first Euro-American to record the character of the Summit Valley, reported that, when he rode through on horseback in 1856, the bunch grass in the valley came up to his horse's knees. He described the Butte hill with scattered tall trees about its upper regions and a grove of trees in the upper basin of what would become Town Gulch, where the first prospectors would cluster a few years later. The first few hearty souls to spend a winter on the Butte hill in 1864–1865 chose the protection of that grove of trees. It provided their crude cabins additional shelter, and the south-facing slope of the Butte hill provided ample winter grazing for their stock. Equally important, it was close by their placer diggings.[4]

The men and their companies who established the modes of production that came to embody the copper industry in Montana, however, had an inorganic focus to their actions. They first sought precious metals and then primarily copper from an environment that came to be known as the Butte hill. To extract those metals, the miners employed processes not at all organic. The miners did not seek to exploit the food chain to produce food or fiber. They did not seek to replenish soil fertility or to optimize forest growth or to find animal species that could thrive at the relatively high altitude and in the semiarid climate of southwest Montana. Unlike persons engaged in logging, fishing, farming, or ranching elsewhere in southwest Montana, who were at some level concerned with and dependent upon the organic processes of photosynthesis, the food chain, and the carbon cycle, miners, metallurgists, and their companies were concerned with and dependent upon processes belonging to the geologic realm, that is, physics and chemistry, but not biology.[5]

Over the next fifteen years, Butte underwent a transition that paralleled a transition taking place throughout the U.S., from a society and economy characterized by production taking place in households, powered by organic sources of energy, to production taking place in factories and powered by coal-fired steam. In the U.S., to be sure, the transition was already well underway by the Civil War; nevertheless, the decades immediately after the Civil War were the tipping point after which the U.S. plunged headlong into the uncharted territory characterized by industrialization and urbanization, with their associated social, economic, and political turmoil. The lives of Americans in the coming decades would undergo more profound change than ever before experienced and, arguably, have experienced since.[6] An energy-related threshold related to this transition concerns the fuel that Americans used: 1885 marks the first year that Americans burned more coal than firewood.[7]

The first settlers in Butte could not have known the changes that would unfold over the next fifteen years, but they possessed an optimism that their work would lead to wealth. The indicators in Butte's rock outcrops promised wealth, but it would take hard work to unlock that wealth. Those first prospectors used their own muscles to pan for gold or to wield picks and shovels to wrest rocks from the Earth, proving only that there was wealth to be had. Extracting it in a profitable way, however, would require improved tools and equipment, which in turn required capital investment. A lucky few prospectors had sufficient wealth to invest in capital improvements, but soon the needs for capital exceeded their abilities, and local bankers often gained control of mining properties. Mine developers tried to make their properties appear as promising as possible to attract outside investment; sometimes the mineral deposits in a mining camp yielded returns on investments, but as often the investments ended in failure. Butte was one of those rare locations that kept appearing and then proving to be richer and richer. By 1880, as the Butte mines proved ever more productive at depth, a few of those local prospectors and bankers were still profiting from their investments, but control of most properties had transferred to distant capital, based in Salt Lake City, San Francisco, Boston, and New York. By 1880, that capital was being invested in ever more mechanized productive capacity, powered by steam and fired by local timber resources; within a few more years, Butte would be linked to the national economy by rail and thereby linked to outside sources of coal, and with that fuel the scale of production at Butte was able to grow at rates that outside observers found almost unbelievable.[8] One consequence of Butte's production would be profound environmental impacts, but that was far from the minds of Butte's early promoters. This chapter summarizes the fifteen years through which Butte moved from uncertainty to apparent permanence.

Butte's landscape attracted men—and Butte was for several generations a very male-centric place—who hoped to prosper from the quartz rock.[9] In time, those men came to derive their primary livelihoods not from working on Earth's sunlit surface, but from toiling deep underground. Men employed fiery, subterranean methods to change rock into metal. Out of men's workplaces streamed water opaque with yellow and gray slimes, flowed molten slag that solidified lavalike on the landscape, and billowed sulfurous smoke that could choke man or beast. The tension between the volcanic nature of life in Butte and the biological nature of the lives of the city's people eventually spilled over into civic affairs. Tension led to conflict between groups engaged in the exploitation of that volcanic nature and groups who wanted to manipulate biologic nature for their purposes.

This chapter sets the stage for that conflict, describing the environment early prospectors found as they searched for gold along the headwaters of the Clark Fork. During its early years, Butte went through a series of transitions not unlike those of other Rocky Mountain mining camps. There were booms and busts, and there was interest in other minerals, even though it was gold that first lured miners to Butte. And during Butte's early years, as in other mining boomtowns, there was little concern for the impact mining might have on the environment. Miners were more concerned with finding mineral wealth and trying to find the right technologies to unlock that wealth from the ores, or to ship their minerals to places where others could unlock that wealth.

Prospecting for Placer Gold

Stories of the old-timers vary concerning who may have comprised the first party to strike gold along Silver Bow Creek in 1864. Butterworth H. "Bud" Barker, Frank Ruff, Peter "Seven-Up-Pete" McMahon, William Allison, and Joseph and James Heister are the names usually recited, but the size of the party varies from three to four. While camped overnight along Silver Bow Creek, several miles below the creek's exit from the Summit Valley, they stuck their pans in the stream's gravels and found colors. Their discovery attracted attention in Virginia City, Montana Territory, where men were constantly on the alert for word of the next bonanza. The first small wave of prospectors arrived in the Summit Valley in the summer and fall of 1864 and began staking placer claims above and below the first discovery. Most gold-seekers wanted placer gold, so they looked for promising gravels along Silver Bow Creek and its tributaries, including the gulches carved into the Butte hill. Other prospectors hoped to find a mother lode, the ore body from which the gold in the gravels had eroded. They dug test pits into the sides of the gulches and across the hillside, finding much of the decomposed rock to be easily worked and some bearing gold.[10]

As prospectors combed the Butte hill and neighboring areas, they could see in the black-stained quartz reefs that many of the rock features contained manganese, and therefore probably silver. Silver was not so desirable as gold, so the black ledges and quartz reefs did not attract as much immediate interest as the placer deposits, because placer gold could be relatively easily won, while extracting values from quartz required additional investment to develop underground workings and the facilities needed to process ores. Some determined miners, however, did fix on the promising quartz outcroppings. William Allison had gone back to Virginia City to bring his friend George O. Humphreys to the Summit Valley. After staking a few placer claims, they began prospecting along the face of the Butte hill. They took quartz samples to be assayed at Virginia City, and the excitement spread. Dennis Leary, Henry Porter, and Herbert Madison followed Allison's and Humphreys's trail back to the Butte hill and found the two men digging a discovery hole on what would become the Original claim. Leary, Porter, and a few other miners also filed some important lode claims that first year. The most important was the Parrot. Charles Murphy located another of Butte's early lode claims, the Black Chief, in the summer of 1864 on a quartz outcropping just north of Silver Bow Creek near Missoula Gulch. Without the means to extract values from the rock, however, his claim, like many others in Butte, remained largely inactive for more than a decade.[11]

Hardrock Mining

> An *ore* is a natural aggregation of minerals from which a metal or metallic compound can be recovered with profit on a large scale. When the percent of metal is too low for profitable extraction the rock ceases to be an ore.
>
> —Robert Richards, 1909[12]

Richards makes it clear that a rock's mineral composition is important in determining whether it qualifies as an ore, but the overriding determinant is economic: Can the mineral content be converted into wealth on a commercial scale and at a profit? Most of Butte's early miners had answered this question about the hill's obviously mineralized outcroppings in the negative. The question nevertheless intrigued a few individuals that first winter in Town Gulch. In February 1865, Virginia City's *Montana Post* reported that a piece of ore weighing less than two pounds and taken from the Original lode had been smelted in a blacksmith's fire to produce a small brick of silver and copper. Such an act did not qualify as "profitable on a large scale," so the minerals locked in Butte's veins had yet to be proven as ore by an economic measure. Despite being ridiculed by their peers for foolishly picking away at more ledges, a few "quartz cranks" kept locating more mining claims and filing on them in the Recorder's Office at the newly established seat of Deer Lodge County in the town of Silver Bow, several miles west of Butte. These men were determined to find a way to derive a profitable return from mineral wealth locked within the Butte hill's geological structure. Assays on samples from their lode claims showed the potential wealth was there. The question was how to get it out.[13]

There were essentially two ways of profitably gaining metal from Butte ores: build reduction works at Butte capable of milling and/or smelting the ore, or transport the material to distant reduction works where it could be treated. Butte was still a remote mining camp high in the Rocky Mountains, so primitive transportation options stymied both options. When gold at Butte was first discovered, the most effective means of transporting equipment over long distances to the Montana mining camps was by steamboat from St. Louis, or other shipping points on the lower Missouri River, to Fort Benton, the head of navigation. Above Fort Benton, falls and rapids made the river impassable, and during some summer months steamboats could not even make it as far as Fort Benton. From Fort Benton, freighters typically hauled goods by mule- or bull-train about 150 miles to Helena, from where several roads branched to the mining camps of southwest Montana. Butte was another sixty-five miles from Helena. As the transcontinental railroad neared completion in Utah in 1869, a less expensive route opened to the Montana goldfields. The main road linked Corinne, Utah, to Virginia City, from where branches extended to other Montana mining camps. Mormons had opened the route in the early 1860s and began providing stiff competition to the Missouri River route well before the completion of the railroad. The cost of shipping goods to Montana was a major reason for the high cost of living in the gold camps. For example, for miners who spent the winter of 1864–1865 along Silver Bow Creek, a sack of flour sold for as much as $30, and coffee and sugar each sold for $1.10 per pound. Freighters offered low rates to anyone wishing to ship minerals back to Fort Benton or Corinne, but even that inducement made it difficult to profitably ship ore from Butte elsewhere.[14]

William Farlin was one of the first of Butte's quartz miners to attempt shipping ore profitably. In 1865, he shipped ore from his Asteroid claim overland to Fort Benton, from there by steamboat down the Missouri River to a railhead, and then cross-country to a smelter at Newark, New Jersey. Farlin's effort was not profitable, and he soon left Butte. When he returned to the camp about ten years later, he made a more significant contribution to the development of mining with the construction of Butte's first successful stamp mill for treating ore locally. (A stamp mill follows the pattern of many early industrial facilities: the building takes the name of the main piece of equipment it houses. The mechanized stamps at the heart of the building operate like a mortar and pestle to crush gold or silver ore to a fineness sufficient to be able to recover the gold and/or silver by amalgamation with mercury. Stamp mills were powered either by waterpower or steam). Farlin's distinction as the first successful operator of an ore-treating facility is not because others had not tried earlier.[15]

During the winter of 1866–1867, the Porter brothers, along with Joe Ramsdell and W. J. Parks, built Butte's first furnace intended to smelt copper ore from a shaft on the Parrot lode, which by then was said to be eighty feet deep. The simple stack furnace was located in Town Gulch. Little is known of this first smelter other than that it was built of stone, that the men tried to use a simple blacksmith's bellows to drive the blast, and that the effort failed. That same year, Charles Hendrie and Harvey Ray began construction of a stamp mill, announcing intentions to build a furnace for smelting ore as well. In 1868, Dennis Leary and the Porters built another furnace about a quarter mile to the east near the Parrot mine. It had an improved blast, driven by a fan commonly used on threshing machines, but the operators still could not find the proper flux (material charged to the furnace with the ore to achieve a good separation of slag from metal) to smelt Parrot ore economically. In 1868, Hendrie and Ray were still trying to get their mill and furnace operating, promising to treat ore from the Gem mine, but they never succeeded. These early Butte pioneers did produce some copper matte (the product of a copper smelter that needs further refining to make pure metal), but the effective recovery of base and precious metals from the Butte hill's complex ores would have to wait about a decade before mine developers could bring the proper mix of capital, transportation, and technology to bear on unlocking Butte's mineral wealth.[16]

In the 1850s, most copper production in the United States had come from Michigan's Keweenaw Peninsula, where the mines yielded native copper (pure metallic copper existing in nature). Found near the shores of Lake Superior, it was called "lake copper." The companies extracting copper from the mines on the Keweenaw Peninsula employed melting furnaces to separate the already-pure copper from the gangue, but they did not have to rely on the smelters, as does every other copper-mining district in the world, to separate the copper from other elements, such as sulfur, to which the copper was bound chemically. Smelters treating lake copper in Michigan and at Cleveland and Pittsburgh employed reverberatory furnaces. Elsewhere in the U.S., where smelting companies had to extract copper from more complex sulfide minerals, metallurgists used furnace methods with German and English/Welsh origins.[17]

The first successful, large-scale copper-smelting operation in the West was Nathanial P. Hill's works at Black Hawk, Colorado. In 1867, Hill and his financial backers formed the Boston and Colorado Smelting Company to treat ores from the mines of Gilpin County, Colorado. Although his enterprise was launching at about the same time as the Porters and others were attempting to smelt Butte's copper ores, and although Black Hawk was similarly remote from convenient transportation, Hill had several advantages over Butte's quartz cranks. He was a young professor of chemistry at Brown University and so brought a scientific understanding to his endeavor. He used his resources at Brown to research available smelting technologies, settled on the Swansea method, and traveled to Wales to observe it in practice. Then he had some ore from Gilpin County freighted east to the Missouri River and from there shipped to Swansea. Hill traveled to Swansea so he could watch his ore being smelted. Finally, Hill hired a trained German metallurgist with experience at Swansea to supervise his Black Hawk operations. With financing, technology, and skill, Hill began to have the necessary equipment manufactured at St. Louis and freighted westward in 1867. The Black Hawk smelter was ready to begin trial operations in February 1868. In June, Hill made his first shipment of copper matte to Swansea for refining. In the mid-1870s, Hill built his own refinery so that he could separate gold and silver from the copper in his matte, thus obviating the need to ship the matte to Swansea. The fact that copper amounted to only about 5 percent of Hill's sales demonstrates that he was mainly interested in the precious metals; copper was important primarily because, in smelting, it effectively collected the precious metals in the ore, and the precious metals could be separated from the copper later in the refining process.[18]

Butte's initial boom was drawing to a close, as recovery of placer gold dwindled, yet the ore deposits buried beneath the surface were attracting notice outside Montana Territory. J. Ross Browne, an agent reporting to the federal government in 1868 on mining and related activities in the American West, wrote about Butte:

> At Silver bar [*sic*] and Butte City the mines have done well. Placers only have been worked; but in every gulch where good placer mines are found, gold-bearing quartz veins are found also, many of which contain silver, copper, antimony, arsenic, and manganese, and are rich but very refractory.
>
> At Butte City some copper mines have been discovered, and a furnace erected for smelting. Owing to a defect in the blast it was not successful. The ore, which is quite abundant, is composed of oxides and carbonates in a concentrated form. It contains gold and silver, and with a well-regulated furnace there would be no difficulty smelting it.
>
> These veins are found crossing a belt about one mile wide and four or five long, and show evidence of being deep and permanent.[19]

Despite such observations, Butte's boom was following closely the typical pattern for western mining camps as the placers played out. After a peak population of perhaps 500 in 1867, Butte's population in the 1870 U.S. Census stood at 241 persons, a third of whom were Chinese. The presence of so many Chinese miners often signaled that a camp was on its last legs. Most white miners had left in search of easier, quicker wealth. The Chinese were willing to exert the hard work necessary to extract what gold remained after the easy pickings had been removed. During the early years of the 1870s, the population of Butte dropped to only about 100 year-round residents.[20]

Interest in Butte's quartz lodes did not revive to any significant degree until 1875, when several quartz cranks gathered the financing needed to build small stamp mills to crush silver ore and recover silver and small amounts of gold by amalgamation with mercury. The person most widely credited with spearheading the revival was William Farlin, who had staked the Asteroid and other Butte lode claims in the mid-1860s. He had taken some samples of Butte quartz with him to Idaho, where he worked in the mines and learned more about milling. From assays of his samples, he concluded that Butte ores were more promising for their silver and copper than for gold. Returning to Butte in late 1874, he prepared to refile on a dozen claims under a revision of the federal mining law (the Mining Act of 1872), which took effect on January 1, 1875. Those claims included the Asteroid, located on southwest portion of the Butte hill. He renamed it the Travona on January 1. He also secured financing from a Deer Lodge banker, W. A. Clark, to build a stamp mill. Farlin named his mill the Dexter.[21]

Farlin ordered a ten-stamp mill so that he could work the silver-bearing quartz from his Travona claim. He had the equipment freighted overland from Corinne to a millsite located near the Travona mine, on the east side of Missoula Gulch, about a half mile above the mouth of the gulch on Silver Bow Creek. By June 1876, the Dexter mill was crushing ore and recovering silver using the Washoe process, in which salt and other chemicals were ground into the pulp of finely crushed ore and water, in order to make the silver amenable to amalgamation with mercury. Farlin was so deeply indebted to Clark that the banker actually controlled the mill. When the mill treated a lot of ore from a different mine, the mill showed poor recovery of silver. In June, Clark decided to complete the construction of reverberatory furnaces Farlin had started. The furnaces would be used to roast silver ores with salt using the Reese River process to chloridize the silver-sulfide minerals (silver sulfide will not amalgamate with mercury, but silver chloride will) and also to oxidize the other minerals so they would not hamper the amalgamation process. By July, Clark had the mill and some of the furnaces operating.[22]

The return of spring weather in 1876 had brought a return of active mining and allowed several other construction projects to get underway as well. Despite Farlin's setbacks in getting his mill operational, there was new enthusiasm for quartz mining in Butte. By April, there were more than one hundred miners working for about thirty partnerships and small companies in as many small silver mines. There were still about sixty men working placer mines in the area. Copper was drawing considerable attention as well. Butte had fifteen small companies employing over sixty miners and working veins in more than a dozen copper mines.[23]

The miner credited with initiating Butte's rise as an important copper producer was Bill Parks, who also achieved his long-awaited success in 1876. He had continued throughout the early 1870s to sink his shaft on the Parrot lode, hoping to strike copper ore rich enough to ship or treat. In early 1876, by which time Parks's shaft was about 150 feet deep, he drove a crosscut 200 feet from the shaft and struck a vein of copper glance (chalcocite, Cu_2S) 4 feet wide. Using a simple windlass, Parks and two men were able daily to hoist a ton or more of ore containing 40–60 percent copper. Parks placed the ore in sacks and freighted them by wagon to Corinne. Railcars then took Parks's ore to Baltimore for smelting. He claimed that lack of freight wagons heading toward Utah was the main factor limiting the amount of ore he produced.[24]

As the nation's July 4 centennial approached, the community was convinced it was truly embarking on a new era of prosperity. Butte citizens decided to stage an elaborate Independence Day celebration, complete with a parade, fireworks, and speeches.[25] The main speaker was Charles S. Warren, who had worked in the County Recorder's Office during Butte's first boom. He was now a prominent citizen, and he was called upon to recite the camp's history. After crediting Farlin with bringing Butte back to life and recounting how Parks's persistence had paid off, Warren named the other mines that likewise showed promise. He concluded with these remarks:

> The future is indeed promising, and within a short time will be heard the music of a hundred whistles of quartz mills—the horizon will be clouded with the smoke of scores of furnaces. Freyer and his process will be firmly established in the camp, and the fondest dreams of the most excitable will be thrown in the shade. Never before in any part of our mining country has such rich mines [*sic*] been discovered for the work done. Take a view of the camp. For four miles around Butte is but a net work [*sic*] of rich lodes; gold, silver, copper and lead abound and in such vast quantities that we cannot help but be the greatest mining center in the west, if not the world. Virginia City and the mines of Nevada, now producing millions of precious metals, will be eclipsed. All we need is transportation and works for the reduction of our ores. To you of faint heart and who feel discouraged, who for years have worked patiently and hard without a murmur, we would say, be of good cheer, hold your grip, for day is dawning. Butte will come out, and with flying banners. Here will be such a community in the near bye-and-bye as you do not dream of. The camp is now getting known throughout the world, and were it not for our isolation, a few weeks would see the capital of the world seeking investment here.[26]

Warren's speech is in many ways typical of the boosterism that sprang across the mining frontier.[27] The quest for wealth was a key factor behind the two-century movement of significant portions of the nation's population westward, and many nascent communities promised that they were the El Dorado where that wealth could be claimed. Few could fulfill their promises. What sets Warren's boosterism off from so much of its kind is that his predictions largely came true. All Butte lacked at the time was transportation and reduction works. Soon thereafter such infrastructure arrived, and Butte indeed became the greatest mining center in the nation and arguably the world for a time. And once the railroads ended Butte's isolation, some of the world's most powerful capitalists sought investments there. Try as they might, then, most of Butte's early mine owners lost their grip—on their valuable mining properties at least.

Warren's speech is noteworthy for another of its predictions: Butte's horizon would soon be clouded with smoke from scores of furnaces. Indeed, the smoke grew so thick on some days that streetlights burned at noon, folks lost their way in the city during daytime, and deaths from respiratory ailments skyrocketed. But that problem was more than a decade away. Before Butte became engulfed in clouds of sulfurous and arsenical fumes, the hazy image in the imagination only foretold prosperity. Warren and his listeners were inspired by images of a smoke-filled horizon, and they spent the rest of 1876 toiling on and under the Butte hill to realize their dreams.

During the 1876 freighting season, Butte's copper miners shipped about 650 tons of copper ore and about 350 tons of silver ore. For hauling that combined 1,000 tons of ore, Butte miners paid freighting companies about $35,000, or about $35 per ton. For the ore they shipped, Butte miners received about $150,000, meaning they paid nearly a quarter of their gross revenue merely to haul their product out of Butte's remote location. With the remaining 75 percent of their gross earnings, miners had to pay for railroad transportation and smelting costs as well as for their own mining costs back in Butte. In many cases, miners could do little more than break even.[28]

It was obvious, then, that to earn a profit in mining at Butte, effective means had to be developed to treat those ores locally. Individuals like W. A. Clark were determined to do just that. By the end of 1877, Clark had taken complete possession of both the Travona mine and the Dexter mill. He continued to operate the mill at a profit for several years, treating his own ore as well as that of other miners. The roasting furnaces were not effective, however, likely because reverberatory furnaces used for roasting are very labor intensive, and other roasting furnaces were being developed that were mechanized and less costly to operate. The Dexter operated for most of its history using the wet process, also known as the Washoe process.[29]

Butte as a Silver-Mining Camp

The Dexter was often said to have been the first successful stamp mill in Butte, but John How's Centennial mill actually started treating silver ore on a somewhat regular basis about six months before the Dexter. John How was a grocer who had owned and operated outlets in Virginia City and Helena before building a ten-stamp mill in Brown's Gulch, northwest of Butte, in 1868. It only ran about three months and crushed only about seventy tons of ore. Eight years later, after incorporating the Centennial Gold and Silver Mining Company, How moved his mill to the north bank of Silver Bow Creek, just south of Butte City. He renamed his works the Centennial mill in honor of the nation's approaching one hundredth birthday. The mill first went into service with five stamps dropping in February 1876, but mechanical troubles kept the mill from operating full time.[30]

The third silver mill to herald the revival of quartz mining in Butte was that of Andrew Jackson Davis, a shrewd banker and merchant based in Helena. In 1868, A. J. Davis had bought a half-interest in the Allie Brown lode, high on the Butte hill. That same year, Charles Hendrie and Harvey Ray built Butte's first quartz mill with financing from Davis. Located near what would become the southeast corner of Broadway and Arizona streets, the ten-stamp mill had primarily been intended to treat gold ore, but it failed to recover enough gold or silver to be profitable. In 1872, Davis acquired Hendrie's assets, including the dormant mill at Butte. For a few years, Davis did little with his new property other than sell a battery of five stamps to another party. By 1876, Davis had acquired the Lexington claim, located on the same lode as the Allie Brown. Both mines promised to be rich silver producers. Davis began to convert the old Hendrie mill to process silver ore, equipping it with ten new stamps and with the pans (cylindrical vats in which the crushed ore could be further ground with mercury, salt, and bluestone, or copper sulfate) necessary for the Washoe process of pan amalgamation. The rehabilitated mill, known variously as the Lexington mill and the Davis mill, began dropping stamps on ore in February 1877.[31]

During its operating life, the Centennial mill probably treated less than 10,000 tons of ore, producing a like volume of tailings. The Dexter mill probably treated about 50,000 tons of ore, and the Davis mill somewhat more than that.[32] Although the 100,000 tons of tailings or more discharged by these early amalgamating mills may seem to be a large volume of solid material to have been released into Silver Bow Creek, it would soon be dwarfed by the pollution of the upper Clark Fork from the later silver mills and especially the copper concentrators.

Soon after Butte's three initial silver mills began operating, several other small entrepreneurs were emboldened to try their hands at the Washoe process. They built the Burlington, Clipper, Grove Gulch, and Gagnon mills. As with the Centennial, Dexter, and Davis mills, these other small mills treated relatively little ore. The Burlington may have treated more than 15,000 tons of ore, but the other three certainly treated less than 10,000 tons each and were short-lived.[33] Although tailings from these four mills undoubtedly polluted Silver Bow Creek as well, their contributions to any downstream damage to property or ecosystems was relatively insignificant compared with what was to come.

Of greater significance was the outflow from Butte's second wave of silver mills, facilities that had greater capacities, were more technologically advanced, and operated for longer periods. The first of Butte's large silver mills, and certainly the one with greatest output of tailings, was the Alice. Rolla Butcher had located the Alice claim on the northern part of the Butte hill in January 1875. The claim was situated along the Rainbow lode, which would prove to be one of the richest silver veins beneath the Butte hill. As was so often the case, however, Butcher did not have sufficient capital to develop a mine, so he offered it for sale. The Walker brothers, prominent mining investors in Salt Lake City, bought the Alice and some adjoining claims in 1876 on the recommendation of Marcus Daly. Manager of some of the Walkers' profitable mining properties in Utah's Ophir mining district, Daly was an Irish immigrant who was quickly proving himself one of the most skilled practical mining men in the West. The Walkers had been impressed with some of the Butte ore that had been shipped to Salt Lake City, so they sent Daly to investigate the Montana camp in August 1876. Liking what he saw, Daly negotiated the purchase of the Alice and some other Butte claims on behalf of the Walkers the following month. They placed him in charge of developing their new Butte properties.[34]

The Walkers freighted equipment for a fifteen-stamp mill from Utah to Butte in July 1877, and it went into operation in October. Initially, the Walkers operated the mill without roasting the Alice ore, because it was extracted from the shallower depths of the mine and was therefore already oxidized. By mid-1878, the mill had produced one hundred bars of bullion worth a total of about $250,000. The Alice added another five stamps that year. In 1880, when the shaft reached below the water table and was encountering silver-sulfide ores, the Alice company installed two roasting furnaces to improve the recovery of silver from the unoxidized sulfide ores found at depth.[35]

Below the water table, Butte's silver ores had not been oxidized through weathering. They therefore remained as sulfides of silver. Mercury does not amalgamate effectively with silver sulfides, whereas mercury amalgamates readily with silver chloride. Therefore, to achieve effective recovery of silver, the silver-sulfide minerals had to be chloridized. To accomplish chloridization, ore in a late nineteenth-century mill was crushed and then roasted with common salt (NaCl). Roasting brought the ore or pulp to a very high temperature, but not hot enough to melt the material. The chlorine in the salt swapped places with the sulfur in the silver sulfide to yield silver chloride, with the sulfur going up the stack as sulfur dioxide.[36]

Operators of early roasting mills employed "craft" knowledge, as opposed to "scientific" knowledge. Rather than measuring furnace temperature in degrees, they observed the temperature of the pulp by the cherry-red color it assumed. Early methods of chloridizing crushed pulp included the Freiberg process, named for the German city where it was first scientifically studied, and the Reese River process, named for the mining district in Nevada where the Freiberg process was adapted to American conditions. To accomplish the roasting, both the Freiberg and Reese River processes relied on hand-rabbled (stirred with handheld rakes) reverberatory furnaces, which were very labor intensive. Several American metallurgists became the namesakes for the improved roasting furnaces they devised. The Howell, Brückner, and White furnaces featured revolving horizontal cylinders through which combustion gases passed to heat crushed ore to the roasting temperature. The revolving motion obviated the need for hand-rabbling. The Stetefeldt furnace was a vertical shaft furnace, similar to a cupola used for smelting, but it did not bring the ore to the melting point. Ore was charged through the top and, after sufficient roasting, drawn out the bottom. It was to these improved roasting appliances that the Alice and Butte's other successful silver mills soon turned.[37]

The Alice mine and A. J. Davis's Lexington mine were both located in Walkerville, a small camp north of Butte City and near the top of the Butte hill. In June 1878, faced with high costs for equipment and supplies because of high transportation charges, and with labor costs high as well, managers of the two Walkerville mines decided to reduce wages from $3.50 to $3.00 per day for miners and from $3.00 to $2.50 for surface workers, including men working in the mills. In response, workers at the Alice and the Lexington quit their jobs, gathered supporters, and marched behind a brass band from Walkerville down to Butte's business district. There were about four hundred marchers in the parade. Within a few days, sixty-five men formed the Butte Workingmen's Union, writing a constitution patterned after those of the pioneering unions on Nevada's Comstock lode. Workers at the Alice and the Lexington held out until the end of July, when the employers acceded to the workers' demands and returned the wage scale to previous levels. The union, with a membership by 1881 of more than eight hundred, changed its name to the Miners' Union of Butte City. Three years later, with the growth of copper mines in Butte, the union's membership reached 1,800, surpassing the size of the miners' union at Virginia City, Nevada, and becoming the largest in the West. By the late 1880s it had established a tacit closed shop in all the Butte mines. The Miners' Union looked out not only for the welfare of its own members; together with the Knights of Labor it helped in 1887 to form a federation of other Butte labor unions called the Silver Bow Trades and Labor Assembly.[38]

In 1878, four local miners, doing business as the LaPlatta Company, built a twenty-stamp silver mill along the west side of Silver Bow Creek, at the east end of the Butte hill. In 1880, they added roasting capacity to their facility, called the Silver Bow mill.[39] In the early 1880s, the Alice and the Silver Bow mills would be joined in the Butte vicinity by two other large silver mills that roasted ores: the Moulton and the new Lexington mill, which Davis built next to his Lexington mine in Walkerville. The combined tailings output of these four mills, estimated at about 1 million tons, was considerably greater than that of the smaller mills using the Washoe process.[40] Yet even that large volume would soon be eclipsed by Butte's copper concentrators, some of which would individually produce more than a million tons of tailings.

Butte as a Copper-Mining Camp

Tailings from treating copper ores had barely begun to foul the waters of Silver Bow Creek by the close of the 1870s. Butte's copper industry was beginning to awaken simultaneously with the silver industry during the mid-1870s, but most of the copper ores were shipped elsewhere without initial treatment in Butte. One of the important early developers of copper properties was W. A. Clark, who had capitalized William Farlin's Dexter mill and soon came to own it. A native of Connellsville, Pennsylvania, Clark had caught the national mining fever in 1862, traveling to the Colorado camps. The following year, he joined the gold rush to Bannack, Montana, where he worked a placer claim. Rather than simply spending his earnings, he invested them in mules and a wagon so he could freight supplies from Salt Lake City and sell to fellow miners in Montana. Clark soon left active mining to become a full-time businessman. In addition to his mercantile business, he started a stagecoach line and gained a contract to haul the U.S. Mail. Clark formed a partnership in 1868 with R. W. Donnell, a prosperous grocer with stores in Helena and Deer Lodge. With their profits, Clark and Donnell established a bank in Deer Lodge in 1870. As president of the bank, Clark turned his attention to making loans to small mining entrepreneurs, acquiring a variety of properties in the manner by which he would acquire Farlin's mill. In 1872, he traveled to the still sleepy mining camp of Butte and began purchasing copper mines, including the Original, Colusa, and Gambetta, each of which eventually became a large copper producer.[41]

To expand his understanding of the mining and smelting of copper ores, Clark took some ore samples from his new copper mines and set off for New York City, where he enrolled in the nation's leading mining school at the time, Columbia University's School of Mines. He did not seek a degree but rather enough technical understanding to be able to develop his mines profitably and to converse knowledgeably with the smelting companies to which he would be shipping ore. He claimed to have been the first to ship large quantities of copper ore from Butte. It cost him about sixty-five dollars per ton to freight ore by wagon to Utah and then by rail to the East Coast. In 1877, Clark visited N. P. Hill's works at Black Hawk, Colorado, ore samples in hand, to see if Hill would offer more attractive smelting rates. Hill saw that Clark's ore, rich in both copper and silver, would complement his Colorado supplies nicely. In 1878, Clark began shipping Butte ore to Black Hawk, inaugurating a business relationship lucrative for both men.[42]

Bill Parks had struck exceptionally rich copper at the Parrot in 1876, but he and several other miners were also finding an abundance of ore bearing less than 30 percent copper, considered the threshold below which it was not remunerative to ship. Such ore would eventually be considered very valuable, once effective reduction works were built locally, but ore with less than 30 percent copper did not return enough payment from distant smelters to cover what it cost to ship it from Butte. One possible solution to the problem was to concentrate the ore, discard tailings bearing relatively little mineral value, and ship the concentrates to smelters elsewhere.

Concentrating copper ores required the three factors involved in industrialization: mechanization, effective organization of a production worksite, and steam engines for powering the operation. (As with many frontier applications of steam power, the initial fuel in Butte was firewood, but in time coal would be the preferred fuel.) Giles Olin was the first in Butte to attempt to concentrate copper ore, announcing his building plans in April 1876.[43]

Olin had lost his legs to amputation after they were frozen one winter. Nevertheless, he acquired a millsite on the east side of Silver Bow Creek near the base of the Butte hill. In 1876, he ordered concentrating equipment from Salt Lake City. Struggling on artificial limbs that he made himself, Olin built a concentrator, hoping that, if his works were successful, he could erect a smelter on the site as well. After numerous delays caused by the difficulty of freighting equipment from Utah, Olin started his concentrator in July 1876. He treated copper ore from several mines, including from Parks's Parrot mine. The Olin concentrator employed a steam-powered rock crusher, Cornish rolls, two hand jigs, and concentrating tables.[44] Water was supplied to the jigs from Silver Bow Creek. Each hand-operated jig could treat about a ton of ore in twenty-four hours. Initially, concentrates were sacked and sent to Baltimore. In 1877, by which time the capacity of the Olin concentrator had increased to about 18 tons of ore per day (producing about 9 tons of concentrates daily), much of the product was being freighted about forty miles to Glendale, Montana, where the Hecla Mining Company's short-lived reduction works was smelting ores and concentrates from Butte. Olin's operation did not last long, however. The Olin concentrator closed for the winter in December 1877, at which time it was being leased and operated by another party. It apparently never reopened.[45]

The first of Butte's smelters to offer a realistic alternative to shipping copper ore to distant smelters, or over the Continental Divide to the Glendale smelter, was known as the Colorado smelter. The inaugural local smelter to successfully treat Butte's complex copper-silver ores, it was located along the south side of Silver Bow Creek southwest of Butte and northwest of Timber Butte. The Colorado smelter was built in 1879 on a site where R. K. Williams had tried to operate a smelting furnace. By March 1879, however, his effort was a recognized failure because the brick he used to build his furnace was not suitable for smelting. He announced his intention to try again, using a better grade of brick, but in June 1879, the Colorado and Montana Mining Company bought his smelter, making plans to replace his smelting furnace with two 12-ton reverberatory furnaces built of brick equal to the task.[46] The Colorado and Montana derived its name from N. P. Hill's operation in Colorado.

The new Butte smelter had greater chances for success because of its association with Hill's Boston and Colorado Smelting Company, which had been successfully smelting base ores at Black Hawk since 1868. Two of Hill's metallurgists were Richard Pearce and Henry Williams (no known relation to R. K. Williams), Cornishmen who had first moved to Colorado to work for British smelters. Soon after Clark had begun regular shipments of rich Butte ore to Black Hawk in 1878, increasing shipments of nearby Colorado ores, coupled with Clark's deliveries from Butte, caused Hill to relocate his smelter from Black Hawk to Argo, near Denver. Construction began that year. About the same time, Hill sent Henry Williams to Butte to explore the possibility of starting a smelter there.[47]

Henry Williams was born at Cornwall in 1840. He studied at the Truro School of Mines and at the Clausthal Academy in Germany before going to work for Vivian and Sons, a renowned smelting concern at Swansea. About the same time Hill was starting the Boston and Colorado, Williams and others formed an English corporation to smelt copper in Colorado. In the face of stiff competition, though, Williams's operation ceased, and he went to work for Hill in 1875.[48]

Hill's new Argo works began smelting at the beginning of 1879. Upon Williams's favorable report on the situation in Butte, Hill sent him back to Montana in 1879 to purchase the R. K. Williams smelter and rebuild it. Hill, Clark, Richard Pearce, Henry Williams, and another Hill associate incorporated the Colorado and Montana Smelting Company under the laws of Colorado in June 1879. The purpose of the new Colorado and Montana smelter was to reduce Butte copper ores to matte containing 50 percent or more copper and rich in silver. Matte from Butte could then be shipped to the Argo works for further treatment at much less cost than Butte ores of lower grade could be shipped. Clark continued to ship high-grade Butte ores directly to Colorado. Both the Boston and Colorado at Argo and the Colorado and Montana at Butte produced copper matte primarily as a vehicle for recovering gold and silver.[49]

The Cornish background of Hill's metallurgists was evident in the choice of a smelting method for the new Colorado smelter at Butte; they built two reverberatory calcining (roasting) furnaces and two reverberatory smelting furnaces. Rock crushers and Cornish rolls crushed the ore to a size suitable for roasting and smelting, but the plant did not concentrate the ore prior to roasting. Thus, the Colorado smelter did not produce any tailings in this early period. The principal product was a silver-rich copper matte suitable to be ground, sacked, and shipped to Colorado. Byproducts included smoke from the roasting and smelting furnaces and slag from the smelting furnaces. The smoke carried sulfur and arsenic from the ore, volatile materials given off in the combustion of fuel, and fine dust consisting of many of the same elements originally bound in the ore. Slag contained primarily silica and iron, but also some amounts of the other elements originally present in the ore, including copper.[50] Less than a decade after the Colorado smelter opened, these byproducts, especially smoke, would become the source of great controversy in Butte. In 1879, however, they were yet another sign that Charles Warren's vision of Butte's future might well be realized.

When the Colorado and Montana smelter began operating in August 1879, it had a capacity to treat 20–25 tons a day. It was strictly a custom smelter, purchasing base-metal ores from miners and paying them according to the silver content. The lowest grade it would buy was ore with 50 oz. of silver per ton, for which the company paid $8.50/ton delivered to the works. For ore with 100 oz./ton, the smelter paid $60/ton delivered, and for ore with 200 oz./ton it paid $160/ton delivered. The smelter also paid miners for the value of the copper in the ore. In addition to copper ores, the smelter purchased argentiferous (silver-bearing) manganese ores and galena (lead sulfide) ores. While the smelter treated ore from several mines, including the Twin Stevens, Mountain Boy, Star West, and Morning Star, its largest supplier by far was the Gagnon mine, John C. C. Thornton's property just north of the county courthouse in Butte. A former Confederate officer, Thornton had moved to Deer Lodge in 1866 and to Butte in 1875. After practicing law for three years, he turned his attention to mining and acquired the Gagnon mine.[51]

In 1880, the Gagnon was producing 50 tons of ore daily, about 30 tons of which went to the Colorado smelter. This amount greatly exceeded the smelter's capacity. Not all the ore was smelted at the Colorado smelter, however. Much of it, the high-grade material, was simply crushed there before being sacked for shipment to smelters elsewhere. Some of the ore produced near the surface at the Gagnon was free-milling ore, meaning it yielded its silver by amalgamation with mercury without roasting or other treatment. In the spring of 1879, Thornton and his Gagnon Company built the Gagnon mill near the Gagnon mine, but as already described, the Gagnon

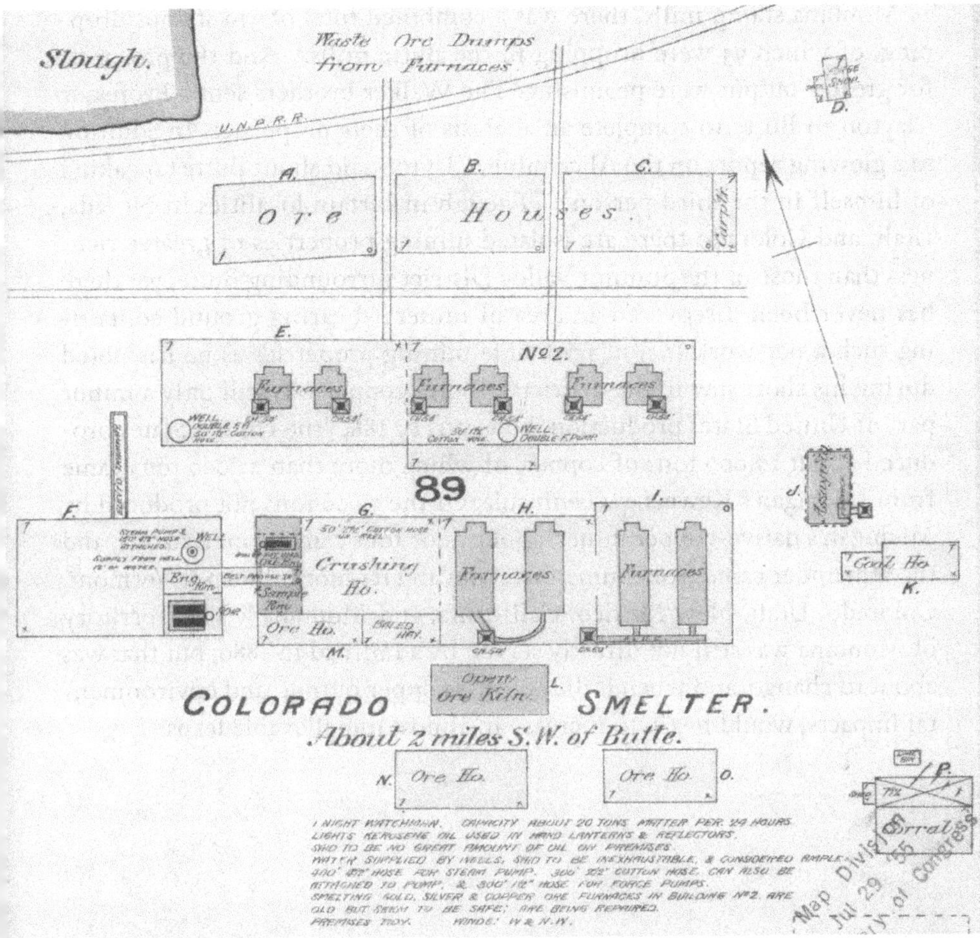

FIG. 3.1. Colorado Smelter. This image from the 1884 Sanborn Fire Insurance Map for Butte shows the Colorado smelter as it stood five years after opening in 1879. Note the locations of the furnace stacks, each of which is represented by a square with an X inside. Sanborn Fire Insurance Map, Butte City, Montana, 1884, sheet 9, Library of Congress, Geography and Map Division, Sanborn Maps Collection.

mill was short-lived. After the Gagnon mill closed, all Gagnon ores went to the Colorado smelter.[52]

By the close of the decade in 1880, Butte had emerged as Montana's largest silver-mining camp. Montana had twenty-six stamp mills operating that year, of which eight were in and around Butte. Only one other mining district in the territory had as many as three stamp mills. At those

26 Montana stamp mills, there was a combined total of 319 stamps dropping, of which 93 were dropping in the Butte mills.[53] And the prospects for greater output were promising. The Walker brothers sent a Professor Clayton to Butte to complete an analysis of their properties. In addition to a glowing report on the Alice mine, Clayton said about Butte (speaking of himself in the third person): "Though in certain localities in Nevada, Utah, and Colorado there are isolated mining properties of greater richness than those of the Summit Valley District surrounding Butte, yet there has never been discovered an area of mineral-bearing ground containing such a net-work [*sic*] of profitable mining properties as he has noted during his short stay in the district."[54] Butte copper was still only a minor part of United States production, however. In 1880, the United States produced about 27,000 tons of copper, of which more than 22,000 tons came from Michigan's Keweenaw Peninsula. Of the 5,000 tons not produced by Michigan's native-copper mines, about 1,000 tons came from Arizona, and the remainder came from numerous states and territories, such as Vermont, Colorado, Utah, New Mexico, California, and Montana.[55] The Territory of Montana was still not directly served by a railroad in 1880, but that was about to change, and when it did, Butte's copper output, and environmental impacts, would begin to increase in almost unbelievable leaps.

CHAPTER 4

Amazing Growth of the 1880s

> While the concentration works now in operation in Butte, as well as those in process of erection, are by no means model establishments, they still do very good average work, and are perhaps quite as well adapted to the peculiar conditions as more elaborate works. The loss of copper in the tailings is considerable, and under eastern conditions could not be tolerated; but it must be considered that in Butte the ore is the cheapest thing we have, while fuel, labor, and machinery are extremely expensive, and that the copper contained in the ore bins has cost so little to mine that it does not acquire any considerable value until a certain amount of labor has been expended on it.
>
> While I do not deny that a more perfect sizing, a graded system of crushing, and a more systematic effort to concentrate the finer classes of ore might even now prove remunerative, I doubt very much if it would pay expenses to greatly extend the present system of slime treatment, in which department the greatest losses [of copper] occur. I can, however, assert positively that at least one of the concentrating mills in coarse of erection is so arranged as to carry out a considerable extent the improvements just discussed.[1]
>
> —E. D. Peters, 1884

Butte's copper industry sprang to prominence in the 1880s, growing from minor output, which was barely noticeable to any but the most scrupulous observers in 1880, to production greater than that of any other mining district in the world in 1887. Butte's fabulously rich copper ores were the basis of its copper boom, and those ores helped the camp overcome its disadvantages: remote location and consequent high transportation costs for fuel, equipment, and output; high labor costs; and available technologies that led to acknowledged waste of potential product (waste that would come to be a source of environmental damage as well). As E. D. Peters reported, Butte's metallurgical engineers were working on those problems, but the surest way to profit, he asserted, was to continue mining and processing ore in the meantime. By the end of the decade, the volume of waste, in the form of tailings, would become a bit of a problem for Butte's mining companies, but the tremendous wealth of Butte's ore deposits, and the potential they heralded for seemingly unlimited growth, captured the imagination of Butte's boosters more than the emerging problems with tailings.

To understand how the mining and metallurgical engineers working for the various Butte smelters (and later Anaconda and Great Falls smelters) approached their work, it is helpful to understand their concept of "waste."[2] Industrial waste, in the mineral industries and elsewhere in the U.S. economy at the time, generally referred to the inefficient utilization of natural resources. Mining and metallurgical engineers set the elimination of waste as one of their major tasks soon after the formation of their professional organization in 1871. Three Pennsylvanians—E. B. Coxe, R. P. Rothwell, and Martin Coryell—initiated the first meeting of the group they proposed to call the American Institute of Mining Engineers (AIME). Coxe, Rothwell, and Coryell suggested the organization's principal objective would be to exchange information about two essential topics: "1st. The more economical production of the useful minerals and metals. 2nd. The greater safety and welfare of those employed in these industries."[3] They understood that one way of maintaining economical production was to eliminate waste by more effectively utilizing resources. An ore, for example, might hold 4 percent copper. If processing the ore recovered 75 percent of the available copper, that meant 25 percent of the copper was discarded, and the discard was called waste because it could no longer be recovered and therefore contribute to the enterprise's revenue stream.

The first exposition of a set of values concerning mining and mineral wealth given by an AIME member before his peers was Abram S. Hewitt's presidential address, delivered in 1876, the year of the United States' centennial celebration. Titled "A Century of Mining and Metallurgy in the United States," the address included a description of significant events in U.S. history related to sources of the nation's mineral wealth. Hewitt also made recommendations to his peers for responsible actions they should take to further advance the nation and for the moral improvement of mine workers. He was firmly opposed to government participating in the mining enterprise to any extent beyond conducting surveys to discover where minerals were located. He cautioned that strict reliance on the capitalist system, however, would not automatically move society toward perfection;

it required the responsible and conscientious efforts of those in positions of management to work toward the common welfare. Such, for example, was the case with the waste of mineral resources, one of Hewitt's major concerns. Government regulation to prevent waste, while justified under reasoning comparable to that prohibiting the waste of timber or the hunting of game during certain seasons, should be a last resort, he argued. He hoped that the waste of mineral resources was merely a symptom of the nation's youth and that, by transferring mines from the public domain to private hands, selfish interest would impel miners to use their resources efficiently.[4] This is the same engineering ideal of efficiency that Hugh Gorman describes being used by the early oil industry to argue against government regulation to prevent pollution caused by that industry.

Early professional mining engineers' concerns over the waste of resources presaged the major resource conservation theme of the Progressive Era at the turn of the twentieth century. Progressives believed in the efficacy of experts solving society's ills. In the realm of natural resource extraction, expert mining engineers would have the responsibility to rid the nation of needless waste.[5] This theme was frequently revisited in presentations given by AIME members and officers. For example, Eckley B. Coxe, one of the founders of the AIME, discussed the matter of waste in his presidential address of 1879: "One of the greatest triumphs of science, one of the points in which modern civilization distinguishes itself most from the ancient, is *the utilization of waste products* [emphasis in the original]—in turning that which formerly was either useless or hurtful to mankind into a source of wealth or a thing of beauty. If he who causes two blades of grass to grow where one grew before is to be praised, how much more he who converts some useless and fever producing swamp, by drainage and cultivation, into a beautiful meadow?"[6] Because their practice was embedded in capitalist enterprise, however, engineers were seldom able to implement techniques for achieving their ideal regarding waste unless those techniques proved profitable.

With profitability as the driving force in determining how mining and metallurgical operations were conducted, engineers would sometimes knowingly waste desired resources in order to achieve greater economies. Such was generally the case at Butte (as described by Peters at the beginning of this chapter) and elsewhere when the decision was made to use gravity concentration on ores before smelting. Smelting recovers a very large percentage of the metal in ore, but the cost of smelting is dependent on the volume of material treated, not on the output of product. As the percentage of copper in ore decreases, it becomes prohibitively expensive to smelt the ore to extract copper, unless much of the gangue can be removed before smelting. Gravity concentration relies on the fact that more dense mineral-bearing particles will behave differently in water than less dense particles of gangue. Gravity concentration is relatively inexpensive, per ton of material treated, and separates relatively high-value minerals from low-value rock (gangue), which is then discarded as tailings. Gravity concentration is not 100 percent effective, however, so some of the copper originally present in the ore remains in the tailings. The resulting high-grade concentrates can be profitably smelted, even though more copper is wasted in the tailings than would have been wasted had all the ore been smelted directly. This concept has been understood since the earliest of texts described the metallurgy of copper.[7]

As will be seen, when early concentrator operators in Butte discharged tailings, they knew they were discarding copper. The first successful copper concentrator at Butte was opened by the Montana Copper Company in 1881 next to its Colusa smelter. The plant later became known as the Boston and Montana upper works. In the concentrator's early years of operation, Montana Copper Company ores were so rich in copper minerals, while available concentrating techniques were ineffective enough in separating the copper minerals from gangue, that Montana Copper was able to operate profitably despite tailings from the concentrator assaying about 4 percent copper (a grade that in a few years would be considered relatively high-grade ore in Butte). As was typical in the Butte district, the Montana Copper Company expanded the capacity and upgraded the machinery of its concentrator in the years immediately following its construction to both increase production and increase the percentage of copper recovered. Knowing that its old tailings still had the potential to yield values, the Montana Copper Company nevertheless continued to focus attention on treating newly mined ore, because there was more money to be made from doing that than from reworking old tailings. Meanwhile, those copper-laden tailings simply got in the way of continuing to mill fresh ores, so the concentrator operators made an effort to have Silver Bow Creek wash the tailings away.[8]

The Boston and Montana would later close its upper works in 1893 and begin shipping ores to a new smelter along the Missouri River at Great Falls. The Great Falls site was chosen, as described in the next chapter, because the asset of an ample water supply for power, ore dressing, and tailings disposal more than offset the added cost of shipping ore 150 miles for smelting. When the Boston and Montana closed its upper works at Butte, there was a sizeable tailings dump next to the creek. The dump existed, though, not because the Montana Copper Company and the Boston and Montana had been trying to save tailings but because there was insufficient water flowing in Silver Bow Creek to carry all the tailings away.[9]

Butte's copper output continued to grow in the 1880s, and more concentrators and smelters opened, despite warnings by many in the industry that Butte's mines could not sustain the accelerating production. A simple economic explanation for Butte's amazing growth might be that growth in demand for copper drove up the price of the metal, and Butte's miners were merely benefiting from the high prices. The actual history, however, is more complicated, and more interesting. Across the decade, as noted by historian Kent Curtis, there was a dramatic drop in the average price of copper from well over twenty cents per pound in 1880 to about fifteen cents per pound in 1890, with the price dropping below twelve cents per pound in 1885. He argues, accurately, that this falling price of copper helped stimulate the growth in electrification in the U.S., rather than electrification stimulating growth in copper production.[10]

Despite the falling price of copper, Montana's annual copper production (meaning Butte copper production) grew by an order of magnitude from 9,058,284 pounds in 1882 to 112,980,896 pounds in 1890, and not because Michigan production was falling. Across those same years, Michigan copper production almost doubled, from 56,982,765 pounds in 1882 to 101,410,277 pounds in 1890, and total U.S. production grew more than fivefold, from 90,646,232 pounds to 259,763,092. In 1890, Montana's copper production stood at 43.5 percent of the national total, and Michigan accounted for 38.9 percent. Those Michigan companies were investing capital in mechanization just as the Butte companies were, and their investments allowed them to continue to be profitable even as the price of copper was going down.[11]

Through the middle of the decade, observers tried to understand how the Butte copper companies could stay in business in the face of declining prices, especially in light of Butte's high costs of labor, fuel, and transportation of copper matte or high-grade copper ores to refineries in England and the eastern U.S. Some observers thought that the high silver content of Butte ores might be the secret to Butte's success, but inspection of annual reporting of the Butte companies showed that only two of them enjoyed argentiferous ores. One development that improved the prospects for Butte copper was the introduction of converters at the Parrot Silver and Copper Company's smelter in 1884, which allowed for the bessemerization of copper matte (described below). The new technology allowed the Parrot to ship converter copper to refineries. Converter copper was more than 99 percent pure, whereas copper matte was only about 60 percent copper.[12]

At the same time, low copper prices were having a pronounced effect on other facets of the U.S. and global economies, notably that the consumption of the red metal skyrocketed. In 1885, U.S. fabricators of finished products of copper, brass, and bronze used about as much copper (84,917,170 pounds) as the U.S. exported to other countries (86,619,951 pounds). The market for copper output varied by region. In 1885, most of Michigan's output went to U.S. consumers, whereas most of Montana's production was exported. One segment of the fabricating industry that was growing rapidly included the plants in the U.S. that drew copper wire, supplying the electrification of the U.S. economy, a development that began in earnest when Thomas Edison commenced operations at his central generating station in New York City in 1882.[13]

The 1880s was a decade of impressive growth in American industry and the economy. Railroad systems expanded tremendously. After laying about 63,000 miles of track in the 1870s, the nation added another 92,500 miles in the 1880s, as several other companies joined the Union Pacific with transcontinental lines. The railroads' expansion spurred comparable growth in other sectors of the economy, as manufacturers received new orders for locomotives, and steelmakers enlarged their works to respond to unprecedented demand for rails. During this period, more than half the nation's iron and steel production was devoted to rails, and by 1890 the United States' output of pig iron surpassed that of England. The 1880s were also the dawn of the electrification of the American economy. As American businesses grew, they also reorganized, and the country began to witness an accelerating rate of corporate mergers, one of the most noteworthy of which was the formation of the Standard Oil Trust in 1882. In response to the trend in mergers, New Jersey passed legislation enabling the incorporation of holding companies and setting the stage for further corporate consolidations, ripples of which would soon have a major impact on the Montana copper industry. Completion of the transcontinental railroads helped further the agricultural settlement of the American West, culminating at the close of the 1880s with six new states joining the Union: North Dakota, South Dakota, Montana, Wyoming, Idaho, and Washington.[14] In Montana, the 1880s was not only the decade of statehood; it was the decade during which Butte became the mining district with the world's largest annual copper output. It was also the decade in which the discharge of metallurgical wastes began to assume problematic proportions.

Prior to the railroad era, remote mining districts in the United States had little chance of developing unless they were close to large bodies of water necessary for transportation of metals to markets. The nation's first two important lead-mining districts—the Upper Mississippi District of northeast Iowa, northwest Illinois, and southwest Wisconsin, and the Disseminated Lead District of southeast Missouri (the Old Lead Belt)—had begun to be exploited during the colonial period, and they benefited by proximity to the Mississippi River. The nation's first major western copper-mining district, on Michigan's Keweenaw Peninsula, could begin to be profitable in the late 1840s only because transportation was available from Lake Superior down through the Great Lakes to the Erie Canal. When railroads began to offer an alternative to waterborne transportation, mineral resources were often an attraction; for example, Galena, in the heart of northwest Illinois' lead-mining district, was the proposed western terminus of Chicago's first railroad, chartered in 1836 to construct a line between Chicago and Galena. Similarly, when mining commenced in Montana in the 1860s, steamboats on the Missouri River provided the main link to the United States east of the Mississippi. It took two decades for railroads to reach Montana, although completion of the Union Pacific's transcontinental line in 1869 began to provide an alternative to the Missouri River system for the mining camps in Montana. Both routes, nevertheless, required considerable overland travel by freight wagons.[15]

William Cronon's *Nature's Metropolis* delineates the crucial role that railroads played during the United States' development in the second half of the nineteenth century in connecting the metropolis (especially Chicago, in his narrative) with the hinterlands of the North American continent, for linking manufacturers and processers of the metropolis with the raw materials and dispersed markets of the hinterlands. The vast mid-continent hinterlands of Cronon's narrative are the fields of grain that farmers established on the prairies, the rangeland for raising cattle on the high plains, and the vast timberlands of the upper Midwest. William Robbins, in *Colony and Empire*, shows how the Twin Cities in Minnesota, Denver, San Francisco, and Portland developed along the Chicago model as metropolises for the hinterlands of the American West.[16] Butte was part of that western hinterland, but in time Butte's wealth potential would be so great that it became a magnet for railroad development, somewhat as a large metropolis would, even though Butte never developed its own hinterland, other than serving as a market center for a very small surrounding agricultural area.

The Utah and Northern Railway, a subsidiary of the Union Pacific, was the first to lay its tracks into Montana. After a tentative start in the early 1870s, construction northward from Utah resumed in 1878, and the tracks crossed Idaho and into Montana in 1880. By the time the Utah and Northern reached Butte in 1881, Northern Pacific Railway tracks had entered eastern Montana, and construction was proceeding westward up the Yellowstone Valley. Northern Pacific crews were also at work in western Montana on the line extending eastward from the Puget Sound. The Northern Pacific tracks were completed at Gold Creek, northwest of Deer Lodge, in 1883. The event, however, did not bring competitive freight rates to Butte and southwest Montana, as many had hoped, because the Northern Pacific and the Union Pacific had formed a pooling agreement in early 1882. The former agreed not to route its tracks through Butte, instead running tracks through Helena and passing about forty miles north of Butte. The Union Pacific agreed not to extend the Utah and Northern north to Helena. The Union Pacific and the Northern Pacific then formed a jointly owned subsidiary called the Montana Union, which was a short-line railroad forming a bridge from the Union Pacific tracks at Butte to the Northern Pacific at Garrison Junction. Through joint ownership of the Montana Union, the two transcontinental railroads could cooperate to set freight rates favorable to both lines and to coordinate the shipment of freight to and from Montana's mining districts. Nevertheless, with rail transportation assured, Butte began to attract new capitalists to compete with the likes of W. A. Clark, A. J. Davis, and the Walker brothers. Many of the new investors were more interested in copper.[17]

Although silver was probably foremost in the minds of many who saw Butte's promise, because its value was measured in ounces and not pounds, copper soon became the main fuel for Butte's economic engine. The nation was embarking on the process of widespread industrialization, creating an unprecedented demand for the red metal. At the beginning of the nineteenth century, world production of copper stood at about 9,000 tons per year, with the British Isles being the largest source and Swansea the leading smelting center. By 1850, world copper production was more than five times that of 1800. Swansea was still an important center for smelting, but the world leader in the supply of copper had passed first to Spain, then Chile, and then the United States, with the U.S. surpassing Chile for the first time in 1883, 51,574 long tons to 41,099 long tons. World production that year stood at about 200,000 long tons. Michigan's Keweenaw Peninsula accounted for about half the U.S. production.[18]

The large increase in worldwide copper usage during the nineteenth century prior to the 1880s had been in such traditional markets as ornamentation, the military, and cookware, and in new markets like clocks, piping, valves, and sheet copper for roofing and ships' hulls, all spurred by the Industrial and Scientific Revolutions. Until the 1880s, however, electrical applications had not played a significant role in copper's expanding market. The metal was known for its superior conductivity, but fabricators had yet to devise means for producing copper wire strong enough for widespread application. Use of the telegraph, for example, had been growing throughout most of the century, but copper wire did not seem to possess sufficient tensile strength for long spans between supports. Iron was the preferred material for most systems of telegraph wire. Iron, however, offered insufficient conductivity for effective telephony or high-speed telegraphy. Coincident with the advent of those technologies in the 1880s, fabricators learned how to draw copper wire with better tensile strength, and copper quickly became the material of choice for new, rapidly expanding communications systems.[19]

The other significant technological development occurring in the early 1880s was Thomas Edison's new central-station lighting system in New York City. On September 4, 1882, one of Edison's electricians sent electricity from the Pearl Street generating station to a system of incandescent streetlights in the surrounding district. Use of electricity spread rapidly from that date to other commercial-district street-lighting schemes, to electric street railways beginning in the late 1880s, and to factory illumination in the 1890s. Each of the new applications for electricity stimulated tremendous demand in copper for switches, transmission wire, and windings for motors and generators.[20] And the falling price of copper in the 1880s certainly helped to expand the market and uses of copper.

The tremendous growth in U.S. copper consumption due to electrification is readily apparent in the following table:

TABLE 4.1. UNITED STATES PER CAPITA COPPER CONSUMPTION, 1850–1890

Year	Per Capita Copper Consumption (lb.)
1850	0.55
1860	0.405
1870	0.603
1880	1.006
1890	3.043

Source: Richard P. Rothwell, ed. *The Mineral Industry, 1892*, Vol. I (New York: The Scientific Publishing Company, 1893), 117.

GROWTH IN BUTTE COPPER PRODUCTION

Investors in Butte's copper mines in 1880 could not have foreseen, of course, that electrification was about to transform American society and the demand for copper. They may have known that, both worldwide and in the U.S., copper production was beginning to increase and, along with the growing supply, copper prices were declining from more than thirty cents per pound in 1872 to about twenty cents per pound in 1879. By 1880, Michigan producers, led by the giant Calumet and Hecla Mining Company, dominated the U.S. market and cooperated as a pool to put a break on production in an effort to keep copper prices from falling further.[21] Capitalists outside the Michigan pool must have seen places like Arizona and Montana as opportunities to act independently of the restraints imposed on Michigan production and to pour copper into an expanding market, reaping high prices. Railroads were beginning to link the copper mines of the Rocky Mountains to the world market, but the question remained whether copper could be extracted from the rich ores of the western U.S. at a cost, relative to the price of copper on the world market, that would leave a profit. The answer to that question depended on whether available metallurgical methods were applicable to the task and whether new metallurgical methods could be found that would make Butte ores even more profitable. Speculation that the answer to that question would be in the affirmative attracted large investments to Butte from major centers of capital in the U.S. and elsewhere in the world.[22]

Such a setting promised to be fertile ground for fast-paced technological development as well as for investment. Both happened in Butte during the

1880s. Several groups of capitalists followed on the heels of the Clark/Hill endeavor, forming companies, acquiring copper mines, assembling technical experts, and building reduction works at Butte. At the close of the decade, Clark had built his own smelter independent of the Colorado, five other companies had established new smelters at Butte, and Marcus Daly had built two smelters twenty-six miles west of Butte at a town he named Anaconda. Another enterprise, the Boston and Montana Consolidated Copper and Silver Mining Company, was beginning construction of a new smelter along the Great Falls of the Missouri River, more than 150 miles northeast of Butte. The metallurgical history of Butte's copper industry presented a rich tapestry of rapid change, growth, failure, and a ripening awareness of the significance of the metallurgical waste products being discharged into the air, land, and waters of the upper Clark Fork basin.

After Henry Williams had set the Colorado smelter operating for Clark and Hill in 1879, the next company to build a reduction works at Butte was the Montana Copper Company, which focused its efforts on the rich ores of the Colusa mining properties, located on the back side of the Butte hill not far from upper Silver Bow Creek. The principals in the Montana Copper Company were the Lewisohn brothers of New York, who were brokers in metals markets, principally lead and copper. Leonard Lewisohn, the most active of the brothers in the Montana Copper Company, was born in Hamburg, Germany, in 1847. Realizing that they had more orders for copper than they could fill from Michigan copper sources, the Lewisohns decided to try developing their own source of copper at Butte. The man the Lewisohns had in Butte to buy new properties was Charles T. Meader. Born on Nantucket Island in 1817, Meader had gone to California in 1849 in search of gold. He learned about smelting copper at the blast furnace of the Union mine in Calaveras County in the 1860s. Moving to Utah in 1874, he met the Lewisohns, who sent him to Butte in 1878. Meader quickly recognized that the Silverite mine and some adjoining claims would have more value in copper than in silver, and he bought them for the Lewisohns, renaming the Silverite the West Colusa and a neighboring claim the East Colusa.[23]

After the Lewisohns incorporated the Montana Copper Company in October 1879, and the Colusa claims were consolidated under the company's ownership, Meader began erecting a smelter on the east side of Silver Bow Creek, across from the Colusa mine. The small community that grew up along Silver Bow Creek near the Colusa mine, more than a mile east of Butte City, came to be known as Meaderville. Unlike the Colorado smelter, which both roasted and smelted ore in reverberatory furnaces, the Montana Copper Company's process was to be based on the water-jacketed cupola furnace. The Colorado's successes notwithstanding, there was still considerable debate among smeltermen concerning the most effective furnace for smelting Butte's sulfide ores. In general, the cupola furnace required less skilled and costly labor to operate than did the reverberatory furnace, but the cupola required greater skill on the part of the metallurgist to develop the proper charge (the mixture of ore, fuel, and flux) and blast pressure. The cupola required a good metallurgical fuel, like anthracite, coke, or charcoal, while the reverberatory could be fired with wood or bituminous coal. Meader, an enthusiast of the cupola, designed the new smelter to have a capacity of 20–35 tons/day. Technical difficulties delayed commencement of smelting operations, however, and in early 1880, the Lewisohns sent Alfred Wartenweiler to Butte to serve as the Montana company's metallurgist.[24]

The Montana Copper Company's new works was known alternately as the Montana smelter or the Colusa smelter. Technical problems, most likely with the water-jacket furnace, kept the works from operating throughout the spring and summer. In October 1880, the company rebuilt the works to include a concentrator and reverberatory furnaces (often called reverbs). Meanwhile, the company continued developing the surface plant and underground workings of its mines, hoisting ore, shipping the material of highest grade elsewhere for smelting, and setting the rest of its ore on a dump awaiting successful operation of the smelter. The reverbs began operating early in 1881, producing about 10 tons of copper-silver matte daily. The blast furnace was finally ignited in June 1881, doubling the smelter's output of copper matte. With the smelter running smoothly, the Montana Copper Company quickly became Butte's largest producer of copper ore, its mines yielding 100 tons/day. Colusa ore assayed at about 70 ounces of silver per ton, which was very rich for copper ore. The company sacked and shipped ore assaying more than 35 percent copper to smelters elsewhere. Ore assaying 20–35 percent copper was smelted at the company's Meaderville works without preparation. Ores of lower grade were concentrated first.[25]

With a capacity to treat 25 tons/day, the concentrator, built under Wartenweiler's supervision, was the first successful copper concentrator in the Butte district. The Montana Copper Company soon added Frue vanners to the concentrator for treating slimes, the very fine particles of rock unavoidably produced by crushing. Slimes were almost impossible to concentrate using conventional jigs. Vanners were concentrating tables developed to treat fine materials. A vanner featured an endless belt that moved around rollers at each end of the table, which was slightly higher at one end than the other. The belt moved in the uphill direction, and the slimes, a slurry of fine solids suspended in water, were introduced near the upper end of the table. A vanner also had rapid shaking motion. This combination of factors was designed to cause the metal-bearing particles to settle, adhere to the moving belt, and be carried to the upper end of the table and collected as concentrate, and to allow the particles of gangue in the slimes to remain suspended in the stream of water to be carried downslope and conveyed away as tailings. The name Frue refers to the inventor of this particular vanner.[26]

Tailings from this early concentrator contained a percentage of copper three times higher than the tailings produced by comparable concentrators two or three decades later, at the turn of the twentieth century.[27] This was due both to ongoing improvements in the techniques used for gravity concentration and to the fact that ores of higher grade were being treated in the early 1880s than around 1900; therefore, the earlier tailings were also of higher grade.[28] The experience of the Montana smelter in its first year demonstrates the skilled methods of trial and error that were necessary in finding an effective technological system of mining, milling, and smelting for Butte ores, made all the more challenging by the absence of a railroad. It also demonstrates the capital that was necessary to build the physical plant, initiate operations, and then remodel the physical plant and operate again, all before returning a profit on the investment. The Montana Copper Company example also demonstrates the waste of copper and other metals that was acceptable to mining companies in that early phase in Butte milling and smelting while trying to generate profits.

The next reduction works for copper ore in the Butte district came on the heels of the Montana smelter. It was associated with the Parrot mine, one of the first copper producers in Butte. Because of the lack of smelting facilities locally and the high cost of shipping ore elsewhere, the mine had sat idle for most of the remainder of the 1870s. Partners Franklin Farrel and Achille F. Migron, of Connecticut, and Samuel T. Hauser and Anton M. Holter, of Helena, formed a corporation in 1880 called the Parrot Silver and Copper Company to build a smelter at Butte. The new Parrot smelter went into operation in July 1881. Its small reverberatory smelting furnace had a capacity to treat 12 tons of ore per day, and it could produce about 3 tons of matte daily. The venture proved immediately successful, and the

Parrot company expanded its smelter in the fall of 1881 with the construction of 4 or 5 more reverbs. The smelter also featured both stalls and hand-operated reverberatories for roasting ore prior to smelting. After the 1881 expansion, the Parrot quickly followed the Montana smelter's lead with the construction of a 50-ton/day concentrator, which opened in April 1882.[29]

By early 1883, the Parrot mine was producing more than 100 tons of ore daily, sending about 50 tons/day to the concentrator and another 50 tons/day directly to the smelter. Some additional ore carried enough copper to be shipped elsewhere for smelting. By 1883, Butte was served by rail, and many of the mining companies continued shipping their highest-grade ore to Baltimore and other eastern smelters despite the growing smelting capacity at Butte. Late in 1883, the Parrot expanded its concentrator to a capacity of 100 tons/day.[30]

When the Parrot smelter first went into operation, it used its reverberatory furnaces to produce blister copper, a relatively pure product. The high cost of producing blister copper at Butte, however, coupled with the fact that the copper market was more receptive to copper matte from western mining districts than it was to blister copper, quickly forced the Parrot company to cease making blister copper. The company then shipped only high-grade ore and matte to eastern markets. One of its principal markets was the Connecticut brass industry, in which Farrel and Migron had interests. To meet the needs of that market, the Parrot company decided in 1884 to build an all-new smelter and equip it with converters, with which the smelter could bessemerize matte using a process for copper pioneered by the French metallurgist Pierre Manhès. Much like the bessemerization of molten iron, the Manhès process entailed pouring molten copper matte (about 45 percent copper, 25 percent iron, 25 percent sulfur, and 5 percent other impurities) into a large vessel lined with silica brick, called a converter, and then blowing air through openings, called tuyeres, into the molten mass to oxidize the ferrous sulfide and copper sulfide. The iron oxide combined with the silica to form a slag, and the nearly pure copper settled to the bottom, with sulfur dioxide leaving the stack. Heat generated by the oxidation helped to keep the molten mass hot, despite the air blowing through it. At the same time as the developments at the Parrot smelter, Farrel and his Connecticut interests developed a copper refinery at their Bridgeport Copper Company to refine copper by recovering gold and silver from the converter copper (similar to blister copper) shipped from the Parrot's converters.[31]

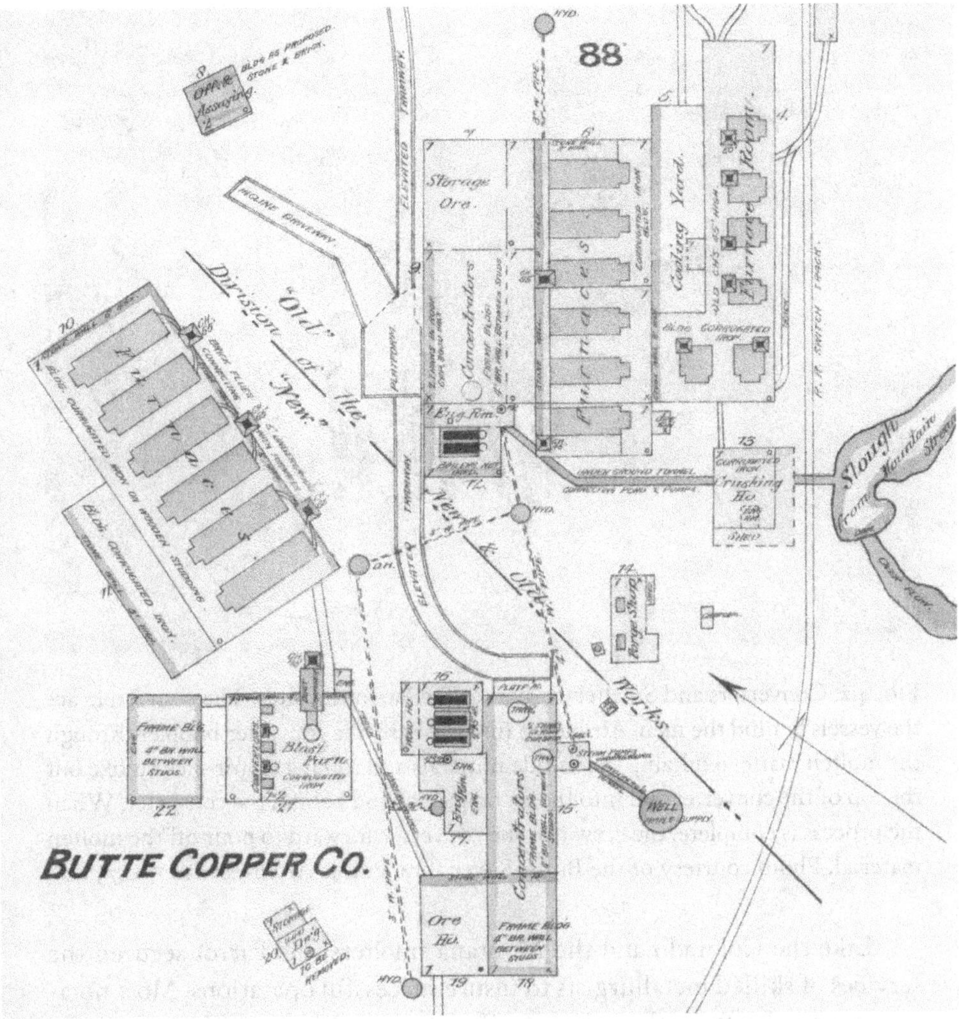

FIG. 4.1. Parrot Silver and Copper Company Smelter. This image from the 1884 Sanborn Fire Insurance Map for Butte shows the Parrot smelter, incorrectly labeled the Butte Copper Co., as it stood shortly after installing converters to bessemerize copper matte. Note the dividing line between the smelter's "Old" works and its "New" works. The building located between the new concentrator and the new calcining furnaces housed the converters installed in 1884. Sanborn Fire Insurance Map, Butte City, Montana, 1884, sheet 9, Library of Congress, Geography and Map Division, Sanborn Maps Collection.

FIG. 4.2. Converters and Smelterworkers at the Parrot Smelter. The converters are the vessels behind the men. Air blown into the base of a converter bubbles through the molten matte, oxidizing the sulfide minerals and forcing sulfur-rich smoke out the top of the converter and into the rectangular hood behind the converter. When the process is complete, the crew tips the converter forward to pour off the molten material. Photo courtesy of the Butte-Silver Bow Public Archives, cat. #1127306.

Like the Colorado and the Montana smelters, the Parrot secured the services of skilled metallurgists to insure successful operations. Most notable among the Parrot's early metallurgical engineers, and perhaps the most notable of all the metallurgists to work in Butte in the 1880s, was Edward Dyer Peters. Born in Dorchester, Massachusetts, in 1849, Peters gained his technical education in Germany at the Royal School of Mines at Freiberg. In 1869, at the age of twenty, he returned to the United States and departed immediately for the mines in Colorado Territory, where he became a highly regarded assayer and mill manager. After five years, however, Peters returned to the East and enrolled in Harvard Medical School, where he graduated first in his class. He practiced medicine for three years in his hometown of Dorchester before returning to the copper industry in 1880. After brief stints in Vermont and New Jersey, Peters went to work in 1883 for Franklin Farrel at Ansonia, Connecticut. In the spring of 1884, Peters accompanied Farrel to Butte, remaining there for most of the year to supervise construction of the new converter department at the Parrot and to work out problems in getting it into operating condition. After leaving Butte, Peters consulted for numerous other mining companies in the U.S. as well as Canada and Australia. His *Modern American Methods of Copper Smelting* was published in 1887. It became the standard text in the field, going through several editions as he kept updating and revising the book through the remainder of the century to reflect rapid technological developments. Peters also wrote numerous reports on the copper industry for the government and the *Engineering & Mining Journal*, and in 1904 he was appointed professor of metallurgy at Harvard.[32]

During his stay in Butte, Peters also supervised the construction of a new concentrator at the Parrot, doubling its capacity to 200 tons/day. It went into operation in June 1884. The old concentrator was remodeled and enlarged to serve as the boiler room, sampling house, and machine shop for the smelter. Technical difficulties with the converters kept the smelter from running full blast during early 1885, but by June the plant was operating smoothly, and by July the converters were producing 15,000 lb. daily of a product called black copper. By the end of the year, the output of black copper had increased to 30,000 lb./day. Through the remainder of the 1880s, the Parrot did little to increase its smelting capacity, but the company did expand its converting department, installing larger converters, in part so it could treat matte from other Butte smelters. Although the Parrot increased its concentrator capacity to about 300 tons/day in 1886, it appears that the concentrator typically treated about 200 tons/day, and the works overall treated about 300 tons/day, the typical production from the Parrot's mines.[33]

Shortly after Peters left Butte, he helped write a government report on the nation's mining industry. Explaining why the Parrot company was willing to discard so much copper in its tailings, he wrote the paragraphs that begin this chapter. Peters was probably voicing a sentiment toward losses of copper in waste discharges typical of the metallurgical engineers working in Butte's mills and smelters at the time. From time to time, engineers found means to increase the percentage of copper they recovered from ore, but they were careful not to let their search interfere with the ongoing treatment of that "cheapest thing" they had: copper ore.

Peters was also highlighting another bit of mineral processing wisdom: metallurgists knew that, along the spectrum of particle sizes discharged in the overall tailings stream, the very fine solids, called slimes, typically had higher concentrations of copper than did the coarser particles. Trying to improve the technology for recovering copper minerals from slimes (or minimizing the loss of copper in slimes) would remain a major challenge for nonferrous metallurgists for the next thirty years, until the introduction of a process called flotation, addressed in a later chapter.

The Colorado, Montana, and Parrot smelters all provided their investors handsome profits after resolving initial technical problems. Not all the smelting ventures of the early 1880s were successful, however. Charles Meader, formerly of the Montana Copper Company, embarked on the single failure of the period. One of the mines Meader had bonded for the Lewisohns in 1880 was the Bell, about three-quarters of a mile west of the East Colusa and atop the main ridge of the Butte hill. When the Lewisohns declined to exercise the bond, Meader decided to sever his association with the Montana Copper Company and start his own operation. At its upper levels, the Bell mine produced free-milling silver ore, which Meader sent to the nearby Clipper mill for treatment. At greater depth, Meader's miners struck ore rich in copper. In mid-1881, he and a group of investors incorporated the Bell Silver and Copper Mining Company and began construction of a smelter in the Summit Valley along Blacktail Creek some two and a half miles from the mine.[34]

The Bell smelter began operating in late 1881 or early 1882, treating about 30 tons of ore and producing about 14 tons of matte daily. The Bell smelter housed two blast furnaces. Ore was first roasted in open heaps, by placing large piles of ore atop a layer of cordwood placed directly on the ground. An ignited heap would be allowed to smolder for weeks, with the heat accelerating the oxidation of the copper sulfide minerals. The company soon built two hand-rabbled reverberatory furnaces for roasting some of the ore. Butte's three existing smelters all used reverberatory furnaces to produce copper matte, although the Montana Copper Company had initially installed only a blast furnace. The Bell was the first and only smelter in Butte to attempt to use blast-furnace technology exclusively for smelting copper. Its sole reliance on the blast furnace may explain some of the company's problems. Although the blast furnaces would have a place in the broad spectrum of appliances used in treating Butte's complex sulfide ores, most metallurgists soon recognized that the reverberatory smelting furnace, with some sort of roasting of the ore prior to smelting, was best suited for treating most of the Butte ores.[35]

As with blast furnaces for other metals, a blast furnace for copper is charged with alternating layers of ore, metallurgical fuel (either charcoal or coke made from suitable coal), and flux (used to facilitate the formation of slag as the mixture approaches and then reaches the molten state and as the impurities in the ore separate from the copper). The ore need not be roasted before charging to a blast furnace; indeed, the heat generated by oxidation of the sulfide ore reduces the amount of charcoal or coke needed as fuel. Air is blown into the blast furnace via openings near the bottom, called tuyeres, to intensify the combustion of the fuel and provide oxygen for the chemical process of unbinding the sulfur from the copper. The molten material settles in the bottom of the furnace, with the less dense slag sitting atop the more dense copper matte (mostly copper, but still consisting of significant portions of impurities). The molten slag can then be drained from an opening at a higher level in the furnace bottom, and the molten matte is drained from an opening at a lower level.[36]

The configuration of a blast furnace is vertical, while that of a reverberatory furnace is horizontal. A reverb has a long floor onto which a bed of calcine (roasted ore or concentrate) is placed, along with the necessary flux. The furnace has a brick, vaulted ceiling over the bed. A firebox at one end of the furnace holds the fuel as it burns (often coal; it need not be a metallurgical fuel). The combustion gases pass from that end, over the bed of calcine and beneath the vault, to a flue at the opposite end. During transit from firebox to flue, the combustion gases heat the quiescent bed of calcine and flux to the molten state (in early centuries of reverberatory smelting, the heat was understood to "reverberate" from the vault down onto the bed of calcine). Once molten, the slag can rise to the upper layer, and the copper matte settles to the lower level. Each molten fluid may then be tapped from furnace through openings at the appropriate level. Because there is no blast of air moving through the charge in a reverberatory furnace, it is more suitable for smelting very fine material that has been crushed and ground in a concentrator, but the material must be roasted before being charged to the reverberatory.[37]

After a troubled period of operation, the Bell smelter closed in the middle of 1882. New investors tried intermittently to reopen the operation during the mid-1880s without success, and the Bell mine was sold to the Chambers syndicate in 1887, eventually to become one of the Anaconda Copper Mining Company's major producers.[38] Meanwhile, the lesson of the Bell smelter is that, while the Butte hill proved to be incredibly rich in copper and other metals, Butte's fabulous mineral deposits did not guarantee success. Profiting on Butte's mineral deposits required developing an effective system of technologies capable of extracting those minerals and putting metals on the market at a cost that was less than copper's price. Meader and his investors were not able to do so with the strategy they pursued.

Not all the development in Butte in the 1880s was in copper. Two new silver mills in Walkerville, W. A. Clark's Moulton mill and a new Lexington mill near the Lexington mine, joined the Alice and Silver Bow mills in the early 1880s in roasting and crushing Butte's silver-sulfide ores. When the new Lexington mill joined the other three in 1882, the four big silver mills had a total of 210 stamps dropping. Of the earlier, smaller silver mills using the Washoe process, only the Dexter mill and the old Davis mill were still in operation. Butte's silver mills were treating about 250 tons of silver ore per day, compared to the approximately 200 tons Butte's copper smelters were treating. After 1882, there was little growth in the capacity of Butte's silver mills. Growth in the capacity of reduction works for copper, however, was phenomenal.[39]

Marcus Daly and the Anaconda Mine

The most incredible development of the period was that of the Anaconda company, offspring of Marcus Daly's vision and huge infusions of capital by three San Francisco capitalists. Daly had managed the Walker brothers' Alice properties in Walkerville until 1880, when he severed his relationship with them after a disagreement over whether shares in the Alice Gold and Silver Mining Company should be traded on the New York Stock Exchange. Looking at other Butte properties, he identified the Anaconda mine as one he would like to purchase. The Anaconda had been located in 1866 by Michael Hickey along a major rock outcropping forming the south rim of the upper Town Gulch basin. Hickey did not file a claim on it until the mid-1870s, when Butte's prospects again were looking bright. His choice of a name for the mine is a favorite part of Butte lore. He claimed not to have known that an anaconda is a kind of snake. He recalled that he first learned of the word as Union soldier in the Civil War, when he read an editorial by Horace Greeley forecasting that the federal forces would wrap Lee's army like an anaconda. Hickey said he liked the sound of it. Hickey's brother Edward filed an adjoining claim, naming it the St. Lawrence for their home county in New York.[40]

By the time Daly bought the Anaconda mine in 1880, Hickey had sold two-thirds interest to Charles X. Larabie, a prosperous Butte mine developer. Daly paid Larabie $20,000 and Hickey $10,000 for the property. He then turned to former associates from Utah to capitalize the development of his mine. While working in Utah, Daly had become acquainted with Robert C. Chambers, superintendent of Park City's Ontario mine, and its owner, George Hearst. By 1881, Hearst had formed a mine-investment syndicate with two other San Franciscans, James Ben Ali Haggin and Lloyd Tevis, lawyers who had roots in Kentucky. The three had already earned fantastic profits from the Ontario mine as well as the famous Homestake mine in the Black Hills of Dakota Territory. Daly traveled to San Francisco in the spring of 1881 to present a proposal to Hearst, Haggin, and Tevis. After Hearst traveled to Butte to inspect Daly's property, the three agreed to invest. They bought the Anaconda mine from Daly for $30,000, leaving him with a share in its ownership, and appointing him superintendent. The syndicate gave Daly a generous budget to develop the mine, build hoisting equipment, and purchase adjoining property, such as the St. Lawrence. He immediately began sinking a shaft on the mine and driving crosscuts to explore for underground ore bodies. Above the water level, he found rich free-milling silver ore. To treat silver ore from his mine, Daly leased the Dexter mill, where he sent more than 7,800 tons. This initial lot of Anaconda ore paid handsomely, returning about $213,000 to the syndicate. As development of the Anaconda mine reached below the water level in late 1882, however, Daly saw that the mine would generate its greatest wealth in copper.[41]

To treat the copper ore, Daly again turned to his San Francisco partners, asking them to build him a smelter. What followed is one of the most unfathomable episodes in early Butte history. Daly, Haggin, Hearst, and Tevis were all experienced in various aspects of financing, operating, and profiting from mines and mills for precious metals. None had experience in base metals.[42] Looking at the copper market, they would have seen that it had been rapidly expanding for years, but they also had to have known that the market was dominated by Michigan's producers of native copper and that the price of copper had been dropping in recent years, despite growing demand. Dropping prices should have made investment in the complex works necessary to reduce Butte's sulfide ores seem risky. The syndicate may have known about electricity, but they could not possibly have foreseen what electrification would mean to the copper market.

In early 1883, as the syndicate considered how to proceed, Daly could have looked around Butte and seen that the Colorado smelter had a daily capacity to treat about 60 tons of ore, the Montana and Parrot smelters each could treat about 100 tons daily, and the Bell smelter was struggling to reopen with a capacity of 30 tons per day. The concentrators at these smelters needed considerable supplies of water to facilitate operation of their jigs and tables and to carry tailings away. Water in Butte was in precious short supply, so it was not surprising that Daly chose to build his new smelter elsewhere, with more abundant water. What amazed all onlookers was the scale of the reduction works he and the syndicate decided to build, reportedly at a cost of $4 million. When the new works opened, it had a capacity of about 500 tons per day, and construction was already underway to increase capacity to 1,000 tons. At a stroke, the largest concentrator in the United States was located in the Territory of Montana.[43]

Records do not survive to give much insight into how Daly convinced his partners to invest so much money in such a large, unproven undertaking. Two of the twentieth-century accounts of negotiations among the members of the Anaconda syndicate, Isaac Marcosson's *Anaconda* and Michael Malone's *The Battle for Butte*, portray the magnitude and the uncertainty of the undertaking, but they cannot explain the basis for the confidence of the men to invest so much money in their Anaconda plans. Local observers were equally baffled, in part because of the secrecy Daly and his colleagues maintained. In 1882, the syndicate began shipping some of its high-grade copper glance to smelters at Baltimore and Swansea. Late that year, Haggin sent technical advisor William McCaskell to Butte to help Daly determine the best location for a new smelter.[44]

Meanwhile, the syndicate was investing huge sums in new hoisting works for the Anaconda and St. Lawrence mines and in developing the underground workings. At every one hundred feet of depth, Daly's men drove crosscuts toward the vein, which was more than fifty feet wide, and, in some areas, showed ore with more than 50 percent copper. This was the material sent to distant smelters. Lower-grade material was set aside on huge dumps in anticipation of the new smelter opening.[45]

Daly announced in May 1883 that he had selected a site for his smelter along Warm Springs Creek, twenty-six miles west of Butte. He had a townsite laid out in June, and construction of the smelter began in July under McCaskell's supervision. The new town took its name from the Anaconda mine. Daly and his syndicate had negotiated an agreement with the Montana Union to haul ore from the mine at Butte to the new smelter. In a foreshadowing of future disputes, the railroad temporarily stopped construction of branch lines to the Anaconda mine in August 1884, responding to a squabble with the syndicate over freight rates. The dispute was quickly resolved, and by the end of the month passenger trains began running between the Anaconda mine and the new smelter site at Anaconda.[46] A measure of the tremendous volume of ore that the Anaconda company moved from Butte to Anaconda is evident in the fact that, within a few years, the daily tonnage of freight the Montana Union carried was more than twice that of the busiest stretch on either of the Montana Union's two parents, the Union Pacific and Northern Pacific.[47]

The first set of smelting furnaces began producing copper matte at the Anaconda works in September 1884. There were two smelter buildings at the new works, each housing reverberatory roasting furnaces and reverberatory smelting furnaces. The complete roster of furnaces at the Anaconda works included twenty-six hand-rabbled reverberatory roasting furnaces (fifty feet long), twenty-six reverberatory smelting furnaces (twenty feet long), and two water-jacketed blast furnaces, which the company decided to build during the fall of 1884 after the reverberatories were fired. Each smelting furnace had its own stack, and the stacks did not even extend above the ridgelines of the buildings.[48] This situation would prove noisome at the smelter, and in a few years the Anaconda company would implement an improved design for discharging smoke.

Although it now had an ample supply of water, the Anaconda company (as it was often called, even though it was not yet legally incorporated) faced most of the same obstacles other Butte companies faced, especially high labor and transportation costs. All the Butte companies were rapidly implementing technological innovations that helped them reduce their costs of production. Observers in the mining industry quickly realized that a key feature of the Anaconda company's strategy to bring its costs down was to treat heretofore unheard of quantities of ore in a concentrator and smelter made more efficient through systematization.[49] Although all of the major Butte copper companies made important contributions to the extractive metallurgy of copper in the nineteenth century, the Anaconda company was to be the leader, as will be seen, in exploiting economies of scale by conceiving of its operations as a large-scale technological system.[50]

Following a brief closure around the beginning of 1885, presumably to make adjustments after trial operation, the new Anaconda works, including the concentrator and both smelter buildings, went into full operation.

FIG. 4.3. Anaconda Upper Works. This photo, looking northeast, shows the first smelter built at Anaconda as it was configured in about 1885. The large group of buildings at the left housed the concentrator, and the trestle from the left delivered ore from the Butte mines. The buildings with the multiple stacks housed the calcining and reverberatory furnaces. Each stack served a single furnace. Photo courtesy of the Montana Historical Society, cat. #Pac 88–19.23.

More substantial changes were also underway, with the Anaconda company trying to achieve the proper capacity in each department so that all equipment could be used at capacity as material moved through the several steps in the reduction process. Such additions included twelve more roasting furnaces, soon thereafter an additional six, and another concentrating mill. Crews also began construction of a building to house vanners for treating slimes, which heretofore had been discharged from the concentrator with the tailings. When the concentrator commenced operations, it only recovered about 65 percent of the copper in the ore, the remaining 35 percent being washed out of the concentrator to the tailings dump and into Warm Springs Creek. Installation of the vanners for treating slimes improved the recovery to about 75 percent, but the vanners reduced the economy of the plant.[51]

By 1886, the capacity of the Anaconda concentrator had grown to 1,200 tons/day, but the Anaconda company had concluded that the system of rolls,

FIG. 4.4. Anaconda Lower Works. This photo shows the Anaconda company's second smelter at Anaconda as it appeared in about 1893. The concentrator steps down the hillside on the left, and the roasting and smelting department extends into the background at center. The long set of buildings on the flat in the foreground house equipment for treating slimes. Note the three stacks on the ridge behind the smelter. Each stack connected to the roasting and smelting department by a flue. Photo from the author's collection.

for crushing ore, and jigs, for separating particles of copper mineral from gangue, was not adequate for processing the huge volumes of material produced in its Butte mines. An addition was built on the concentrator in 1886 to house an experimental array of crushing and concentrating equipment, including a Ball steam stamp of 300 tons/day capacity, hydraulic separators, jigs, and circular tables for treating slimes. The steam stamp, which could pulverize tremendous volumes of material, had by then become the principal mechanism for crushing native-copper ore on Michigan's Keweenaw Peninsula. The steam stamp caused more copper to be lost to tailings because it produced more slimes, and mill engineers had not yet devised effective means for separating valuable minerals from gangue in the slimes. Nevertheless, the new system proved satisfactory, meaning it was more economical, despite the high loss of copper, because of the tremendous volume of ore it could crush at low cost. The company then built a second smelter and concentrator more than a mile downstream along Warm Springs Creek. When the new works went into operation, it was known as the Lower Works and the original Anaconda smelter became the Upper Works. When the Lower Works went into operation in 1889, the combined capacity of the two concentrators was about 3,000 tons/day, six times the capacity the Upper Works had been when it opened only five years earlier.[52]

The Lower Works had a new feature for handling smelter smoke that became a model for an important retrofit of the Upper Works as well. At the Lower Works, three masonry flues ran up the hillside to the west, connecting the smelting departments with brick smokestacks along the ridge. Although flue-and-stack system at the Lower Works was not operating yet in 1888, the Anaconda company recognized such a system would improve conditions at the existing smelter as well. Under some weather conditions, smoke would settle around the works rather than being carried aloft, making it unpleasant for smelterworkers, and if the wind was out of the southwest, smelter smoke would drift into town. The Anaconda company

Fig. 4.5. Anaconda Upper Works. This photo shows the Upper Works in the early 1890s after the individual stacks were removed and the roasting and smelting furnaces were connected to stacks atop nearby hills. The earlier of the two stacks is visible at the upper left, and the second stack and flue are visible at the right. The stacks at the center of the photo serve the powerhouse and discharge coal smoke. Photo courtesy of the Butte-Silver Bow Public Archives, cat. #201112202–17.

built a flue from the roasting department of the Upper Works to a 172-foot brick stack atop the hill to the north of the smelter. In the early 1890s, Anaconda built another flue from the reverberatory department to a 162-foot brick stack atop a small hill immediately east of the smelter. An additional benefit of the flue-and-stack system was that some flue dust would settle in the flue. That dust could be recovered periodically and resmelted to yield metals otherwise lost.[53]

The next chapter describes the beginning of smelter-smoke controversies in Butte in the 1890s, and the subsequent chapter describes the suits brought by farmers in the Deer Lodge Valley against the Anaconda company over damage to their land, crops, and livestock caused by smoke and tailings from a new Anaconda smelter that commenced operations in 1902. Meanwhile, the Anaconda smelters built in the 1880s were so situated that most of their smoke apparently drifted into the hills north of the smelters, where only a few farmers grazed stock and noticed the damage that smelter smoke caused to grazing lands, but the smoke of the 1880s and 1890s did not illicit the widespread outrage that smelter smoke in the early twentieth century would cause.[54]

Continuing Growth in Butte Copper during the Late 1880s

About the time the first Anaconda works commenced operations, W. A. Clark also started his own new smelter along the west bank of Silver Bow Creek in Meaderville. By 1884, one of Clark's mines, known as the Clark's Colusa, was producing 10–15 tons of copper ore daily, but it had been developed to produce as much as 100 tons/day. Clark planned to ship about 15 tons of first-class ore to smelters outside of Butte, to send about 25 tons of second-class ore directly to his new smelter, and to concentrate about 60 tons of lower-grade ore, yielding about 25 tons of concentrates for the smelter. The concentrator started up in October 1884, and the smelter began operating in June 1885. Early in 1886, the Clark's Colusa smelter increased its total capacity to 150 tons/day.[55]

As the Clark's Colusa began smelting, another smelter went into operation near the Colorado smelter. Built by the Butte Smelting Company, the Butte Reduction Works was situated along the south side of Silver Bow Creek just west of Montana Street. Rudolph Siever was the first president of the company. He had helped to build the first blast furnace in June 1885, and then served as foreman of the plant. Late in 1885, the Orford Copper and

Sulphur Company of New Jersey purchased the Butte Smelting Company and made plans to increase the capacity of the smelter. The new blast furnace was of the type known as an Orford brick furnace. It employed water pipes embedded between layers of brick rather than the water jacket typical of conventional blast furnaces at copper smelters.[56] When the Orford Company encountered financial problems in 1887, the opportunistic W. A. Clark bought the Butte Reduction Works and sold his Clark's Colusa smelter to the newly formed Boston and Montana Consolidated Copper and Silver Mining Company. Almost immediately, he began making substantial improvements at the Butte Reduction Works, building a concentrator in 1888, with a capacity of 150 tons/day. The new concentrator began treating low-grade ore in mid-September.[57]

The year Clark bought the Butte Reduction Works, 1887, momentous changes in transportation were underway for Butte: competitive railroad service finally appeared in the form of James J. Hill's new transcontinental line, known then as the St. Paul, Minneapolis, and Manitoba Railroad (and Manitoba Road for short). Hill planned his route across Montana to pass between the Northern Pacific line and Canadian border. To tap the rich freight revenues available at Butte, he planned a major branch line, called the Montana Central, from his main line at Havre and running southwestly through the Northern Pacific's domain in Helena to Butte. The Manitoba / Montana Central route passed by the Great Falls of the Missouri River, a hydropower site of tremendous potential. Coalfields east of Great Falls also held promise for the fuel needs of Montana's mining industry. Hill vigorously solicited the freight-hauling business of Marcus Daly's Anaconda syndicate. Construction of the Manitoba Road into Montana began in 1886, and the Montana Central completed its line from Havre to Butte in 1888. Hill's direct competition to the Union Pacific / Northern Pacific pooling agreement caused its demise, and freight rates in and out of Butte immediately fell, sometimes by nearly 50 percent. In 1889, Hill reorganized his new railroad system as the Great Northern Railway.[58]

The year 1887 was an even more significant milestone in Butte's history as the year the camp surpassed Michigan as the leading copper producer in the nation. Michigan's output of copper had been slowly growing, reaching about eighty million pounds in 1886. Butte produced nearly fifty-eight million pounds of copper that year. In 1887, Michigan's production dropped off slightly to less than seventy-six million pounds, but Butte's output surged to almost seventy-nine million pounds. The state with the next largest output of copper that year was Arizona, with about eighteen million pounds. Montana produced more that 43 percent of the United States' domestic copper in 1887, with virtually the entire Montana product coming from Butte. Observers were amazed at Butte's record for several reasons. First and foremost, all Butte's production was coming from a district of only a few square miles, while Michigan's came from several districts along a narrow mineral belt about eighty miles long. Furthermore, the Michigan companies, led by the dominant Calumet and Hecla, had tried to damage the Butte companies financially by depressing the price of copper to as low as ten cents per pound. Finally, Butte's location far from the nation's industrial centers meant that transportation and labor costs were quite high. Nevertheless, most of Butte's copper-mining companies not only maintained output but increased it during the mid-1880s as Butte forged ahead of Michigan. In 1888, U.S. copper production grew by 26 percent, and Montana's output grew accordingly, maintaining its 43 percent share of the nation's total, outproducing Michigan by twelve million pounds.[59]

Butte companies had to adapt to the trying circumstances of increasing demand for copper even as Michigan producers sought to drive prices down. The Montana Copper Company responded by effecting a bold reorganization. Faced with dropping prices in the mid-1880s, the company reduced its output for a couple years, despite the fact that its mines held some of the richest copper-silver ores in the Butte hill. Then in July 1887, the Lewisohns joined with capitalists from Boston to incorporate the Boston and Montana Consolidated Copper and Silver Mining Company (B&M) in Montana. The Boston capitalists were led by Albert S. Bigelow, who also controlled the Tamarack and the Osceola copper mines in Michigan. Because of the B&M's association with the Michigan operations through Bigelow, the *Engineering & Mining Journal* predicted that the B&M would not produce copper at a rate as high as its mines and reduction works were able to maintain if such high production served to reduce the price of copper, thus hurting the Michigan producers. Nevertheless, observers expected the B&M to be a large producer and that production would exceed the present capacity of the old Montana smelter.[60]

The new Boston and Montana company was intended to consolidate several rich properties into a more efficient operation that could compete in a market characterized by greatly expanding consumption of copper at prices lower than historical levels. The B&M brought together the properties of the

FIG. 4.6. Boston and Montana's Lower Works, Formerly Clark's Colusa Smelter. This view to the west shows the Butte hill in the background. Note the dead trees on the right, no doubt killed by smelter smoke. The thicker stacks served roasting and smelting furnaces and discharged metallurgical smoke laced with sulfur, arsenic, and flue dust containing metals. Photo courtesy of the World Museum of Mining, Butte, cat. #00273.

Montana Copper Company (the reduction works, some placer claims, and several Butte mines, including the East and West Colusa) and the Mountain View properties of Charles Larabie. A Butte mining entrepreneur who once had owned a share of the Anaconda mine, Larabie had developed the Mountain View mine, an extremely promising property, but had yet to produce any ore other than that extracted during development.[61] And early in 1888, the B&M pursued a short-term solution to its need for increased smelting capacity by purchasing the Clark's Colusa mine and smelter from W. A. Clark. Henceforth, the old Montana smelter was known as the B&M's upper works, with the Clark's Colusa smelter becoming the B&M's lower works. In September 1888, stockholders approved B&M managers' plan to increase capitalization of the company so that a new, more efficient smelting plant of sufficient capacity for the company's expected future production could be built. In November 1889, B&M directors and officials visited Great Falls, Montana, where they decided to build the large new smelter so that it could take advantage of the tremendous waterpower potential of the Missouri River. This decision also assured that the B&M's two Butte smelters would soon close.[62]

In August 1888, incorporation papers were filed in Montana for the Butte and Boston Mining Company (B&B). The original trustees included A. S. Bigelow and several other New England entrepreneurs who were also prominent in the Boston and Montana Consolidated Copper and Silver Mining Company. A. J. Davis of Butte was named president, and Charles H. Palmer was named superintendent and general manager of the Butte and Boston's operations. In addition to acquiring the Silver Bow Company's mines and silver mill, the B&B acquired its water rights, including the rights for the old Noyes and Upton ditch, which took water from Elk Park east of the Continental Divide. Several of the Silver Bow Company's mines also appeared to hold promise as copper producers, and the B&B intended to build a smelter to reduce its ores. For nearly the entire first year of its operation, however, the B&B mined and milled only silver ore, producing no copper ore other than that extracted during development of the Gray Rock and the Silver Bow mines. The new B&B smelter opened in early October 1889.[63]

Moving into 1890, the B&B added roasting and smelting furnaces and built a 400-ton concentrator next to the smelter. The new concentrator would be equipped with crushers and rolls—because they required less water—to pulverize the ore. The concentrator went into operation in December 1890.[64]

Tailings Disposal

When the B&B concentrator went into operation in late 1890, it was treating about 335 tons of ore daily and yielding about 135 tons of concentrates and 200 tons of tailings, which it discharged along Silver Bow Creek. The B&B tailings joined those being discharged from Butte's other copper concentrators and silver mills, which discharged a combined 800 tons per day (see table 4.2). Sending such a large volume of waterborne solids into Silver Bow Creek and Missoula Gulch choked the watercourses. There is no evidence, however, that this situation bothered anyone yet, other than the milling and smelting companies themselves. The problem for the companies lay not in contaminating the environment; rather, the problem was simply the matter of physical space for disposal.

To understand the reasons for losses of copper discharged with the tailings, it is important to understand the variables involved in determining the effectiveness of concentrating ores. In the late nineteenth century,

TABLE 4.2. BUTTE TAILINGS PRODUCTION, 1890

Silver Mills

Mill	Ore treated (tons/day)	Tailings discharged (tons/day)
Alice	110	100
Lexington	65	60
Moulton	40	35
Silver Bow	60	55
Total	**275**	**50**

Copper Concentrators

Mill	Ore treated (tons/day)	Tailings discharged (tons/day)
B&M (both)	300	180
B&B	335	200
Butte Reduction Works	65	40
Colorado	60	35
Parrot	185	110
Total	**945**	**565**

Source: Quivik, "Expert Report" (August 1997), B-8, B-60, B-72, B-80, C-11, C-46, C-86, C-109, C-172.

concentration used gravity by exploiting the fact that, given two solid particles of like size, the more dense particles (the particles with greater specific gravity) will sink in water more readily than will the less dense particles. To effectively use this principle, concentrators first classified crushed ore by particle size using apparatus called trommels, which are essentially cylindrical sieves. A series of trommels would separate, or classify, crushed ore into separate streams, each with its own relatively narrow range of particle sizes. Each stream of like-sized particles then moved to a set of jigs, a type of apparatus that agitated the mixture of crushed ore and water with an intensity at which the dense, metal-bearing particles could sink while the less dense, non-metal-bearing particles remained in suspension. Jigs yielded two streams: concentrates, containing elevated levels of metal, and tailings, which were discharged as waste.[65]

Although jigs were not 100 percent effective, they were efficient enough that their tailings were relatively clean, meaning they retained relatively little copper, at least in the coarse particle ranges. Ore crushing, however, did not produce only coarse particles. Crushing inevitably also produced fine solid particles, in the ranges from sand down to fine powder, called slimes, and jigs were relatively ineffective at concentrating crushed ore in the fine particle ranges. Innovators in the late nineteenth century had developed varieties of shaking tables, like the Frue vanners already described, which could effect a certain degree of separation of fine metal-bearing particles from the fine particles of gangue, but they were much less effective at doing so than jigs were at effecting that separation in the coarse particle ranges—hence E. D. Peters's observation that treating fine particles in Butte had reached the limits of cost-effectiveness, and that companies could realize greater profits by treating fresh ores than by endeavoring to achieve ever more perfect treatment of the fines.[66]

There was another, intermediate class of material produced in concentrators, called middlings. These were particles with not enough metallic value to send to the smelter as concentrates but with too much value to be discarded as tailings. Mills typically crushed the middlings to a finer state so they could be sent to the sand and slime tables for further treatment. Even though the tables were not as effective as jigs, they did produce some fine concentrates, and their use was more cost-effective than including middlings with the concentrates or discarding middlings with the other tailings.[67]

Tailings discharged from the Butte mills and concentrators represented the full range of particle sizes produced by the mills' crushing and grinding operations, from relatively coarse sands down through slimes. When discharged, the coarse tailings tended to settle nearer the mills, while the fine tailings tended to remain in suspension longer and thus were more readily carried downstream by flowing water. At the same time, fine tailings tended to have a higher metals content than coarse tailings, because gravity methods of concentration were less effective on fine particle sizes than coarse. These factors would play roles in the history of tailings in the Butte mining district as the waste product became a problem, first for the mills trying to dispose of ever-larger volumes of tailings, and then for the farmers downstream who owned riparian lands and who used water, now laden with tailings, to irrigate crops.

In the 1880s, the companies simply wanted to be rid of their tailings so there would be room to discharge more tailings in the continuous process of milling ore and generating profits. Had Silver Bow Creek provided sufficient flow, the stream could have obliged the companies by simply washing all their tailings downstream, out of site and out of mind. Silver Bow Creek and its tributaries, however, had insufficient flow to carry all the tailings away, so several companies had to develop alternative strategies for handling their tailings. This was especially true for the concentrators located directly along Silver Bow Creek. The problems the mining companies faced in Butte in the 1880s foreshadowed the problems that would be imposed on farmers, in subsequent years, dozens of miles downstream of Butte.

The silver mills in Walkerville did not face such a difficult problem of tailings disposal, but it was not for lack of volume. In 1890, the three Walkerville mills (Alice, Moulton, and Lexington) were discharging nearly 200 tons of tailings daily. Although they were located at the head of Missoula Gulch, the mills were not situated down in the gulch. Each was located well up on a hillside with a generous slope down to the gulch. Thus, even though the gulch did not usually run enough water to carry the tailings away, it did serve as a large basin, which never filled to an elevation close to the millsites, a situation that would have hampered operations. Moreover, the head of Missoula Gulch never filled with tailings because spring runoff and cloudbursts periodically eroded the deposits, washing tailings down the gulch to Silver Bow Creek. By the late 1880s, a large bank of tailings four or five feet deep had built up on the floodplain along the north bank of the creek adjacent to the mouth of Missoula Gulch. Mill operators described the silver-mill tailings as having a dark, roasted appearance, reddish-purple in color. They had the consistency of fine beach sand. Silver-mill tailings were finer than tailings from copper concentrators, so they were readily displaced by the flow of the creek and eventually by the deposition of Butte Reduction Works tailings once that concentrator began operating in 1888.[68]

Butte's copper concentrators faced a more serious problem of tailings disposal because they were all situated along Silver Bow Creek, which in most seasons did not provide sufficient flow to carry their tailings away. Furthermore, their streamside locations meant there was no downslope gulch into which they could dump their tailings, as the Walkerville mills could. Problems were compounded for the concentrators downstream of the Montana Copper Company/B&M upper works—the upstream-most concentrator along Silver Bow Creek—because tailings from each successive mill could foul the works of the next concentrator downstream. These conditions led to several innovative techniques for managing tailings.

As with other concentrators in the Butte district, the Montana Copper Company expanded the capacity and upgraded the machinery of its concentrator in the years immediately following its construction to both increase production and increase the percentage of copper recovered. W. A. Kidney reported that while he was superintendent of the B&M's upper works concentrator, he was only able to recover about 60 percent of the copper in the ore. At a 1905 trial in a Butte courtroom, Kidney testified that in 1888 the upper works concentrator was treating ore of about 7–8 percent copper. When it closed it was treating ore about 4 percent copper. If Kidney recalled correctly that the concentrator was indeed only recovering 60 percent of the available copper, then the tailings may still have been as rich as 2.67 percent copper when the upper works closed. Incremental expansions and improvements also increased the volume of tailings produced.[69]

Maps showing the site plan of the works and the topography of the site show that tailings from the Montana Copper Company/B&M upper

FIG. 4.7. Boston and Montana's Upper Works. This view to the northeast shows the tramway trestle for conveying ore from the mine (on the Butte hill behind the camera) to the ore bins at concentrator. Silver Bow Creek flows from left to right in the foreground. The light-colored area is accumulated tailings. Smelter smoke obscures Rampart Mountain along the Continental Divide in the background. Photo courtesy of the World Museum of Mining, Butte, cat. #11308.

works concentrator were discharged toward the southwest along Silver Bow Creek as it flowed south toward Meaderville. Tailings would easily have eroded downstream during freshets. Most times of the year, however, little enough water flowed down the creek that most of the tailings could not be carried away. When the tailings dump reached eighteen to twenty feet in depth, tailings would no longer simply flow away from the concentrator by gravity. Thus, the B&M built an elevator so the tailings could be laundered across to the west side of the creek, farther from the concentrator.[70]

Like the other Butte concentrators, the Parrot discharged tailings still bearing considerable copper. E. D. Peters's explanation of the economic rationale for letting so much copper go to waste has already been quoted above. When Peters had written his report for the U.S. Geological Survey, tailings leaving the Parrot concentrator contained about 4 percent copper. Not until 1888, when the Parrot had a new concentrator and its capacity was again increased, could the company treat middlings. After the company installed equipment necessary to crush and treat middlings, assays of the tailings leaving the concentrator showed about 1.5 percent copper. The concentrator also lost a significant amount of copper in its slimes (extremely fine material resulting from crushing the ore). Sampling of the materials flowing through the concentrator indicated that as much as 3 percent of the copper in the Parrot ores was lost in slimes that could not be treated, and that the slimes carried as much as 10 percent copper (about the same concentration of copper as was in the ore itself).[71]

The Parrot faced an additional problem not experienced by the Silver Bow mill or the copper concentrators upstream. The other plants had sources of water from above the sources of tailings. The Parrot, on the other hand, took water for its concentrator directly from Silver Bow Creek adjacent to the works. The creek at that point carried a significant load of solids from upstream tailings dumps. To clarify the water before using it, the Parrot built dams, eventually three in number. The upper dam was an east-west embankment just upstream (northeast) of the Parrot works. This dam backed water up into a slough northeast of the smelter. From this slough, the reduction works obtained much of the water needed for concentrating ore. Impounding the water allowed slimes flowing downstream from the Silver Bow mill and the three upstream copper concentrators to settle. The Parrot then pumped relatively clear water from the impoundment. Not well engineered, the upper dam occasionally gave way. It burst in April 1891, for example, under pressure of heavy spring runoff, forcing temporary closure of both mine and reduction works.[72]

The Parrot's new 1884 concentrator was built on the creek side of the complex. Tailings from the new concentrator were deposited on a dump south of the concentrator, extending south and east toward the creek. For the first few months of operation, the new concentrator discharged its tailings onto the bank of Silver Bow Creek, and much of the material ran into the creek. In 1885, the Parrot built a dam extending about a thousand feet south of the works to impound the tailings, forming a reservoir in which Parrot slimes and tailings could settle, allowing the concentrator, during periods of low flow in the creek, to reuse water it had already discharged. The impoundment caused a sizeable tailings dump to accumulate adjacent to the Parrot concentrator. It existed, however, not because the company was attempting to save tailings but rather because it wanted to save water for potential reuse. In 1890, the Parrot needed about 750,000 gallons of water

FIG. 4.8. Old Parrot Tailings. This photo was given in evidence at the Magone trial in 1905. The view is to the northeast, with Rampart Mountain in the background at right. Silver Bow Creek flows from right to left in the foreground, and the Parrot tailings are the light material sitting between the smelter and the creek. Photo courtesy of the National Archives at Denver, cat. # NAID 298153.

each day to operate its concentrator, and Silver Bow Creek did not always supply enough.[73] As the dump grew to the south, it also grew in height, so the company had to install a tailings elevator and an elevated launder to transport tailings from the jig room to the dump.[74]

The Butte Reduction Works was located a little more than a mile downstream of the Parrot. Prior to construction of its concentrator, the Butte Reduction Works had drawn water needed for smelting from a nearby well. When the concentrator began operating in late 1888, the mill drew its water from Silver Bow Creek, which ran along the north side of the works. A simple dam built in the bed of Silver Bow Creek just north of the concentrator diverted the flow through the mill. In about 1890 or 1891, the Butte Reduction Works began construction of a wall in the bed of Silver Bow Creek upstream of the smelter grounds. The wall effectively kept the relatively clear water of Blacktail Deer Creek separate from Silver Bow Creek's water, which had been fouled by the numerous concentrators, mining operations, and Butte's municipal sewer drains tributary to it. In this way the Butte Reduction Works could convey only good water from Blacktail Deer Creek to its concentrator.

Segregating the two streams became especially important after the St. Lawrence mine fire in 1889. The fire in the mine had acted on unextracted sulfide minerals, much as the roasting furnaces in a smelter acted on the sulfide ores and concentrates charged to the smelter, which is to say that fire oxidized the sulfide minerals. The unoxidized sulfides deep in Earth's crust were not water-soluble, but oxidation converted them to soluble minerals. As a result, copper and other metals in the rock could be taken into solution in the mine water. So much copper-bearing water was pumped from the St. Lawrence mine and into Silver Bow Creek that, when the Butte Reduction Works used water from Silver Bow Creek for concentrating ores, copper precipitated onto the iron and steel parts of the concentrator's equipment, damaging the pumps and machinery. (In a process called precipitation, copper ions drop out of solution, replacing iron, which goes into solution.) Iron parts are strong and suitable for machinery parts; copper is soft and not suitable for machinery parts. Separating the waters of the two streams also made it possible for the Colorado smelter, even further downstream, to draw water accorded it by right from Blacktail Deer and Basin Creeks without mixing with Silver Bow Creek.[75]

The Butte Reduction Works began producing tailings in 1888. When the first concentrator was built, tailings, both coarse sands and slimes, were simply discharged to the west into Silver Bow Creek. The company made no effort to store the tailings and in the early years, in fact, made an effort to have the creek carry away as much of the tailings as possible. Even before Rudolph Siever began construction of the first furnace at the Butte Reduction Works in 1885, the area adjacent to the site, especially along the north side of Silver Bow Creek near the mouth of Missoula Gulch, was already covered with several feet of tailings that had washed down Missoula Gulch from the Walkerville silver mills. In 1890, the silver mills used 30-mesh or 40-mesh screens on their stamp batteries, while the finest screens used at the concentrator of the Butte Reduction Works were 16-mesh. The silver-mill tailings were readily displaced as the tailings of the Butte Reduction Works expanded northward, because the silver-mill tailings were much finer, so the creek preferentially eroded the silver-mill tailings.[76]

As 1890 drew to a close, Butte's copper concentrators and silver mills were daily discharging about 800 tons of tailings into the upper basin of Silver Bow Creek. This was a much greater volume of solids than the creek could carry away, so large tailings deposits accumulated at some locations. The margins of these dumps were carried away during cloudbursts or spring runoff, but year after year the volume of tailings produced continued to be greater than high water could transport. The tailings that were being transported downstream proved problematic mainly to downstream concentrators, which required relatively clear water for efficient operations. There was as yet little conflict with property owners outside the industry for fouling water and riparian land resources. Members of the mining industry seemed to respect each other's water rights, and some impoundments were built merely to make the best use of the scarce water available. Other sorts of enterprises, however, like farming, had not tried to establish themselves along the upper stem of Silver Bow Creek, so there was minimal potential for conflict over the fouling of the creek in the immediate vicinity of Butte. That was not the case, however, as tailings from Butte made their way downstream to the Deer Lodge Valley.

By 1890, however, Butte was not the only source of tailings in the upper (southern) Deer Lodge Valley. At Anaconda, the Upper Works and Lower Works had a combined capacity to treat about 3,000 tons of ore daily and were concentrating an average of more than 1,400 tons of ore per day, at least 200 tons per day more than all the Butte mills and concentrators combined, and the two Anaconda works were discharging on average more than 600 tons of tailings daily. Almost immediately after the concentrator had begun discharging tailings in 1884, downstream residents began to complain of damage to Warm Springs Creek. Gravels were turning green, fish were dying, and the owners of the Warm Springs Hotel and Asylum had to dig a well for fear that water from the creek had become poisonous and was no longer suitable for drinking. The Anaconda company issued assurances that tailings would be diverted from the creek and that the creek would remain pure.[77] It is questionable whether the company complied with its promise.

The Anaconda company was more concerned with increasing throughput, reducing costs, and thereby maximizing profits. Daly's syndicate hired Charles Rueger, who had studied mining and metallurgy in Germany, to manage the concentrator in late 1884, with a special charge to find means for reducing the losses of copper in the slimes. His two years at Anaconda are indicative of the changes the company was willing to engage as it sought to balance the objectives of running greater volumes of material through the concentrator, improving the recovery of copper at the concentrator, and keeping costs down, so the company could remain profitable. When Rueger arrived, he found a concentrator designed to process 500 tons of ore per day but which was handling less than half that. Moreover, the concentrator was recovering only about 65 percent of the copper in the ore, and tailings sometimes ran as high as 3–4 percent copper. Rueger immediately set about improving the operation of the jigs, increasing the recovery to nearly 70 percent, while he began the installation of Frue and Embry vanners and round concentrating tables for treating slimes. With the additional equipment, he was able to get the recovery rate as high as 76 percent of the copper in the ore, but operating the concentrator in that manner reduced the throughput of ore, thereby reducing the overall profitability of the operation.[78]

As the output of the Anaconda company's Butte mines increased and exceeded the capacity of the works at Anaconda, Daly and his investors decided to build the second smelter about two miles downstream (east) of the first smelter. To design the new concentrator, Daly replaced Rueger with Otto Stalmann, who at the time was a draftsman for the giant Calumet and Hecla operation in Michigan. By the 1880s, Michigan copper producers had settled on the steam stamp as the most effective crushing technology for the Keweenaw district's native-copper ores. A single steam-powered stamp could process hundreds of tons of ore daily (depending on the character of the ore). Stalmann, began by experimenting with steam stamps at the Upper Works. Based on those tests, Daly and Stalmann opted for steam stamps at Anaconda, despite their propensity to produce a higher proportion of slimes, because steam stamps enabled tremendous throughput at low cost. Over time, Daly reasoned, his engineers could work to improve the recovery rates of his overall operation, including the recovery of copper values in the slimes, but for the present, maximum throughput was the objective, conducive with increased profits.[79]

Although news reports and maps depict the efforts made by Anaconda to recover as much value as was economically feasible from the tailings, and especially the slimes, there is little detailed information concerning how the company disposed of its tailings once they could economically be treated no more. There is no record of tailings from the Lower Works being discharged directly into Warm Springs Creek, as had initially been done at the Upper Works. The Upper Works processed a huge volume of ore during the 1880s and simply discharged its tailings by means of launders (channels or sluices built to hold a stream of flowing water as a means of conveying solids) out onto the floodplain adjacent to Warm Springs Creek. Those tailings must eventually have taken up all the available disposal area, necessitating the implementation of methods to carry tailings farther from the smelter area. In later litigation, company officials acknowledged that a considerable but unspecified portion of the tailings made its way into the creek. As the volume of tailings from the two works grew, the Anaconda Copper Mining Company continued to develop its disposal infrastructure, digging launders to carry tailings farther eastward toward the marsh lying between Silver Bow and Warm Springs Creeks.[80]

Despite the tremendous volume of tailings being discharged into the upper Clark Fork in the 1880s and 1890s, however, the first significant controversy concerning the impact of tailings from Montana's copper industry would not arise until late in the century. The first metallurgical waste to generate controversy would be smelter smoke, and that issue grew to prominence in Butte in 1890.

CHAPTER 5

Beginnings of Conflict over Smelter Smoke in the 1890s

> I must say that the ladies are very fond of this smoky city, as it is sometimes called, because there is just enough arsenic there to give them a beautiful complexion, and that is the reason the ladies of Butte are renowned wherever they go for their beautiful complexions. . . . I say it would be a great deal better for other cities in the territory if they had more smoke and less diphtheria and other diseases. It has been believed by all the physicians of Butte that the smoke that sometimes prevails there is a disinfectant, and destroys the microbes that constitute the germs of disease. . . . it would be a great advantage for other cities, to have a little more smoke and business activity and less disease.
>
> —William Andrews Clark (1889)[1]

Such were the benefits of smoke as W. A. Clark described them to the delegates of the Montana Constitutional Convention, over which he presided shortly before the territory became a state in 1889. Many in Butte shared Clark's rosy outlook on smoke, but by Montana's year of statehood the emanations from Butte's smelting furnaces were becoming a nuisance in some minds. In 1885 a group of Butte women had begun complaining that the thick smoke from the smelters was damaging their flowers and other plants, a complaint that was largely ignored. Most Butte residents still saw smoke as a sign of prosperity.[2] This view may have helped Thomas Couch, one of the Butte smelter managers, to see benefits more readily in the clouds of smoke. In a January 1889 letter to a corporate official in Boston, Couch observed how helpful Butte's smoke had been in helping him view a solar eclipse on New Year's Day. The smoke was so thick that he could look directly at the eclipse without hurting his eyes.[3]

A favorable attitude toward smoke persisted among some segments of society well into the twentieth century, but in industrial communities where smoke pollution was worst, the populace had long agitated against being subjected to the ill effects of smoke, scientists and engineers had long struggled to understand and control the causes of smoke's harmful effects, and government officials had sought acceptable methods to regulate sources of smoke. In the nineteenth century, several American cities, such as Pittsburgh, Chicago, and Cleveland, as well as numerous European industrial cities, like Manchester, Glasgow, London, Berlin, and Paris, undertook extensive studies of their smoke problems and passed ordinances aimed at the sources of smoke.[4]

When smoke first became a source of conflict between Butte's mining industry and its neighbors in 1890, the conflict was a minor annoyance for Butte's mining companies compared to their major objectives of increasing their capacities, meeting the ever-growing market for copper, and reducing costs in order to remain profitable even in the face of dropping copper prices. In 1890, Montana mines produced 112,980,896 pounds of copper, almost all of it coming from Butte, and Butte's smoke controversy was just getting underway. Montana's production represented about 43 percent of total copper production in the U.S. that year. In 1895, Montana's 190,172,150 pounds of copper was half of the national output. And in 1900, Montana mines produced about 45 percent of the nation's copper, putting 270,738,489 pounds on the market. By then, smoke was still a big problem in Butte, tailings were becoming an issue, significant strides in mechanizing operations were continuing, and a wave of major corporate consolidation on the Butte hill was beginning.[5]

Smelter smoke began to appear less benign to many residents of Butte in late 1890, when thick smoke, coupled with a series of winter inversions, typical in high mountain valleys, brought about a noticeable increase in deaths from respiratory disease. The anti-smoke movement in Butte would develop differently from that in many other American industrial cities, where the obvious culprit was smoke from burning coal. As environmental historian David Stradling shows, the Progressive movement took up the anti-smoke fight as one of the means of improving life in cities, and Progressives identified the source of the problem as the inefficient burning of coal. Thus, engineers who could devise more efficient means of burning coal would bring solutions. Stradling notes, however, that improved efficiency would not address the smoke problem in cities like Butte, where the problem was metallurgical smoke, not coal smoke filled with particulate matter that represented the incomplete combustion of coal.[6]

Fig. 5.1. Birdseye View of the Great Mining Camp, Butte City, Mont. In this 1890s view to the northeast, the Butte hill is barely visible in the background through the smelter smoke. The Colorado smelter is on the left, the Butte Reduction Works is just right of center, and the Parrot smelter is producing the smoke in the background at right. If one looks closely, one can see smoke emanating from the Meaderville smelters upstream (left) from the Parrot smelter. Photo courtesy of the World Museum of Mining, Butte, cat. #03181.

By 1890, there were six copper smelters operating along the base of the Butte hill, ranging from the Boston and Montana's upper works north of Meaderville to the Colorado smelter southwest of Butte. The Butte and Boston was just finishing a major expansion of its new smelter, raising its capacity to nearly 400 tons per day. On any given day in late 1890, the six Butte smelters together treated as much as 1,150 tons of ore, smelting perhaps 200 tons directly and sending 950 tons to their concentrators, which lost perhaps 25 percent of available copper minerals to the 550 tons of tailings they discharged daily. The remaining 400 tons of concentrates were roasted and then smelted, meaning that, on a typical day in late 1890, about 600 tons of cupreous material was being charged to the roast-heaps and smelting furnaces at Butte. These materials derived from Butte ores might contain as much as 35 percent sulfur. Not only was copper bound up in sulfide minerals; iron, too, was present in Butte ores in generous portions of the sulfide called pyrite, which concentrated well along with the copper minerals and which was therefore charged to the furnaces along with the copper. As much as 75 percent of the sulfur was given off in the smoke during roasting, and much of the rest was eliminated with the smoke during smelting. Thus, as much as 200 tons of sulfur may have been discharged daily into Butte's airshed in late 1890.[7]

Sulfurous smoke was particularly problematic around smelters employing the heap-roasting technique, whereby coarse ore and cordwood were piled on the ground in heaps covering an acre or more, ignited, and allowed to smolder for as long as six weeks or more to drive sulfur from the ore. During the first few hours after a pile was ignited, the wood was the main fuel burning, but after the fire had spread throughout the pile and the sulfide ore began to burn (oxidize), the pile would begin to give off dense yellow smoke with the characteristic smell of sulfur.[8] After about three days of burning, an attendant had to mount the pile to check the covering material to insure no parts of the pile were burning too hot. E. D. Peters had this to say about the work: "In doing this, a disagreeable obstacle is encountered in the clouds of sulphurous [*sic*] gas, which, to one unaccustomed to the task, seems absolutely stifling. By taking advantage of their momentary dispersion by currents of air, and retreating when they become too thick, no difficulty need be experienced in covering the upper surface of the heap as thoroughly and carefully as any other part of it."[9] The heap would continue to smolder for its allotted time, giving off combustion gases throughout. The resulting smoke would envelope the surrounding area in a thick shroud unless the wind was sufficient to carry it away. Peters acknowledged that the sulfurous fumes were very injurious to vegetation: "In the case of smelting establishments of such capacity that not more than twenty-five tons daily of sulphur [*sic*] are oxidized and poured into the atmosphere, it is probable that all vegetation outside of a circle of four miles in diameter may, under ordinary circumstances, be considered safe from the effects of the fumes."[10]

If the wind blew adequately, it would carry that sulfurous smoke away from the community as well, but on calm days, especially during winter inversions, when stagnant air was trapped in the Summit Valley, Butte could become enshrouded in dense smoke for days. Citizens came to recognize a link between thick smoke and high death rates from respiratory ailments after Dr. Heber Robarts, the city health officer, began a campaign to educate the city council and the public about that link in 1890. In reports publicized in detail by the *Anaconda Standard*, he compared death rates in Butte to those in larger industrial cities, finding that Butte's rates exceeded those of

Chicago and London and were about equal to those of New York and Paris. Robarts also asserted that persons suffering from pneumonia and typhoid fever, who might otherwise recover, were dying in Butte for lack of pure air.[11]

The idea that smelter smoke was bad for human health was not new. Robarts was merely the first public official in Butte to press the issue. Nor was word of Butte's smoke troubles surprising to observers in the copper industry. The industry's leading periodical, *Engineering & Mining Journal*, had carried a report in 1887 that the Spanish government had concurred with local officials at Rio Tinto that the practice of roasting copper ore in open heaps was a public nuisance. N. P. Hill's smelter at Black Hawk, Colorado, was notorious for the pollution it caused, located as it was in a narrow mountain canyon. Most attention during the nineteenth century, however, was on the damage smelter smoke caused to vegetation, including both forests and crops. The first edition of E. D. Peters's *Modern Copper Smelting* (1887) included a section describing the damage sulfurous fumes from heap-roasting could do to surrounding vegetation, but he seemed little concerned with harm to humans or animals. Peters cited research conducted in Saxony around the mining center of Freiberg since the mid-nineteenth century. Several German scientists had performed extensive experiments to gain a scientific understanding of how sulfurous fumes and other pollutants caused damage to vegetation, how weather influenced the intensity of damage, what species of plant were most susceptible to damage, and how to reduce the harm done by a smelter to its vicinity.[12]

Observers in the copper industry probably felt confident that complaints in Butte would not threaten to diminish the camp as a supplier of copper. As in the Rio Tinto district in Spain, where mining represented the principal source of employment, those active in the American copper market could expect that local government actions in Butte would result in little more than modification of the way Butte ore was smelted. The primacy of mining in the employment of Butte workers may be seen in the following table:

Of the 3,000 men employed by the Anaconda company, perhaps 1,000 worked in Anaconda, and about 2,000 worked in Butte. A total of about 6,000 men worked at the mines and reduction works of Butte. In the 1890 U.S. Census, Butte recorded a population of 10,723. Many people living in greater Butte, however, lived outside the city limits in the town of Walkerville or the unincorporated suburbs of Meaderville, east of Butte, and Centerville, located on the hill between Butte's northern boundary and

TABLE 5.1. EMPLOYMENT AND PAYROLLS FOR BUTTE MINING COMPANIES, SEPTEMBER 1890

Company Name	Men Employed	Monthly Payroll
Anaconda (including smelter)	3,000	$300,000
Boston & Montana	900	90,000
Parrot	500	50,000
Butte & Boston	400	40,000
Colorado	400	40,000
Bluebird	350	35,000
Lexington	350	35,000
Alice	300	30,000
Butte Reduction Works	200	20,000
Moulton	100	10,000
Cora, Wabash, Volunteer, Stevens, Ramsdell Parrot, Star West, Clear Grit, etc.	600	60,000
Total	**7,100**	**$710,000**

Source: Engineering & Mining Journal 50 (13 September 1890): 318.

Walkerville. The population of Silver Bow County in 1890 was 23,744, and most of those people lived in Butte and its immediate vicinity. Thus, about one-quarter of the county's total population found employment with the mining companies.[13]

The most extensive newspaper coverage of Butte's emerging smoke controversy was in the *Anaconda Standard*. Although the *Standard* had the largest circulation of all the dailies serving Butte, it was owned by the Anaconda syndicate and published in Anaconda. Marcus Daly had hired John Durston, former university professor and editor of the *Syracuse Standard*, to serve as the *Anaconda Standard*'s first editor when the newspaper launched in 1889. Clearly, the *Standard*'s owners had little to lose and much to gain in public controversy over smoke in Butte caused by the Anaconda company's competitors. Whether it limited the competitors' production or added to their cost of production by installing smoke-abatement devices, local government action would enhance the Anaconda company's relative position in the market.[14]

In late 1890, Butte's local board of trade formed a special committee, with Robarts at its head, to investigate possible technical solutions to the smoke problem. The committee sent letters to *The Scientific American* and *Engineering & Mining Journal* asking readers to convey recommendations for smoke abatement to the committee. Butte's city council instructed the city attorney to draft an ordinance regulating the roasting of ore in open heaps. The ordinance passed on December 17, 1890.[15] Meanwhile, officials for the smelting companies let it be known that technical solutions were available; they just were not profitable to implement. Both the B&B's C. H. Palmer and the Parrot's J. E. Gaylord told the *Anaconda Standard* that most of the sulfur in Butte's ores could be recovered in the form of sulfuric acid. There was no nearby market, however, for all the acid that could be produced. Furthermore, there were already ample supplies of sulfuric acid where there were markets, negating the possibility of making acid at Butte and shipping it elsewhere.[16]

Inventors responding to the Smoke Committee's letters offered other technical solutions. A Chicagoan named William Hutchinson was the first of these "smoke messiahs" to be invited to Butte to see if his invention could abate the city's smoke nuisance. He claimed to have helped Chicago reduce smoke problems stemming from the burning of coal. Characterizing himself as a practical man who knew little theory, he now offered to destroy the arsenic and sulfur emanating from the stacks of Butte's smelters. Although dubious, the smelter managers agreed to cooperate with the Smoke Committee in testing Hutchinson's device. In mid-January 1891, while the city was inundated by some of the thickest smoke it had seen, "Old Hutch" brought his invention to Butte and installed it on one of the smelting furnaces at the Parrot. By injecting steam into his special firebox, Hutchinson claimed he could rid the smoke of noxious substances. His first test received mixed reviews, with the mayor, many members of the city council, and the *Anaconda Standard* proclaiming the trial a promising success, but with many smelter people seeing little benefit from the device. The city council agreed to pay for a larger test, this time involving a special stack being built at the Parrot to treat the smoke from two furnaces. A large crowd gathered to watch the test, but the smeltermen declared it a failure and Hutchinson soon left town. Two months later, W. A. Clark brought his own "smoke messiah," Jacob Storer, to Butte from New York. Within a week of his arrival in Butte, however, Storer was stricken with illness and nearly died. Forced to return to New York, he never tested his invention. No smoke messiah ever found a technical fix for Butte's smoke nuisance.[17]

Meanwhile, the mining companies responded in a variety of ways to the smoke ordinance. One closed its mines and smelter, claiming it was forced to do so by the ordinance, but observers asserted that the closure had more to do with the current low price of copper. Two smelters agreed to let their roasting heaps burn out, but they continued to operate in general. The B&B and B&M continued to build new heaps and, demonstrating their contempt for Butte's new smoke ordinance, ignited them in 1891. As a consequence, early 1891 was a deadly period in Butte. During calendar year 1891, Butte officially recorded 620 deaths, of which 246 (40 percent) occurred in the first three months of the year. January and March were particularly deadly months, and 71 percent of the deaths during those months were from pulmonary causes.[18]

The manager of the Boston and Montana who gave the order to ignite the company's roasting heaps was Thomas Couch, the man who had appreciated the benefits of smoke for eclipse gazing. He was a prominent western mining expert who had been appointed superintendent of B&M operations when the company was formed in 1887. Born in Cornwall, England, in 1843, Couch, like so many Cornish mine supervisors of copper mines on Michigan's Keweenaw Peninsula, took on the title of captain, which meant he combined the practical skills of manager, geologist, and mining engineer while lacking a university education. Like the Michigan copper companies, the B&M's corporate officials relied on Couch's practical sense, rather than on an expert with formal scientific or technical training, to find and develop copper resources in Butte. When Couch had arrived in the United States at the age of twenty, he went immediately to the Keweenaw. Gold soon lured him to California, where he spent about five years in the mines of Grass Valley. From there, he moved to Nevada and then Utah, gaining appointments as manager of various mines and, as an autodidact, eventually gaining a reputation as an expert in mineralogy and metallurgy. Eastern capitalists began hiring him to inspect prospective mining investments throughout western North America, and such travels eventually took him to Butte in 1883. Upon Couch's recommendation, the new Boston and Montana and its eastern investors assembled the Colusa, Mountain View, and other Butte properties under its single ownership.[19]

During the respiratory epidemic of 1891, Couch reported to B&M

president A. S. Bigelow that mining and smelting operations had to be curtailed because so many men were sick at home and he could not fill his shifts with workers. Yet he reported no connection between the epidemic and the smoke, noting that he had read in the newspapers that the disease was afflicting cities throughout the country. In a separate paragraph to Bigelow suggesting no connection between the epidemic and the anti-smoke ordinance, Couch assured Bigelow that the Boston and Montana would be able to roast ore at the same cost as in heaps if the ordinance forced the company to utilize roasting furnaces, flues, and stacks. When the epidemic finally caught up with Couch, he did allow that he would like to get out of Butte for a couple of days while he recovered, because the smoke gave him a headache.[20]

In 1891, Couch had greater issues on his mind than problems with smoke in Butte. When the company had decided in 1889 to build its new smelter at Great Falls, the directors determined to continue smelting at Butte until the new smelter was completed, thereby ensuring continuing profits to pay for construction. Couch was instructed to keep the Butte smelters in good working order, but not to make any major improvements. Late in 1890, the B&M did add several mechanical roasting furnaces, both Brückners and O'Haras, to its Butte smelters, but the company continued to roast large volumes of ore on open heaps.[21] The installation of the new roasting furnaces at Butte may have been mainly to test their effectiveness and to help the company determine what appliances to install at the Great Falls smelter.

On December 10, 1891, the new city ordinance notwithstanding, Couch gave instructions to crews to set fire to the B&M's roast-heaps again, and that same day he left town (because Butte's mayor had threatened to arrest him if he resumed heap-roasting), igniting the ire of the anti-smoke forces. Reluctant to take action against a private company's operations, Butte's mayor wired Bigelow in Boston, asking him to authorize someone in Butte to act in Couch's absence to comply with the ordinance and extinguish the heaps. As smoke and tempers flared, Butte citizens, reportedly numbering about two hundred, staged a demonstration outside City Hall, demanding that the government extinguish the B&M's heaps. Although there were no suggestions of violence, city officials were likely aware that, in early 1888, riots stemming from unrest over smoke from roasting ore on heaps at Rio Tinto, Spain, had led to a call for assistance from the military, which in turn led to twenty deaths. Butte's mayor finally hired contractors to cover the heaps with earth, taking three days to smother the fires. According to

FIG. 5.2. Brückner Roasting Furnace. This was one of the first mechanized roasting furnaces. The flue for receiving combustion gases is the brick structure at right-center. The cylinder to the left of the flue is brick-lined and holds the charge, either concentrate or crushed ore. A rolling firebox is connected to the other end of the cylinder, out of view in this photo. Combustion gases from burning coal in the firebox pass into a rotating cylinder, heating the charge until the heat from the oxidizing sulfur sustains the roasting process, at which time the firebox is rolled to another furnace. Photo courtesy of the Butte–Silver Bow Public Archives, cat. #201112202–8.

reports in the *Anaconda Standard*, eleven people had died in Butte during the first nine days of December; in the nine days from when the B&M set fire to its roasting heaps to when the heaps were finally put out, the lives of forty more people were extinguished.[22]

When Couch returned a few days later, he calmly asserted that his order to ignite the heaps was not in blatant disregard for the ordinance but rather an effort to demonstrate that the Boston and Montana's heaps were not the source of Butte's smoke. He also claimed that his departure from Butte on the day workers ignited the heaps had nothing to do with the smoke controversy. Indeed, he claimed to have been planning since November to travel to Arizona on that date to inspect some mining properties there for the Boston officials. In a report to Bigelow about the incident, Bigelow downplayed the significance of smoke in the controversy, characterizing the anti-smoke agitation instead as an Irish mob action brought against him and the company because of its non-Irish makeup. To be sure, Butte in the 1880s and 1890s was rife with conflict between nativist, anti-Irish factions and Irish immigrants, Butte's dominant ethnic group; this ethnic conflict may well have complicated the smoke issue.[23]

After the Boston and Montana smoke episode, Butte's ordinance effectively ended the open roasting of ores in and around town. Historians Don MacMillan and Duane Smith assert that this Butte episode is the only nineteenth-century example in the U.S. of a community successfully limiting the practices of smelters.[24] Butte's smoke nuisance nevertheless continued relatively unabated through the 1890s. Meanwhile, Butte's mining industry continued its rapid growth and change. This, coupled with a lack of any technical solutions to the smoke problem—and possibly an absence of severe inversions for a couple winters—diverted the attentions of the populace and its local government away from smoke issue.

Rather than thinking about smoke, many of Butte's miners turned greater attention to issues of class. By the mid-1880s, the Butte Miners' Union had grown to be the largest union of miners in the American West, and it had helped establish unions in other Montana mining camps. In 1893, in response to mining companies' efforts to reduce wages throughout the West and in the wake of intense conflict between miners and mining companies in the Coeur d'Alene mining district of northern Idaho, the Butte Miners' Union, then with a membership of 4,600, issued a call to form an alliance of miners' organizations and hosted the founding convention of the Western Federation of Miners (WFM) in May of that year. Seventeen locals from throughout Rocky Mountains and South Dakota were represented, and the WFM went on to become a major force in the western labor movement until well into the twentieth century.[25]

New Smelters and a New Railroad

Another ongoing development that directly affected Butte was the changing geography of smelting in Montana. Although the largest share of those employed by the copper industry at Butte worked in the mines, the smelters still accounted for an important part of the community's payroll. Already about half of the copper ore hoisted from the Butte mines was smelted at Anaconda, so the smelter under construction at Great Falls could in some minds have foreshadowed the demise of Butte's smelter industry. Observers in Butte therefore likely took great interest in the announcement that the Great Falls smelter was to be enlarged even before construction was complete. To investigate potential changes in the new smelter's design and construction, the B&M appointed a four-man committee to visit Great Falls in early 1891. The committee consisted of Couch and the B&B's Palmer from Butte and two experts from Bigelow's Tamarack mine in Michigan.[26]

In June 1891, the B&M decided to expand the capacity of its smelter to more than 1,000 tons per day, to increase the construction budget from $750,000 to $1,000,000, and to place Frank Klepetko, one of the committee members from the Tamarack mine, in charge of construction. Born in Austria in 1856 and educated at Columbia University's School of Mines, Klepetko took employment in Michigan's Copper Country after completing a master's degree in 1880, making his way to superintendent of the smelting department for the Tamarack-Osceola smelting works in 1890. When he got to Great Falls, one of Klepetko's first tasks was to devise a means to use the coal available from the nearby Sand Coulee mines for smelting. Sand Coulee coal too was high in ash content to be burned effectively in the reverberatory furnaces. After a period of experimentation, Klepetko installed coal-gas producers at the smelter so he could fire the reverbs with gas. The Great Falls smelter began smelting on a limited basis in March 1892. At about the same time, the B&M board of directors authorized selling $600,000 in bonds to finance construction a new electrolytic refinery to take advantage of the hydroelectric potential at Great Falls. Construction of the refinery began in the summer.[27]

One of the most important reasons for the B&M locating its new smelter at Great Falls had been its waterpower potential. In anticipation of the construction of its metallurgical plant, the B&M had entered into an agreement in 1889 for waterpower with the Great Falls Water Power and Townsite Company. Under the agreement, the Water Power and Townsite Company would build and operate a dam at Black Eagle Falls and sell waterpower to various entities, who would be responsible for building their own plants for converting the waterpower to mechanical or electrical power. Although James J. Hill and his associates had had a dam designed for Black Eagle Falls when the Montana Central Railroad arrived in 1887, they did not have cause to build the dam until the deal was struck with the

B&M. Construction of a timber-crib, rock-filled dam at the crest of the falls was completed in 1891. The fourteen-foot-high dam, in combination with the falls, provided an operating head of forty feet. The B&M built its powerhouse below the north end of the dam and near the smelter in 1891–1892. About half of the power it generated was transmitted mechanically by manila ropes to the nearby smelter to drive such equipment as the crushers at the concentrator and blowing engines to supply air to the converters. The dam also served to produce mechanical power for a flour mill on the south side of the river and to produce electricity for city lights and a street railway system.[28]

The Great Falls smelter went into full operation in early 1893, and by September the B&M was shipping 800 tons of ore to Great Falls daily. That amount was nearly twice what the company's two old Butte smelters could treat together. The new smelter was one of the finest in the country, featuring very low energy costs, up-to-date equipment, and a converter department and electrolytic refinery so that the company could produce marketable, pure electrolytic copper. The B&M had hoped that its new smelter would save a penny per pound of copper in operating costs. By the end of the summer 1893, the Great Falls smelter demonstrated that it could actually produce copper for two cents less per pound than it had cost at the B&M's old Butte works. The price of copper averaged about eleven cents per pound in 1893, so this was a significant cost savings. With production estimated at about 30,000,000 pounds of copper per year, that would mean an additional $600,000 in net revenue each year.[29] The B&M closed its Butte smelters as soon the Great Falls facility was in full operation, with the B&M's upper works closing in January 1893, and the lower works closing at the beginning of July 1893.[30]

Closure of the B&M smelters did not significantly diminish the quantity of ore being smelted in Butte, however, because that same year a new reduction works, that of the Montana Ore Purchasing Company (MOP), was opened by F. Augustus Heinze. Heinze was the third, after W. A. Clark and Marcus Daly, of the principal combatants in Butte's fabled War of the Copper Kings. Heinze had first arrived in Butte in 1889 and worked briefly for the Boston and Montana as a mining engineer. After working for the *Engineering & Mining Journal* in New York for a year, he returned Butte in 1891, by which time his father had died. Heinze was able to convince his two brothers to invest their father's estate in Butte properties. After traveling to Germany to study geology and mining at the University of Freiberg and find creditors for his new enterprise, Heinze returned to Butte and, in early 1893, incorporated the Montana Ore Purchasing Company (MOP), a Montana corporation of which he owned 51 percent of the stock, his brothers owning the rest.[31] The MOP built its smelter in 1893 to treat custom ores (ores produced by other mining enterprises) in addition to ores from mines Heinze leased and operated. By the beginning of 1898, the MOP smelter was treating 700–800 tons of ore daily from Heinze's mines (primarily the Rarus and Glengarry) and was still treating considerable custom ore.[32]

At about the time the B&M was moving its smelting operations to Great Falls and the MOP was putting its new smelter into operation, Marcus Daly and his associates were expanding the embrace of their industrial

FIG. 5.3. Converters at the MOP Smelter. This photo is similar to the one above showing the first converters at the Parrot smelter. This photo shows a converter being tipped to pour either molten slag or molten converter copper into a ladle. The converter in the background is in the upright position, and smoke is visible discharging into that converter's hood. Photo courtesy of the Montana Historical Society, cat. # Lot 005 B5 F12 02.

system in Montana by building a wholly owned subsidiary railroad to link their mines at Butte with their smelters at Anaconda. The existing rail link between Butte and Anaconda was an important part of the area's industrial infrastructure but, being owned by the Montana Union, it was a component Daly did not control. Echoing the dispute that had delayed inauguration of Montana Union service between Butte and Anaconda in 1884, a dispute arose between Daly and the Montana Union in 1889 over the rate the railroad was charging to haul ore from Butte to Anaconda. The new confrontation continued for a couple years, culminating in a seven-month suspension of the smelter operations in 1891, when the railroad refused to haul ore pending resolution of the conflict. With the promise of help from his friend, James J. Hill of the Great Northern Railway, Daly decided to form his own railroad in 1892. The Butte, Anaconda and Pacific Railway (BA&P) was incorporated in September 1892 and began service in December 1893. The company hauled passengers and other freight between Butte and Anaconda, but its most important function was to haul ore.[33] Cognizant of the fact that he was building a technological system, Daly considered the railroad integral to his copper mining and smelting operation, writing that "the road from the Butte Mines to the smelter is as much a part and parcel of the works as the Bruckner furnaces or the jigs in the Concentrator."[34]

As Anaconda developed its own railroad, it also developed a method for the electrolytic refining of the converter copper its Anaconda smelters were producing. Heretofore, copper refineries had used furnaces to remove the last of the impurities and produce a product of pure copper. Refining by electrolysis used a different set of principles. Converter copper was cast into anodes (a positive electrode for an electric current) in the shape of a thick shape with tabs at two corners, which allowed the cathode to be suspended in a tank, or cell. Cathodes (the negative electrode), which were similarly shaped starter sheets, were suspended between the anodes, and the tank was filled with electrolytic fluid. After wires were connected to the anodes and cathodes and a current applied, atoms of copper would dislodge from the anodes, pass through the electrolytic fluid along with the current, and attach to the cathodes. The cathodes would be removed from the tank periodically and the pure copper removed. Impurities, largely gold and silver, would settle to the bottom of the tank as a slime or sludge from which the precious metals could be refined, adding to the company's profits. Anaconda's electrolytic refinery went into operation in 1891. It was the second one in the United States and preceded the B&M's electrolytic refinery at Great Falls by a year. Unlike the B&M's plant, however, the Anaconda refinery operated on electricity generated with coal-fired steam. Anaconda enlarged its refinery through the 1890s, but it continued to send about half its converter copper to an electrolytic refinery in Baltimore because labor costs there were about half what they were at Anaconda.[35]

The first half of the 1890s also saw Daly's Anaconda enterprise undergo some corporate changes. Prior to 1891, the enterprise had been owned by the syndicate of Marcus Daly and his investors, James Ben-Ali Haggin, George Hearst, and Lloyd Tevis. They referred to their business variously as the Anaconda company, the Anaconda Gold and Silver Company, and the like, all names linked to the Anaconda mine. In 1891, the partners incorporated their business as the Anaconda Mining Company, which assumed title to all the properties previously owned by the syndicate in the name of Daly, Haggin, or another individual. Hearst died shortly thereafter. In 1895, the principals incorporated a new company, the Anaconda Copper Mining Company (ACM), with Haggin as president and Daly as superintendent of operations. The ACM took title to most of the Anaconda Mining Company's assets, including all its Butte mines and the two Anaconda smelters.[36]

This evolution of the Anaconda enterprise occurred during the period 1890–1916, which business historian Martin Sklar has called "the corporate reconstruction of American capitalism," when industrial businesses made a transition from being proprietary entities to being corporate entities featuring new sets of social relations among owners, managers, workers, customers, communities, and governments.[37] Entities like Anaconda made a transition from being what Alfred D. Chandler Jr. calls "entrepreneurial enterprises," which were managed by their owners (in the case of Anaconda, by Daly, Haggin, etc.) to being "managerial enterprises," which were owned by stockholders and managed by skilled executives who may or may not have owned significant portions of the company's stock. Early in the transitional process, the initial entrepreneurs typically held prominent positions as executives, and they still thought of the company as their property, as Daly did in Anaconda, but in time the top executives held powerful positions in which they were accountable to the company's stockholders, who owned the company. For Butte, this meant that the community went through a transition from being dominated by powerful mining entrepreneurs, like Clark and Daly, to being dominated by a corporate entity, the Anaconda Copper Mining Company.[38]

The year 1893 marked another turning point; it was the year Butte became unmistakably a copper camp. During the early 1890s, the declining price of silver continued a pattern that had started in 1873, when Congress eliminated the silver standard for U.S. currency. As silver production in the western states and territories increased through the 1870s and 1880s, miners and those dependent on the mining industry had pressured the Congress to support the price of silver by buying large quantities of the metal. Silver production continued to outstrip demand; however, and the price of silver continued to fall, dropping below $1.00 per ounce in 1888. When an economic crisis hit the national economy in early 1893, President Grover Cleveland announced his intention to call a special session of Congress to repeal the Sherman Silver Purchase Act of 1890. In late June 1893, England's colonial government of India announced that it would halt the free coinage of silver, causing the global price of silver to plummet further. Silver-mining operations came to a halt throughout the West, and many silver camps became almost instant ghost towns. The financial panic and silver slump of 1893 affected Butte as well. By mid-July, all the silver mills in Butte were closed. The Butte and Boston never reopened the Silver Bow mill. The Moulton and the Lexington mills resumed operations only briefly at the end of 1893 and early 1894. The Alice was the exception to the general permanent closure of the silver mills in Butte. After a brief closure in 1893, the Alice company was able to keep its sixty-stamp mill operating nearly continuously through the remainder of the decade. Thereafter, Butte remained one of the major silver producers in the United States, but most of the silver produced in Butte was a byproduct of the copper industry.[39]

Smoke and Tailings at Butte in the Late 1890s

By the mid-1890s, the amounts of ore being concentrated and cupreous material being roasted and smelted in Butte was again what it had been before the two Boston and Montana smelters closed in 1893. Production of copper in the United States had been hurt little by the financial panic that had been so devastating to silver producers that year. U.S. copper producers put 287,000,000 pounds of copper on the market in 1891. That number had increased to 323,000,000 pounds in 1892, continuing a long-term increase in domestic output. There was little increase in 1893, as U.S. mines produced 327,000,000 pounds, but the upward climb continued in 1894, with U.S. copper producers putting 354,000,000 pounds on the market. In 1894, Montana maintained a commanding lead among U.S. copper-mining districts, accounting for 52 percent of domestic production, at 183,000,000 pounds. As usual, virtually all of Montana's copper came from the Butte district, but by 1894 more than half of Butte's copper, 96,000,000 pounds, was being smelted at Anaconda. It is difficult to determine how much of Butte's copper was smelted at Great Falls that year, because B&M and B&B production were reported together at 58,000,000 pounds. Assuming that, of the Boston companies' combined output, 40,000,000 pounds were smelted at Great Falls and 18,000,000 pounds were smelted at Butte, then a total of about 47,000,000 pounds of copper were smelted at Butte in 1894, including that produced by the B&B, MOP, Parrot, Butte Reduction Works, and Colorado smelters. That is the same amount as had been smelted there in 1890.[40]

Copper output in the United States continued to grow throughout the 1890s, surpassing 500,000,000 pounds for the first time in 1897. Montana's share of domestic production was 237,000,000 pounds (47 percent). The Anaconda smelters continued to treat more than half the copper being produced from the Butte mines. About 50,000 pounds of copper was smelted in Butte in 1897, and a similar amount was smelted at Great Falls.[41]

It should be remembered that the richest copper ores in the Butte district were at a depth of several hundred feet. As those rich ores were exhausted and the mines grew deeper, the percentage of copper in the ore being sent to the mills and smelters declined, so it took more ore to produce 50,000,000 pounds of copper in 1895 than it had taken in 1885. The grade of the ores being concentrated and smelted continued to decline into the twentieth century. It was profitable to treat ores of declining assay values, however, because the technologies available for mining copper ore and for extracting copper from the ore were growing ever more efficient and effective, especially because of cost saving through coal-fired, steam-powered mechanization.[42]

Even though only about one-quarter of the copper being produced by the Butte district in the late 1890s was being smelted at Butte, as much or more ore was being treated at Butte at the end of the decade than was being treated at the beginning, because copper production had increased so much through the decade, and because it took greater amounts of ore to produce a given amount of copper as the years progressed. As a result, more pollutants were being discharged into Butte's immediate environment in the late 1890s than had been the case in 1890. Tables 5.2 and 5.3 show estimates of the amounts of copper tailings discharged daily into the Silver Bow Creek watershed and the amounts of sulfur discharged daily into Butte's air shed during the mid- and late 1890s.

TABLE 5.2. COPPER TAILINGS DISCHARGED DAILY AT BUTTE, 1895–1898 (AVERAGE TONS/DAY)

Smelter	1895	1896	1897	1898
B&B	220	150	300	300
Butte Red. Wks.	55	80	155	190
Colorado	165	190	205	175
MOP	170	195	170	190
Parrot	95	115	275	235
Total	**705**	**730**	**1,105**	**1,090**

Source: Quivik, "Expert Report" (August 1997), C-46, C-86, C-109, C-143, C-172.

TABLE 5.3. SULFUR DISCHARGED DAILY IN THE AIR AT BUTTE, 1895-1898 (AVERAGE TONS/DAY)

Smelter	1895	1896	1897	1898
B&B	52	31	60	60
Butte Red. Wks.	15	21	39	49
Colorado	26	31	34	30
MOP	56	65	55	63
Parrot	37	38	72	61
Total	**186**	**186**	**260**	**263**

Source: Quivik, "Expert Report" (August 1997), C-46, C-86, C-109, C-143, C-172. To generate conservative estimates, it is assumed that 30 percent of the weight of the materials smelted was sulfur, even though records from the period indicate that often more that 34 percent of the material smelted was sulfur.

As the tables show, in the mid-1890s, copper concentrators alone were discharging nearly as much tailings each day as Butte concentrators and silver mills combined were daily discharging in 1890. By the late 1890s, the concentrators were discharging considerably more. Butte's smelters were spewing just about as much sulfur dioxide into the air in the mid-1890s as they were in 1890, at the beginning of Butte's smoke controversy. By the late 1890s, Butte's smelters were sending considerably more sulfur into the air.

The main difference was that heap-roasting was no longer an obvious culprit. All of Butte's smelters were now fitted with smokestacks one hundred feet or in height. Under the right atmospheric conditions, the stacks would help carry the smelters' sulfurous fumes aloft. When cold, still winter air settled into the Summit Valley, however, it often led to an inversion. Emissions from the furnaces during inversions hovered in the Summit Valley, shrouding Butte in a giant, choking cloud, just as if the smoke had been given off by open heaps.[43]

So long as smoke poured from Butte's smelter stacks, however, the city was prospering in a certain sense. The mixed blessing of smoke in the air was described by one A. C. Snow in a letter to cousins in Moscow, Idaho, shortly after his arrival in Butte in October 1897. Snow earlier had been a miner in the silver mines of Aspen, Colorado. Like so many silver camps in the West, Aspen had prospered until the silver crisis of 1893. A few hearty souls had stayed, hoping to find ways to continue earning a living from Aspen's minerals, which no longer qualified as ores.[44] After visiting his cousins, Snow took the train from Moscow through Pullman to Spokane, and then boarded the Northern Pacific eastbound for Butte. Arriving at 10:00 PM, he immediately took stock of the place:

> I was much surprised for the city is much larger than I had expected. I fount [*sic*] the largest smokiest wickedest mining camp on the face of the earth, and if one is fond of Boom business hurry bustle noise excitement and crowds of people he can have it here to his heart's content. And I was much surprised to find the city building very rapidly and some of the largest buildings I ever saw. And the large amount of work going on in the city and mines and smelters is simply indescribable and wages are away up $3.50 a day 8 hours in the mines [illegible] out side.

Although Butte was flooded with job seekers, Snow knew the manager of one of the mines and felt confident he could find work. After describing other aspects of the city, he returned to the theme of smoke: "Butte City is situated on the south slope of a low flat mountain. . . . the mines are all in a bunch in the north part of the city, the smelters are located all around and smoke is no name for it. The country around looks desolate on the account of the copper smoke killing all vegetation."[45]

By January 1898, Snow was gainfully employed as a miner, but working conditions in winter were not healthful. He had such a bad cold that he had missed two shifts. He worked night shift. Since it was quite warm in the mine, the exertion made him very sweaty. Then at 4:00 AM, at the end of a shift, he was hoisted to the surface and into the cold night air, where he had to walk to his rooming house.[46] Winter temperatures in Butte often drop below zero. But it was the smoke that caused the most hardship. For example, in early February 1898, a man named Murdock McLeod was visiting his sister in Butte. McLeod suffered from asthma, which made him susceptible to Butte's smoke. When he was stricken with respiratory problems, his sister called a doctor, but McLeod died that same evening.[47]

As the volume of smoke increased in the late 1890s, winter inversions again plagued Butte with serious smoke problems. In mid-December 1898, a thick pall settled over the city, and the newspapers reported increases in the number of people checking into the hospital with pneumonia and other respiratory ailments. Some died. Conditions became so dire that on December 12, a committee comprised of the mayor, the chief of police, and the chairman of the county commissioners visited all the smelters, asking each manager to close until the inversion lifted. The committee was met courteously at each smelter and shown the operations. What amazed the committee was the apparent absence of smoke at each of the works. Likewise, the smelter managers expressed their inability to explain the source of the smoke troubling the city, although R. D. Grant at the Parrot was certain it did come from the smelters. The managers all stated that they would be willing to temporarily close if the other Butte smelters did. A mass public meeting was called for the next night at the city hall.[48] Meanwhile, the Centennial Brewery had its own solution to the problem: "In these days of smoke and trouble when the undertaker gives you a pleasant greeting and mentally asks how much you will be worth to him dead, it is necessary that one should take the greatest precautions to overcome disease. The best beverage during the smoky season is Centennial beer. It will, in a measure, counteract the smoke and strengthen your lungs. Old time residents of Butte who have tried it say that beer gives better relief than any other stimulant."[49] Other people who could afford it had a more certain solution to the problem: leave town. The well-to-do booked passage on trains to neighboring cities and stayed in hotels until the smoke cleared.[50]

At the public meeting, a committee of five was appointed to study solutions to the problem. Representatives of the smelter companies generally agreed that they would be willing to close if that was the public will, but not all residents were sanguine about closing the smelters. T. T. Updegraff, who had been a smelterworker for twenty years, said that "his wife and children had been buttering their bread with smelter smoke and he did not want to take this bread out of their mouths."[51] The *Daily Inter Mountain* editorialized that there were two solutions to the smoke problem: close the smelters (and eliminate many payrolls), or connect each of the smelters to a giant flue system and convey the smoke to stacks on the neighboring hills. The paper advocated the latter. The *Daily Inter Mountain* also suggested Butte's citizens would have to be willing to put up with some unpleasantness, given its dominant industry: "Butte would be a pleasant place to live if it had no discomforts; so would be hades."[52]

A few days later, District Judge John Lindsay ordered the county attorney to investigate the cause of the smoke problem and make recommendations for a solution. County Attorney C. P. Connolly wrote each of the smelter managers, asking for a formal opinion on the source of the problem and ideas for solving it. Each manager denied that his smelter was the source of the smoke troubling Butte, with the Meaderville smelters claiming they were too far east of town and the smelters south of town saying there was only a problem when there was a light wind from the east, meaning it must be the fault of the Meaderville smelters. Several pointed to the likelihood that the combustion of soft coal throughout the city contributed to the problem. Frank Klepetko, by then manager of the Butte and Boston smelter, was the only one who thought there might be merit to the idea of building a system of flues to carry fumes to hilltop stacks. He cautiously predicted that such a solution might pay for itself by allowing the smelters to more effectively capture flue dust, which could be resmelted to recover copper and silver.[53]

On the last day of 1898, the county attorney submitted a report to Judge Lindsay laying out the possible legal action that could be taken against the smoke nuisance and presenting the plan favored by the smelter managers. They had met and jointly agreed that building a flue system would be the best solution. They asked that a committee of three engineers not affiliated with any of the smelter companies be appointed to recommend a design for the flue system. The same day, the five-man committee appointed at the mass meeting recommended a similar flue system to the mayor.[54]

In early January 1899, Judge Lindsay ordered Butte's five smelter companies to take steps necessary to prevent their smoke "from deluging the said city of Butte and smothering the inhabitants thereof."[55] In his order, the judge gave the companies until May 1 to abate their smoke pollution. He would enjoin any company that failed to comply by that date from operating its smelter. The judge declined to take up the county attorney's recommendation that a panel of engineers be appointed to devise a plan to abate the smoke nuisance, instead placing that responsibility with the smelter companies. As May 1 approached, the *Butte Miner* received a letter from another "smoke messiah," who claimed to have devised a solution to the smoke problem much less expensive than an extensive flue system promised to be. The newspaper recommended that the community continue to investigate such solutions, predicting that the only other real alternative for solving the issue was to close the smelters. The *Miner* opined that closing the smelters would not be good for Butte, benefiting only the railroads and the Anaconda and Great Falls smelters. Evidently, the smelter companies believed that the prevailing public opinion with the *Miner*, for when May 1 arrived they had done nothing to abate the problem. Judge Lindsay ordered the county attorney to investigate.[56]

Ten days later, County Attorney Connolly presented the judge with responses from the companies. The Colorado and the B&B reported that they had conducted extensive engineering studies and that they were willing to share the results of those studies with the judge. Heinze told the judge that the MOP had developed cost estimates for building a flue to a stack atop Anaconda hill, but he also reported that the MOP was exploring experiments being done by an International Smoke and Vapor Condensation Company, which hoped to offer a means of "condensing" the MOP's smelter fumes. The Butte Reduction Works reported that it was already contemplating moving its smelter to another location, where water would be more abundant and labor less expensive. A. H. Wethey, the superintendent, stated that his company would probably move rather than spend as much money as it would likely take to abate the smoke problem from its Butte works. Instead, Wethey said the Butte Reduction Works would close during periods of inversion when smoke choked the city. R. D. Grant of the Parrot smelter reported that, as of April 4, the Parrot Silver and Copper Company no longer owned the Parrot properties, which had been sold to the Amalgamated Copper Company. Grant predicted that the Parrot smelter would no longer be a cause of Butte's smoke problem.[57]

The Amalgamated Copper Company

Grant's report to the judge may have included the most salient information, that the Amalgamated Copper Company's appearance on the Butte scene would shape Butte's future as a smelter city. Amalgamated was the culmination of several years of behind-the-scenes dealings among some of the leading capitalists in the copper industry and the most successful trust builders in the nation. The idea of consolidating the copper industry is said to have been that of Thomas Lawson, a Boston stock speculator. He claimed to have hatched the idea while speculating in Butte and Boston stock and observing the way its principal owners, who also owned most of the Boston and Montana stock, manipulated the transfer of valuable Butte mining properties between the two companies to make the B&M more profitable as an operating company and to profit from the manipulation of public trading of B&B stock. Lawson presented his idea of consolidating not only the Butte copper industry, but the Michigan and Arizona mines as well, to Henry H. Rogers of the Standard Oil trust. Rogers quickly grasped the meaning of Lawson's proposal and invited his Standard Oil associates, William Rockefeller (John D. Rockefeller's brother), A. C. Burrage, and James Stillman, into the deal. The B&M and the B&B were to be the first to be brought into the new copper trust, and in 1897 Rogers and associates began buying stock in the two Boston companies. Meanwhile, unbeknownst to Lawson, Rogers and his group began buying Anaconda stock as well. By early 1899, the New York and Boston stock exchanges were rife with rumors of an impending consolidation of several of the Butte companies.[58]

Marcus Daly was interviewed in Chicago about the rumored copper trust late on April 25, 1899. In saying he knew little about a possible deal, he coyly reported, "I heard of this talk in New York on Saturday, and made some inquiries among some people whom I thought would know something about it if the deal had been carried through, but none of them knew anything more than I did."[59]

Questions were answered on April 27, 1899, when Rogers, Rockefeller, and their associates incorporated the Amalgamated Copper Company in New Jersey. The new company announced that it had acquired major interests in the Anaconda Copper Mining Company, Washoe Copper Company, Parrot Silver and Copper Company, and Colorado Smelting and Mining Company. Marcus Daly and J. B. Haggin had formed the Washoe Copper Company in 1894 and began using the new company, a corporate entity distinct from the ACM, to acquire and develop new Butte mining properties, like the Moonlight and Poulin claims, which now became part of the Amalgamated inventory. Daly was named president of Amalgamated, Henry Rogers vice president, and William Rockefeller secretary and treasurer. Some investors in the B&M and the B&B expressed surprise that the two Boston companies were not part of the deal, but B&M president A. S. Bigelow informed the press that he was pleased with the deal. He said that the amalgamation of the first Butte companies was only the first part of a longer-term plan. Bigelow expected that the two Boston companies would later be added to the Amalgamated Copper Company. The purpose for gaining control of the Butte mining companies was to exercise overall control of mining operations in Butte, thereby eliminating the costly litigation and other conflicts among the various companies over subsurface mineral rights. About two years after its incorporation in 1899, Amalgamated gained control of a majority of both B&M and B&B stock.[60]

At first glance, it would appear that R. D. Grant was somewhat inaccurate in stating that Amalgamated had acquired the Parrot interests on April 4, because Amalgamated was not announced until April 27. Looking behind the public announcement, however, it appears that Grant's statement is likely correct. In Thomas Lawson's exposé of the methods Rogers and others used to gather the Amalgamated companies together and to manipulate stock values, Lawson recalled that, on Rogers's behalf, he had acquired control of the Parrot company's stock well in advance of the incorporation of Amalgamated and the announcement of what companies it had acquired. Early in April, the Parrot company had held its annual meeting in New York, electing several of Rogers's associates to the board of directors and publicly announcing that almost three-quarters of the company's stock was now owned by Standard Oil people. The more important aspect of Grant's report to Judge Lindsay, at least as far as the Butte smoke nuisance was concerned, was his prediction that the Parrot smelter would no longer be a source of the problem. Shortly after Amalgamated was formed, the Parrot company announced that the smelter would close and henceforth ore from the Parrot mine would be smelted at the Colorado and Anaconda works.[61]

Smoky Butte at the Turn of the Twentieth Century

The Parrot smelter's imminent closing notwithstanding, it was apparent to Judge Lindsay that the other smelters had not really done anything to abate their smoke pollution. Rather than taking action against the smelter companies, he ordered the county attorney on May 27 to make another report on November 1, addressing two questions: 1) whether the smoke nuisance had been abated, and 2) what ought to be done if the nuisance was continuing.

In early November, County Attorney Connolly reported to the judge that the smelter companies had not even partially addressed the problem. The companies continued to describe investigations they were pursuing, but they were unable to find a satisfactory technical solution. About the only option they could see would be to close their smelters during the coldest of the winter months, which they complained would place great hardship upon their workers. In his recommendation to Judge Lindsay, Connolly expressed his awareness of the legal principle under which the abatement of a nuisance should not be imposed on a community if the abatement would cause a greater hardship than the nuisance caused. The smelter companies, of course, claimed that permanently closing the smelters would be the greater hardship. In response, Connolly asserted that Butte's primary industry was mining, not smelting. He further asserted that the smelters would not have been built in such proximity to the city if it had already reached a population of 60,000 when they were built. Connolly argued that the time had come for the smelters to be removed from the vicinity of Butte.[62]

Despite Connolly's strong recommendation, Butte's smoke nuisance continued relatively unabated into the twentieth century. Relief came only as the Amalgamated Copper Company continued to close its Butte smelters. Following the Parrot's closure in 1899, Amalgamated closed the Colorado smelter in December 1904, the Butte and Boston smelter in May 1905, and the MOP in July 1906. The latter smelter was actually not yet owned by Amalgamated, but rather by the Red Metal Mining Company, incorporated by parties friendly to the Amalgamated for the purposes of buying and operating Heinze's properties, a transaction that brought an end to the troublesome litigation between Heinze and Amalgamated over apex rights to mining claims. As each of the smelters closed, the mines supplying them began shipping ore to Anaconda and Great Falls, so the smoke abatement for Butte that ensued was due mainly to Amalgamated's consolidation and not any effort to alleviate conditions in Butte. After the MOP smelter closed in 1906, there were only two independent copper smelters operating in Butte. Clark's Butte Reduction Works was the major remaining smelter. In 1905, Clark had increased the capacity of his smelter to 1,000 tons/day and built a new, 352-foot-tall reinforced-concrete stack to carry smoke from all the smelter departments aloft. When completed in November 1905, the stack was believed to be the tallest concrete stack in the world. And a new smelter called the Pittsburgh and Montana, or Pittsmont, had begun operations on the flat east of the Parrot in 1904.[63]

The Butte Reduction Works continued operating until 1910, when it was bought and closed by the Anaconda Copper Mining Company. The Pittsmont smelter operated until August 1924. Although the overall elimination of smelting at Butte alleviated the city's smoke nuisance, it did not end the problem in Montana, as increased smelting at Anaconda and Great Falls transferred the problem elsewhere. Especially at Anaconda, the amount of ore being treated was so immense that an entirely new phase of the conflict over the harmful discharge of metallurgical byproducts arose in the twentieth century.[64]

CHAPTER 6

Corporate Consolidation and the Farmers' Suits

> To show how unfair the law is in its delay [i.e., delay in taking action against the Anaconda smelter for spewing arsenical smelter smoke across the Deer Lodge Valley], if an individual should strew poison on the ranges as this big corporation has done, and even admits, he would be lucky if he went unhung. He would at least be speedily arrested and punished.
>
> —*Silver State* (Deer Lodge, Mont.), April 15, 1903[1]

Across the history of the republic, Americans in the thrall of industrialization altered ways they thought of various aspects of collective life in the United States, including in the legal and political spheres. At its founding the U.S. had adopted a legal system from the British common law, in which many legal matters had been established by precedent in the courts and not by legislation. One tradition in the common law that Americans had adopted was tort law, which protected a person and his or her property from harm by another's actions, whether the other's actions were criminal or not. Tort law held that, if a person was creating a nuisance that prevented another from the enjoyment of his or her property, then the person experiencing the harm could file suit in a court of law to seek relief from the offender. Relief could be in the form of payment for damages caused by the offender and/or in the form of an injunction, legally restraining the offender from continuing the annoying activity.[2]

There could be both private nuisance and public nuisance. As Theodore Steinberg describes in *Nature Incorporated*, courts in the United States began to deviate from traditional understandings of tort law in the early nineteenth century, as industrialization seemed to promise considerable public benefit in the form of economic development. In particular, Steinberg shows, individuals and members of the public, when they went to court seeking relief from corporations that were developing the New England textile industry, began to fail in their legal challenges because the courts ruled that the public benefits of industrial development outweighed the harm that the corporations were causing. Dams may have flooded private property or interfered with spawning fish, for example, but the dams also powered industrial development.[3] Such conflicting popular, legal, and political views on the relative contributions to the common good of protecting private property or promoting economic development came into full conflict in the Deer Lodge Valley in the early twentieth century.

At the dawn of the twentieth century, Butte still rode high atop the list of U.S. copper-producing districts, contributing 38 percent of domestic production in 1901.[4] The Butte hill's tremendous ore reserves meant production could continue well into the future, but an evolving legal crisis threatened to curtail copper output, at least for some companies. The intensifying legal fight between F. A. Heinze and the Amalgamated companies therefore attracted a great deal of attention to Montana's copper industry, yet the sensational court battles masked another kind of conflict looming in the agricultural areas, where most of Montana's copper smelting had moved. Unlike Butte, where mining enterprises had obviously laid claim to most of the important resources before much settlement with contrary economic interests had developed, the areas around Anaconda and Great Falls were already spawning agricultural economies before the smelters arrived.

Marcus Daly's main attraction to Warm Springs Creek in the upper Deer Lodge Valley had been the ample supplies of water and timber nearby. The Boston and Montana's main attraction to Great Falls was the huge waterpower potential the Missouri River offered. Yet both the Anaconda company and the B&M also relied on the vicinities around their new smelter sites as dumping grounds for vast tonnages of tailings, flue dust, and slag discharged as waste products of the metallurgical operations. Although the companies discharged some of these wastes on their own property, large volumes, especially of tailings and smoke, migrated beyond the companies' property lines. For a time, it may have seemed that the wastes dumped beyond the boundaries of the companies' property were conveniently out of the way, but huge quantities of those wastes came to rest on property belonging to others. Herein lay the basis for legal conflict.

Whereas in Butte there was a strong argument that the well-being of the smelters was in the wider community's economic interest, the smelter industry seemed alien to the communities of farmers and ranchers in the vicinities of Anaconda and Great Falls. Agriculturalists relied on the fertility of their land and the purity of their water to produce crops and livestock, and some farmers had been doing so in the Deer Lodge Valley since the 1860s. Indeed, the valley was Montana's first important agricultural district. Prominent cattlemen, like Deer Lodge Valley pioneer Conrad Kohrs, had large holdings of rangeland elsewhere in Montana as well, while smaller landowners took up farming in the Deer Lodge Valley to supply Butte and other nearby mining camps with meat, grain, and produce.[5] The smelters at Anaconda and Great Falls had arrived after the agricultural economies had been established, and now pollutants emanating from the smelters seemed to be rendering the property of some farmers unproductive. And there was another difference between the earlier situation at Butte and the situations at Anaconda and Great Falls: in the former, there were several smelters operating, allowing an individual company to deny culpability and point a finger of blame elsewhere. At both Anaconda and Great Falls, there was a single smelter that was the obvious source of the pollution. This consolidation of responsibility mirrored the corporate consolidation transpiring in Montana's copper industry.

Continuing Corporate Consolidation

The period of the 1890s and the first decade of the twentieth century featured, for businesses, not only the transition from proprietary capitalism to corporate capitalism, described in the previous chapter, but also numerous corporate mergers that in many industries consolidated wealth and power into a shrinking number of giant corporations. These mergers, which occurred during what business historian Naomi Lamoreaux calls "the Great Merger Movement in American Business," resulted in both horizontal and vertical integration of the industries involved.[6] This movement in American business had profound local impacts on the mining and smelting operations of Butte and Anaconda, and it also linked them, through capital and managerial control, more directly to mining and smelting enterprises elsewhere in the Western Hemisphere as well as to sources of supply for the Montana operations and to the fabrication and marketing of finished products made from Montana's raw materials.

The first decade of the twentieth century would conclude with almost all of Butte's copper-mining properties consolidated under the ownership and operation of the Anaconda Copper Mining Company (ACM), but at the beginning of the century, there were several large mining companies operating and competing against each other in Butte. In 1900, when Amalgamated was admitted to the New York Stock Exchange, it owned virtually all the stock of the Colorado Smelting and Mining Company and the Washoe Copper Company, and it owned a majority of shares in the Anaconda Copper Mining Company and the Parrot Silver and Copper Company. A year later, Amalgamated acquired a majority of the shares of the Boston and Montana and the Butte and Boston companies, and their Butte and Great Falls operations were added to the industrial portfolio of facilities for copper mining, smelting, and refining that Amalgamated was developing in Montana. Each of the six mining companies continued to exist as a distinct corporate entity, but each was owned and controlled by Amalgamated.

The formation of the Amalgamated Copper Company propelled Montana into what became, arguably, its greatest political trauma, as a giant corporation, controlled from New York City, suddenly appeared on the scene, and that corporation was so powerful that it could single-handedly manipulate the governor and the legislature and could dictate the economic well-being of several of the state's major communities. It was in this context that the farmers and ranchers of the Deer Lodge Valley began to seek redress of their grievances against the mining industry. They sought relief through the legislative process and through the courts. But even though the damages being sustained by farmers and ranchers were monumental, the threat of domination by Amalgamated loomed even larger to most of the rest of Montana. Without relief through the legal system, agriculturalists in the Deer Lodge Valley were forced to watch their lands become a sacrifice zone for the industrialization and urbanization of American society.[7]

As Amalgamated's growing power became more apparent, F. Augustus Heinze, owner of the Montana Ore Purchasing Company smelter, became a populist champion in Butte and Montana politics. Heinze owned several mining claims on the Butte hill, including the Rarus, the Minnie Healy, and the Nipper, from which his miners were driving workings into the ore bodies beneath claims owned by such Amalgamated companies as the ACM and the B&M. Accompanying that physical work underground, he cleverly used the local courts to fight legal complaints that he was raiding those companies' mining claims. Amalgamated officials were so appalled that they concluded Heinze had bought the judges. Heinze and his candidates for public office, however, were immensely popular with Butte voters, as he roundly criticized all who were associated with Amalgamated as being oily and as smelling of kerosene (obvious references to Standard

Oil).[8] Undaunted, Amalgamated continued to expand its power and consolidate operations in Butte and Anaconda.

At the end of the 1910s, Amalgamated would cause its subsidiary companies to transfer all their properties and operations to the ACM. At the same time, W. A. Clark decided to divest himself of all his copper-mining and -smelting properties in Butte, selling them to the Anaconda company, and the ACM would absorb F. Augustus Heinze's copper-mining and -smelting enterprise as well. This series of transactions made the ACM the world's fifth-largest industrial corporation and by far the largest copper-mining company not only in Butte but also in the world.[9]

The ACM, however, had to traverse a tumultuous decade to get to that point. Much has been written by previous scholars, in distinct accounts, about both the political tumult that convulsed Montana and the legal conflict between the copper industry and the Deer Lodge Valley's agricultural community.[10] This chapter integrates the episodes to show how the political controversy and the ACM's technological prowess put the farmers and ranchers at a legal disadvantage.

Construction of the Washoe Smelter

Among the six major companies Amalgamated had consolidated at the turn of the century, the Washoe Copper Company was the newest on the Butte scene. Toward the end of 1896, word had begun to circulate that Daly and Haggin planned to build a smelter at Anaconda for their new Washoe company, which was expected to be a heavy producer of copper ore. The Washoe took no concrete action toward that end, however, until late in 1898, when it began construction, in the mountains west of Anaconda, of a water-supply system for the proposed smelter. The Butte, Anaconda and Pacific Railway, the ACM subsidiary connecting the Butte mines with the Upper and Lower Works at Anaconda, began constructing new switching yards just east of Anaconda to serve the new smelter.[11]

The corporate separation between the Washoe and Anaconda companies became somewhat moot in 1899, when Daly and the Standard Oil interests formed the Amalgamated Copper Company. Each of the companies comprising the Amalgamated, however, remained a distinct corporate entity. Planning for the Washoe's new smelter at Anaconda proceeded, but now efforts between the Washoe and Anaconda companies were coordinated. For example, even though the existing Anaconda smelters would be rendered obsolete with the opening of the Washoe smelter, they were remodeled in stages late in 1899 and throughout 1900 so that engineers could test concentrating and smelting equipment before making final decisions about processes and machinery to be used in the new works.[12]

In September 1899, the ACM began converting its concentrator at the Upper Works from the system based on steam stamps to one based on the Gates crusher and Cornish rolls, which was similar to the system that had been discarded back in the 1880s. The change was intended to reduce the production of slimes during crushing. The change in crushing equipment accompanied a return to the use of trommels for screening material before sending it to the jigs for concentration, and the installation of Wilfley tables to replace the forty round tables used for treating slimes. The ACM also installed Griffin mills for regrinding tailings from the upper (coarser) jigs. The reground material was then sent to the Wilfley tables to recover additional copper. In other words, the company was testing new equipment to determine whether, in the concentrator at the new works, the savings of copper yielded by the new equipment would merit the costs of the additional treatment. By this time, Amalgamated had determined that the Washoe company would build and own the new smelter and lease it to the ACM, which would be the operator. Construction of the new Washoe smelter was well underway, and ACM engineers were testing other equipment at the Upper and Lower Works for possible use at the Washoe.[13]

In 1899, the ACM demolished its twenty-foot-long reverberatory furnaces at the Upper Works and replaced them with twenty-nine-foot-long reverbs, increasing the smelting capacity of the Upper Works by 50 percent. The ACM learned that longer furnaces produced a matte richer in copper and a slag with less copper going to the slag dump, a trend in longer furnaces that would continue into the new century at the Washoe smelter. The company installed two experimental MacDougall roasting furnaces to supplement the Brückner furnaces and Wethey furnaces already in use. By June 1900, there were two more MacDougalls in place at the Upper Works.[14] The ACM went on to select MacDougall furnaces for roasting concentrates at the Washoe smelter, and MacDougalls would become the standard for the industry after the turn of the century.

Although a few modifications were tried at the Lower Works, it generally did not sustain as many alterations as the Upper Works did during the last few years of its operation. On the other hand, it operated longer than the Upper Works did. The ACM began closing the Upper Works during the fall of 1901 in anticipation of opening the new Washoe smelter. The

FIG. 6.1. Anaconda Concentrator, Jigs East Section. Large concentrators featured huge sections for each step in the process. This view shows two rows of jigs used to separate coarse concentrates from coarse tailings. The machinery visible here, belt-driven from overhead line shafts, agitates a slurry of water and like-sized particles of ground ore, held in tanks beneath the machinery. Photo courtesy of the Montana Historical Society, cat. # PAc 82–62.0128.

Lower Works continued operating at half capacity through the end of 1901, as the ACM's Butte mines reduced their output pending the opening of the Washoe smelter. The new smelter went into operation January 25, 1902, at which time the Lower Works was closed. The Lower Works was reopened again briefly during the summer of 1902, but only to smelt ore and other material already on the site.[15]

Shortly after the Washoe smelter opened, the ACM undertook an exercise to recover additional copper in the vicinity of the Old Works, which illustrates the metals that passed as dust through the flues and stacks of the Old Works and settled on the terrain around the stacks. The superintendent of the new smelter "caused the bare mountain sides back of the plant [the Old Works] to be swept for flue dust in order to re-smelt this dust to secure the copper contents of same." The superintendent claimed to have recovered more than $6,000,000 worth of copper (approximately 60,000,000 pounds) by means of the exercise.[16] At the time, the ACM was only interested in the copper contained in the dust collected, but that dust undoubtedly contained other metals as well, including arsenic, as we shall see.

The man who had spearheaded the amazing rise of the Anaconda Copper Mining Company, Marcus Daly, did not live to see the opening of the new Washoe smelter. He had died in 1900. Upon his death, Amalgamated's directors made William Scallon ACM president. A native of Canada, Scallon had graduated from McGill University and practiced law in Montreal before taking his practice to Michigan's Copper Country. Moving to Butte in 1884, about the time Daly was launching the Anaconda enterprise, Scallon began doing legal work for Daly. He served as counsel for the Anaconda Mining Company and then for the ACM, when it was incorporated in 1895. With the formation of the copper trust, he became Amalgamated's legal counsel as well, and that gave him leadership of much of the apex litigation with Heinze.[17]

When the smelter began operating in January 1902, it had nine departments, which yielded the following products and byproducts (wastes):

A significant feature of the new smelter was that each of the four departments that used heat to treat materials, the roasting, reverberatory, blast furnace, and converter departments, had its own adjacent smokestack to discharge smoke. The roasting department housed forty-eight MacDougall roasting furnaces of the Evans-Klepetko type. Although economical to operate, the MacDougalls, like any roasting furnace, gave off vast quantities of sulfurous smoke laden with flue dust. The five blast furnaces, fourteen reverberatory furnaces, and eight converter stands also gave off considerable smoke. Thick, light-gray smoke emanating from the individual stacks is clearly visible in early photographs of the smelter, and it can easily be distinguished from the black coal smoke from the stacks of the two powerhouses.[18]

As originally built, the concentrator at the Washoe smelter consisted of eight independent sections or units, each with a capacity to treat 600 tons of ore daily, giving the concentrator a total capacity of 4,800 tons/day. The sections of the concentrator were arranged parallel to each other, with ore bins and ore crushing and grinding equipment located at the upper (south) end of each section, various arrays of trommels, jigs, tables, regrinding mills, and settling tanks installed in stages downslope, and concentrates and tailings discharged from the lower end. The concentrator powerhouse was centrally located between two groups of four parallel sections. By arranging

TABLE 6.1. WASTES & BYPRODUCTS DISCHARGED BY THE WASHOE REDUCTION WORKS, 1902

Dept.	Product	Waste or Byproduct
Concentrator	concentrates	tailings (sand & slimes)
Concentrator Powerhouse	power to drive concentrator	clinkers, coal smoke
Sampling Mill	information	negligible
Roasting	calcines	smoke
Reverberatory	matte	slag, smoke
Blast Furnace	matte	slag, smoke
Converter	blister copper	smoke (slag re-smelted)
Smelter Powerhouse	power to drive smelter	clinkers, coal smoke
Briquetting Dept.	briquettes of slimes	negligible

Source: Engineering & Mining Journal 73 (1 February 1902): 191, (1 March 1902): 311–313.

FIG. 6.2. Washoe Smelter Shortly after It Opened in 1902. This view to the north, from farther up smelter hill, shows the individual stacks in the smelter's original configuration. The two stacks discharging black smoke (on the right and third from the right) serve the smelter's powerhouses that burned coal. The four stacks discharging white, metallurgical smoke serve (from left) the reverberatory department, converter department, roasting department, and blast furnace department. Photo courtesy of the Butte-Silver Bow Public Archives, cat. #2021008.

FIG. 6.3. Washoe Smelter Shortly after It Opened in 1902. This view to the south from across Warm Springs Creek, near the Anaconda Lower Works, shows the tremendous volumes of waste materials that the Washoe smelter was discharging into the Deer Lodge Valley. In addition to smelter smoke and coal smoke, the smelter discharged tailings and granulated slag, both of which were conveyed by streams of water. Granulated slag can be seen splashing to the base of smelter hill below the several smelting departments. The stepped roof of the concentrator can be seen at the center of the photo behind the smelter stacks, and a large stream of tailings is spilling off the end of a launder at the left edge of the photo. The wet material in the foreground is tailings from the Lower Works. Photo courtesy of the World Museum of Mining, cat. #11093.

the concentrator this way, the ACM could suspend or modify operations in one or more sections of the concentrator without impinging on the efficient operation of the other sections. Shortly after the concentrator went into service, its capacity was increased to 1,000 tons/day for each section, or a total of 8,000 tons/day. When the concentrator was operating at the 8,000-ton capacity, it used more than forty-four million gallons of water daily, or about five thousand gallons of water per ton of ore treated. The Washoe smelter was by far the largest nonferrous metallurgical plant in the world at the time.[19]

Farmers' Smoke Complaints in the Context of Amalgamated's 1903 Shutdown

The potential political threats posed by corporate consolidation to Montana's citizens and the physical threats posed by the giant new smelter to local farmers merged in the Deer Lodge Valley in 1902. A few months after the Washoe smelter went into operation in early 1902, horses and cattle at the south end of the Deer Lodge Valley began dying in large numbers. In September, several farmers and ranchers in the area petitioned Governor Joseph K. Toole to send the state veterinarian to investigate the cause of their livestock dying. The first notice of the problem to appear in Deer Lodge's weekly newspaper, the *Silver State*, ended with an ominous conclusion: "It's probably the poison from the Amalgamated works, the same outfit that is attempting to get control of Montana. If they succeed, it will be the people that will die instead of the stock."[20]

The *Silver State* was affiliated with the Democratic Party, at a time when local newspapers often made their party affiliations known. The *New North-West*, affiliated with the Republicans, had been the leading newspaper in Deer Lodge since 1869. The *Silver State* started publishing its weekly newspaper in Deer Lodge in 1893. After competing with the *New North-West* for several years, the latter went out of business in 1897, leaving the *Silver State* as Deer Lodge's principal paper. When Marcus Daly had established Anaconda as his smelter city, it was in Deer Lodge County, one of Montana's original counties, of which Deer Lodge was the county seat. As Anaconda's population grew, it was able to prevail in a local election in 1896 to move the county seat to the smelter city. Most of land area of the county, however, was agricultural or mountainous, so the residents of that part of the county petitioned the legislature to divide the county, which occurred in 1901. After the split, Deer Lodge County encompassed a small portion of the southern end of the initial county (741 square miles), with Anaconda remaining as county seat. The newly created Powell County encompassed the rest of the area (2,333 square miles), with Deer Lodge as its county seat.[21]

Noting that for years tailings from the smelters had created serious problems for owners of lands along the river flowing from south to north through the Deer Lodge Valley, observant farmers quickly concluded the new smelter was suddenly wreaking havoc across the valley, not just along the stream. While the Old Works had flues connecting the furnaces to stacks atop nearby hills, thereby capturing poisonous materials in the flues, the new works had no flues and only relatively short stacks adjacent to the smelter buildings. Washoe smelter's smoke was having a severe impact on livestock within five miles of the new works, according to early reports, and by October at least 250 horses and cattle had been killed and numerous others were sickly. Farmers observed that the first symptoms were sores forming in the mouths and nostrils of stock, after which an afflicted animal would lose appetite, lose weight, and finally die. Farmers and ranchers who could do so drove their stock to other pastures, some over the Continental Divide into the Big Hole Basin to the south, some north to the downstream areas of the Deer Lodge Valley. Others who could afford it loaded their stock on railroad cars to be transported to other parts of the state. For example, W. S. Staton, whose place was only a couple miles east of the smelter, shipped four carloads of cattle and two carloads of horses to his brother's ranch near Havre, in north-central Montana.[22]

The first general meeting of farmers and ranchers in the southern part of the valley took place on Saturday night, October 10, 1902, in the Willow Grove School. Called an "indignation meeting" by the *Silver State*, it allowed people to voice their complaints, to develop a preliminary plan and means to finance its implementation, and to organize committees. One committee would convey the community's grievances to the ACM president, William Scallon, and the other committee would visit landowners and ask each to contribute, in proportion to the amount of land he or she owned, to the costs of bringing suit against Amalgamated. The first concrete step the group decided to take in preparing for legal action was to "send east" for a qualified chemist who could analyze the poisons being inflicted upon the valley. Within a month, the farmers had retained the services of Dr. Charles Gresswell, a veterinarian from Denver, not from the "east." Governor Toole agreed to send M. E. Knowles, the state veterinarian, to be present during Gresswell's investigations. By mid-November, the farmers' surveys had documented that the intensely affected area was five miles wide and extended out into the valley about fifteen miles. One hundred ranchers in that zone had suffered losses.[23]

With the governor's decision to send Knowles to the Deer Lodge Valley again, Helena's newspaper took an interest in the problem, describing other symptoms that farmers were reporting. The pads of the feet of dogs living near the smelter were reportedly being eaten by arsenic. When farmers harvested their grain, they found dead rabbits in the fields. Under certain conditions, people as far as ten miles from the smelter reported being able to feel the poisons in the smoke striking their face and eyes with a stinging sensation. Some parts of the valley had seemed immune until three weeks earlier, when the wind shifted. The *Helena Independent* reported that these conditions were similar to those experienced near Cornwall's world-renowned Swansea smelters. The *Independent* also addressed an argument suggesting that the smelter was not the source of the problem in the Deer Lodge Valley, proof of which was the fact that farmers downwind of the Great Falls smelter were not experiencing the same problem. The *Independent* responded that it had a letter from a man in Great Falls describing conditions on the range there. When the grass was green in spring, the problem was not severe, but when rains stopped in summer, materials from the smelter smoke settled on the forage and the condition of the stock worsened. Sheep that got sick in November stayed that way over the winter and died as late as the next June. The health of cattle did not deteriorate as quickly as that of sheep, and horses seemed to be the most hardy in resisting to arsenic's effects, but when they did start to succumb, the first symptom they developed was large sores in their nostrils.[24]

By late December, the *Silver State* reported that farmers in the upper (southern) end of the Deer Lodge Valley could not sell their hay in Butte or Anaconda, and ranchers as far north as Deer Lodge were reporting problems. A local chemist tested hay from the Race Track area, midway between Anaconda and Deer Lodge, and found arsenic. A group in the Deer Lodge area mailed a notice to neighbors and published it in the *Silver State*, announcing a mass meeting in Deer Lodge on December 29. Gathering at the West Side Club in downtown Deer Lodge, an assortment of farmers, ranchers, businessmen, and interested citizens of Powell County elected Conrad Kohrs, who had just been elected as the county's first state senator, to chair the meeting, which then began with landowners describing the losses they had sustained. Among those speaking was John Bielenberg, Kohrs's ranching partner. The meeting appointed a committee of Kohrs, as chair, and four other ranchers to represent the assembly in calling upon officials of Amalgamated to ask what measures the company would take to provide the farmers and ranchers relief, and to ask that the company pay damages for losses. In a demonstration of the impact of new technologies on life in Montana at the beginning of the twentieth century, someone from the meeting then made a telephone call to Scallon and asked for an appointment, which he granted in two days. The meeting also asked those attending to contribute to a fund with which to retain attorneys to file suit against Amalgamated to recover damages and to enjoin the company from continuing to injure residents of the valley. The *Silver State* reported, "A liberal response was had from all interested. These people are very much in earnest, which, however, is only natural, when it is considered ruination stares them squarely in the face."[25]

After meeting with Scallon, the committee called a meeting at the West Side Club for Saturday, January 3, to make their report. Engaging the committee courteously, Scallon assured them that he would settle fairly with farmers and ranchers both for dead livestock and damages to hay, although he wanted time to investigate claims to be assured they were legitimate. He said that he would reimburse stockgrowers for the cost of alternative feed, if they had had to purchase any. Scallon also assured the committee that Amalgamated had experts researching the problem, and they believed they would have a solution by spring. The committee concluded its report optimistically, expressing the belief that Amalgamated was "using every effort to control [the destructive elements contained in the smelter smoke] to the end that the ranchmen and stockgrowers may be protected in their business, and to save the Deer Lodge valley from depopulation."[26]

The winter of 1902–1903 was a particularly bad smoke season, not only in the Deer Lodge Valley but in Butte and Anaconda as well. In mid-December, Butte's city council directed the city health officer to visit each of the four smelters still operating in the city (Colorado, Butte Reduction Works, Butte and Boston, and Heinze's MOP) to investigate rumors that the smoke was so bad because smelters had returned to roasting ore in open heaps. He found that each of the smelters was using approved methods, and none was roasting in heaps. A month later, Silver Bow County's Board of Health met, with the county commissioners and the county physician in attendance. After a discussion of the dangers posed by sulfur and arsenic in Butte's smelter smoke, the board authorized the county physician, under terms of the state health law, to take actions necessary to decrease the smoke nuisance, and the board authorized the county attorney to take legal action against the smelters, if necessary. The board's resolution authorizing the local officials' actions recognized that smelter smoke had "become a deadly nuisance and a grave menace to the public health of the people of Butte and vicinity."[27]

Two days later, the Butte Business Men's Association met in the council chambers at city hall to discuss the smoke problem. Members were following the controversy in the Deer Lodge Valley with great interest, and they thought that the technical solution developed for the Washoe smelter might be applicable at Butte's smelters as well. Members debated the advantages and disadvantages of smelting for Butte. One member asserted that property values in Butte would increase if a solution were found for the smoke nuisance, and he suggested that Butte's notorious smoke problem might be keeping visitors away. On the other hand, he suggested that if Butte officials did not act hastily in trying to curb the smoke problem, the complaints about the issue elsewhere might result in Butte become the smelting center of the state, which might be a good outcome. Butte's streams were already fouled, so no more harm could be done to agricultural interests locally. Meanwhile, consolidating smelting activity in Butte might serve to protect agricultural production and water supplies outside the Summit Valley.[28]

In late January, a large crowd of several hundred people gathered outside Anaconda to watch the end of an era, as the ACM dynamited its brick stack on the hill behind the now-abandoned Upper Works. The spectacle was scheduled for 4:00 PM, but it did not occur until 4:30. Meanwhile, a winter storm blew in from the east, bringing with it snow and a dense cloud of smoke from the Washoe smelter. The combination of snow and smelter smoke made it difficult for the stack to be seen from where the crowd had gathered. Under such unpleasant conditions, there were few spectators remaining when workers finally detonated the explosives.[29]

Smoke during the last two weeks in January was particularly bad; some said the smoke during that period was the worst they had ever experienced. The cloud of smelter emissions in the Deer Lodge Valley was so thick that people compared the darkness to an eclipse. People began to worry that poisons on the ranges might begin to pollute water and air so badly that human health would be threatened as well. Editorializing about the problem, the *Silver State* recognized that closing the smelters, to address the smoke nuisance, would lead to the depopulation of Butte, yet allowing the Washoe smelter to continue as it was currently operating would bring ruin to the Deer Lodge Valley, and its population would eventually move away. Trying to make light of the situation, the editorial closed with reference to the old adage W. A. Clark had used to defend Butte's smoke two decades earlier, that arsenic, taken in proper doses, could be a beautifier.[30]

Seeking legislation that would give citizens of the Deer Lodge Valley the legal means to redress their grievances and give local officials the legal authority to prevent nuisances, Senator Kohrs introduced four bills when the Montana legislature convened at the state capitol in January 1903. Senate

Bill 34 would amend the state's eminent domain law to its original 1889 language. In 1899, the legislature had amended the law to allow private corporations to exercise eminent domain for the movement and storage of tailings, including for lands used in common by several mines for conveying and depositing tailings. The 1899 amendments had declared those uses by mining companies to be public uses eligible for the right of eminent domain. S.B. 35 would authorize the sheriff of a county, when a fine was imposed by a court against a corporation for maintaining a public nuisance, to collect the fine out of the real or personal property of the corporation, and also to collect the cost of abating the nuisance. S.B. 36 would amend the state's penal code so that an officer, director, attorney, or agent of a corporation could be found guilty of a misdemeanor in the event the person aided, consented to, or acquiesced in the corporation's maintaining a public nuisance. And S.B. 37 would amend the state's civil code to enable a private person to take legal action against a public nuisance that was injurious to himself, or on behalf of others who were injured by the nuisance. The senate, acting as a committee of the whole, heard the first reading of the bills in late January and, without opposition, recommended them for passage. Kohrs later introduced a fifth bill, S.B. 51, which clarified that Montana's district courts had jurisdiction over corporations.[31]

Senator Kohrs's bills quickly attracted the attention of Montana's divided press and Amalgamated's lobbyists. One of the newspapers opposing Kohrs's bills was W. A. Clark's *Butte Miner*, which, with Marcus Daly now dead, sided with Amalgamated on most issues. The *Miner* claimed that the true intent of Kohrs's legislation was to force the closure of the Washoe smelter and argued that his bills, if enacted, would go too far in enabling a private person to take legal action against the employees of a corporation. Not only could individuals take action against the Washoe and the Butte smelters; a person could also use the laws to close water companies, streetcar systems, and other companies that served a public purpose. The practical application of Kohrs's legislation, the *Miner* predicted, would lead to the closure of all smelters in Montana, which in turn would force closure of all the mines, leaving the state "a howling wilderness." The *Miner* summarized by asserting that industrial enterprises always pollute, which creates certain unpleasantness for a community and perhaps some small businesses. The remedy proposed in Kohrs's bills, however, would be too extreme. If enacted elsewhere, the bills would destroy every manufacturing center in the U.S.[32]

At the other end of the political spectrum was *The Press*, a Helena weekly that identified itself primarily as opposed to corporate control of government and corporate corruption of democratic processes like campaigns, elections, and legislatures. In addition to its support of Kohrs's bills, *The Press* advocated taxing railroads and other corporate property at the same rates that the property of individuals was taxed, and it advocated for a commission to regulate freight rates and certain other corporate activities on which the public depended. *The Press* took a special interest in the damage being done in the Deer Lodge Valley and in Kohrs's bills. The weekly newspaper argued that S.B. 37 was important, for example, because current Montana law made no provision for the attorney general or a county attorney to take legal action against a corporation that committed a nuisance adversely affecting the public. *The Press* asserted that only those who are part of a corporation could object to the bills. As a practical matter, corporations could not be arrested (being artificial entities); S.B. 36 would allow the law to reach a corporation by holding its directors, officers, and agents accountable. The editor opined that "it would be a good thing if every member of the senate had a ranch in the Deer Lodge valley, so that he could know what it means to be ruined by a corporation that is maintaining a nuisance," and that "members of the Montana senate who are opposed to Senator Kohrs' [*sic*] bills should get some cattle and pasture them in the Deer Lodge Valley."[33]

The *Silver State* in Deer Lodge came out in support of Kohrs's bills, admitting that it initially had been slow to recognize how dire conditions had suddenly become in the valley. By early February, the paper claimed, damage extended as far north as Garrison, at the confluence of the Clark Fork and Little Blackfoot River. If any readers doubted the extent of the damage, the *Silver State* listed two dozen farmers and ranchers who could be asked about the losses they had sustained because of smelter smoke. Kohrs's bills, according to the paper, were intended to put "soulless corporations within reach, if necessary, of the humblest citizen, who may have sustained damage from them." The *Silver State* had confidence that the legislature, the people's representatives, would support Kohrs's bills.[34]

In early February, Amalgamated announced with great fanfare that the Washoe's smoke issue had been solved. As Heinze's newspaper, the *Reveille*, put it, "every kerosene newspaper in the state" carried an article about the new processes the smelter would implement.[35] Experts at the ACM had spent three months investigating the problem and had devised a technical solution, construction of which would begin immediately. The company would build a flue, three thousand feet long and sixty feet wide, which would deliver smoke from the smelter furnaces to a three-hundred-foot-tall stack atop the hill behind the smelter. The broad cross section of the flue would allow the velocity of the gases to slow so they could cool, and the harmful compounds, including arsenic, antimony, and zinc, would then condense and settle to the bottom of the flue, where they could be recovered along with valuable copper in the flue dust. The stack would then discharge the remaining gases at an elevation high enough that no one in the valley would notice them. "Every objection to the present location of the works will be entirely removed."[36]

The same day that Amalgamated announced its planned modifications for the smelter, the senate judiciary committee met to hear testimony concerning the Kohrs bills. ACM president William Scallon, along with ACM attorneys C. F. Kelley and W. B. Rodgers, testified against the bills, and three attorneys presented legal arguments on behalf of Deer Lodge Valley farmers, who also sent a delegation to Helena to protest the damage being done to their property and to describe conditions in the valley, noting that the old Anaconda smelters had not caused so much damage. One member of the judiciary committee asserted that the bills were unconstitutional; if they were, Kohrs responded, then Amalgamated had nothing to fear. The *Silver State* editorialized that the Amalgamated announcement of the planned solution to the program was meant only to convince the legislators that Kohrs's legislation was not necessary. To the contrary, the *Silver State* argued, if the company's technical solution was as good as touted, then the company would have nothing to fear from the legislation, because farmers would not have to avail themselves of the legal remedies it provided. On the other hand, if the legislature did not pass the bills, residents of the valley faced certain ruin. *The Press* concurred: if the company's claims for its technical fix proved accurate, then it had nothing to fear from Kohrs's bills.[37]

When the senate tried to postpone action on Kohrs's bills indefinitely, Kohrs demanded a vote on whether to pass or defeat the legislation, wanting his colleagues to be on the record in their support for the corporations. The legislature defeated Kohrs's bills on the same day the senate committee gave a favorable report to a change-of-venue bill Amalgamated was shepherding through the legislature, to gain relief from the district judges Heinze controlled in the apex litigation in Butte. The lists of senators voting in favor of Kohrs's bills and voting against the change-of-venue

FIG. 6.4. The Legislative Graveyard. This cartoon by Dutton appeared in the *Butte Miner* on March 8, 1903, depicting two important pieces of legislation that were defeated in the just-concluded session of the Montana Legislature: legislation that would have provided for woman's suffrage in Montana and Conrad Kohrs's smoke bills (Kohrs is depicted kneeling, smoking pipe in hand, at the grave of his bills). Image from the *Butte Miner*, 8 March 1903.

legislation were similar (but not identical), leading the *Miner* to crow that the defeat of the smoke bills was a major setback for the Heinze crowd. The fact that one of Heinze's main allies in the legislature, Senator Kennedy of Anaconda (Deer Lodge County), had worked in support of the smoke mills served to link the smoke legislation with the Heinze tactics in some minds. Another factor linking the two otherwise distinct streams of legislation was the assertion that Amalgamated had spent large sums of money lobbying to defeat Kohrs's legislation, just as the trust was lavishing large sums on passage of its change-of-venue bill.[38]

Senator Kohrs's effort to give his constituents the legal means to fight the giant trust and its damaging smelter smoke failed in the 1903 legislature, but the fight between Amalgamated and Heinze continued, in the halls of the state capitol as well as underground. As the 1903 session drew to a close in early March, the legislature had passed relatively few bills, as the Heinze and Amalgamated forces fought skirmishes in anticipation of the showdown concerning S.B. 71, Amlagamated's change-of-venue bill (also called the fair-trial or change-of-judge law), which would allow either party in a civil suit to petition for a change of venue, if the party believed that the court in which the trial was to be held, or the judge before whom the case would be argued, would not provide the party with a fair trial. On March 3, after the legislation had finally passed both houses with veto-proof majorities, Governor Toole sent it to the secretary of state without his signature, as a means of allowing the bill to become law without his approval. Along with the bill, he sent a communication stating that, although S.B. 71 was an improvement over a similar bill he had vetoed two years ago, he still opposed it. Nevertheless, he did not believe that he should use his veto power too often. When S.B. 71 became law, however, Amalgamated's victory was short-lived; a few months later, the Montana Supreme Court voided the law.[39]

Meanwhile, as the ACM was making its changes to the smelter and Amalgamated was lobbying the Montana legislature to pass or defeat laws it wanted, farmers in the Deer Lodge Valley were responding in their own ways to the nuisance. Since William Scallon's meeting with the Deer Lodge committee at the end of December 1902, ACM officials had publicly spoken of a willingness to settle with stockowners who had suffered losses, but it became clear that the company was willing to pay damages only if a farmer signed a contract releasing the company from further damage claims. In early April, the committee appointed at the December meeting called another meeting in Deer Lodge, just prior to another meeting scheduled with Scallon, so they could collect the latest reports on harm from residents of the valley. Several farmers and ranchers reported that the company refused to pay damages when the stockowners, on the advice of attorneys working with the committee, refused to sign the release. By comparing experiences, folks at the meeting realized that the company was offering different release contracts, depending on circumstances, and the company was universally refusing to pay for sheep that had been killed. It seemed that the ACM was anxious to get releases signed before stockgrowers started turning their animals out to pastures in the valley again for the summer. Those in attendance agreed to compile an itemized list of all losses in the valley, including sheep, and present it to the ACM by June 1. If the company would not settle, then they agreed to file suit against the company, and they devised a formula for taxing each other based on the value of their respective properties, in order to cover legal expenses.[40]

Some farmers, however, were not waiting for collective legal action against the company. Back in December 1902, Peter Valiton had filed suit against the ACM in Powell County District Court, seeking $50,000 in damages for discharging tailings into the Deer Lodge River. He claimed that tailings were overflowing lands along the entire stretch between his land and where tailings entered the river, and the ACM denied that it was causing any damage. In February 1903 John and Lizzie Tremblay filed suit in district court seeking an injunction to prevent the ACM from operating its smelter. The plaintiffs argued that, unless the smelter was enjoined, they would have to file suits continuously to claim damages. A month later, Peter Levengood, who farmed west of Anaconda, filed suit seeking $35,000 in damages for harm done by smelter smoke for the previous five years (which would have included smoke discharged by the old works prior to the Washoe smelter being built). Meanwhile, John Bielenberg, Conrad Kohrs's partner in the Kohrs and Bielenberg ranch just north of Deer Lodge, had his men rounding up cattle in mid-May in order to ship the stock to the Milk River country in north-central Montana, where the partners owned large tracts of range land. Bielenberg wanted to get his animals "away from the smelter smoke poison on the pastures of the [Deer Lodge] valley."[41]

The damage being caused by smoke and tailings was not the only physical conflict underway in the upper Clark Fork drainage. The legal fight between Heinze and Amalgamated over ore in the ground assumed physical proportions as Heinze's miners in the Rarus mine moved into workings that had initially been developed by the Boston and Montana's Pennsylvania mine, ground that the Montana Supreme Court had ruled was the B&M's property. First, Heinze's men drove B&M miners from the five-hundred level of the Pennsylvania mine, then the B&M's workers used compressed air to drive Heinze's men back into the Rarus workings. In early March, the Montana Supreme Court enjoined both sides from mining the underground where fighting had occurred over the weekend.[42]

As the contest between Heinze and Amalgamated continued to simmer throughout 1903, changes to the landscape of smoke emissions were underway at Anaconda. Shortly after Amalgamated announced its plans for remodeling the Washoe smelter's emissions system, the company sought about five hundred men to build the flue and new stack, but a labor shortage made that difficult. Yet as the stack grew in height and visibility, Amalgamated's principal newspaper, the *Anaconda Standard*, appeared to be frank and transparent in acknowledging the harm that had flowed from the smelter in its original configuration. The newspaper assured its readers that the new system would eliminate the arsenic, sulfuric acid, and other poisons that caused so much damage. By mid-May, the stack had reached a height of nearly one hundred feet, and the *Standard* predicted that the system would be completed by July 1. In late June, the ACM issued hundreds of invitations to farmers and ranchers, from the south end of the Deer Lodge Valley all the way down to Garrison (twenty miles north of Deer Lodge), to attend an open house at the smelter, where they could inspect the giant stack and new flues that were nearing completion.[43]

The ACM arranged a special train to leave Garrison at 10:00 AM. It stopped at every station between there and the smelter, picking up 450 passengers, and arrived at the Washoe works at noon. Many other property owners from the valley arrived by wagon. ACM president Scallon greeted the guests with a short speech, after which the Montana Hotel served about 1,300 lunches of sandwiches, chipped beef, candy, cake, beer, and lemonade. Following lunch, Scallon gave a longer speech, in which he admitted that the new works had been configured initially so smoke could drift down the valley and cause considerable damage to livestock and vegetation. He then assured the gathering that he believed his experts had devised a system of flues and a stack that would solve the problem. He also said it was the company's policy to pay any just claims for damages. At the end of his speech, he invited Deer Lodge's mayor and several representative farmers to make some remarks; all were appreciative. Amalgamated attorney C. F. Kelley made the concluding remarks, assuring his listeners that the company wanted to settle fully with all who had sustained damages. One guest then joined company officials to the top of the stack to participate in the ceremonial laying of the last brick. The rest of the guests were invited to

Fig. 6.5. Stack under Construction at the Washoe Smelter in 1903 and Excavation Underway in Preparation for Building the Flue. The flue, connecting the smelter, behind the camera, to the new stack atop smelter hill, would be built at grade. The excavation was to make way for the apparatus beneath the flue used to collect flue dust settling out of the smoke. Photo courtesy of the Montana Historical Society, cat. # PAc 82–62.0170.

tour throughout the grounds, inspecting not only the new flue system but such departments as the concentrator and the reverberator furnaces as well.[44]

Much of the good feeling evaporated a few days later, when the ACM announced that operations at the smelter would be suspended for four or five weeks, beginning July 1, so that the flue system could be connected to the furnaces of each of smelting department. The company had completed as much of the work as possible while the smelting furnaces were functioning. About 1,200 smelterworkers would be kept on the payroll to work on the construction, but about 1,000 men would be laid off at the end of June. At the same time, the ACM and the Washoe company ceased production at nine mines in Butte, because there would be nowhere to ship their ore. Development work and repairs would continue in those mines, but about 2,000 Butte miners were laid off because of the smelter closure. The actions corresponded with standard Amalgamated practice of planning major repairs prior to the end of the fiscal year (June 30) and then implementing the repairs at the beginning of July. All told, about three thousand workers in Butte and Anaconda would be without their regular employment for at least a month. The Washoe closure would not affect the other Amalgamated operations in Butte, including those of the Colorado, B&M, and B&B companies, but the Colorado announced its own suspension of smelter and mine for repairs. Clark's mines and Butte Reduction Works, Heinze's mines and MOP smelter, and the mines of other small companies were likewise unaffected by the suspension of operations at the Washoe smelter.[45]

Completion of the flue system took longer than ACM officials had estimated, and the Washoe smelter was not returned to operation until mid-September. After being closed a full eleven weeks, the ACM gradually put its Butte mines back into production, began shipping ore to Anaconda, and began firing its furnaces, just in time to host a tour of the remodeled smelter on Saturday, September 19, for representatives from some of the nation's leading newspapers.[46] The *Standard*'s article describing operations of the new flues and stack proclaimed headlines of "No More Smoke Nuisance," and "Draft of the New Stack Is So Strong That It Carries the Impurities High in the Air and Mixes with the Clouds." The article itself asserted that now the Washoe's smelter smoke was of a different color, almost white, and that farmers were already claiming that now the smoke was harmless to both plant and animal life. The *Standard* predicted that Deer Lodge Valley farmers had "experienced the last of the smoke nuisance."[47]

With the smelter operating again, Amalgamated turned to uniting the spirits of all the workers of its several companies that operated mines in Butte by sponsoring a parade in Uptown Butte and then a barbecue picnic at Columbia Gardens, an amusement park tucked against the base of the East Ridge that W. A. Clark had developed for the people of Butte as a beautifully landscaped refuge from the mining city's barren industrial environs. Seventeen Butte policemen headed the parade, followed by a band of one hundred musicians, officials of the Amalgamated companies, and then 6,500 workers organized into companies by the mines where they worked. The ACM's three main mines, the Anaconda, St. Lawrence, and Neversweat, were first, followed by a band from Helena and then workers from the B&B smelter, the Colorado smelter, two other bands, and miners from the other Amalgamated mines, including other ACM mines. Each worker wore a badge featuring an American flag. After walking several blocks through the business district, the parade proceeded east on Park Street to a location where marchers boarded streetcars. Amalgamated had arranged with the streetcar company to put every car in service to ferry workers to the Gardens, where they were greeted with four roast steers, roast pigs, and mutton, as well as twenty thousand sandwiches, forty-five thousand cigars, and ninety barrels of beer.[48]

After the men had eaten, they were regaled by a series of speakers, headed by Scallon, who said that the event was intended to secure friendly feelings between Amalgamated and its employees. The rivalries that once existed among the ACM, B&M, B&B, Colorado, Washoe, and Parrot companies had ended, and now the interests of their workers, whether from Finland or Ireland, Italy or England, were bound together with Amalgamated's. Scallon wanted the workers to know that Amalgamated relied on them to "demand equal rights and justice for all." He said, "My friends, our interests are bound up together here; if you don't prosper here we can't prosper, and if we don't, you can't either, I am sorry to say."[49] In reporting the speech, the *Miner* recognized Scallon's allusion to a pending judge's ruling in a case in which minority stockholders, friendly to Heinze, objected to Amalgamated's acquiring control of the Boston and Montana and the Parrot companies and asked the judge to deny Amalgamated the right to own B&M or Parrot stock and to derive dividends from them.[50] In the meantime, the plaintiffs sought to have the two companies placed in the hands of a receiver. If that were to happen, Scallon warned, "there would be many long and dark days for Butte." Later in the speech, he again sounded an ominous note: "If certain plans which have been formed should succeed, if the Amalgamated company should be declared an outlaw, if it should be forced to go into liquidation, if it should be forced to go into the hands of a receiver, . . . it would be pretty dull in Butte."[51] Amalgamated would be an outlaw if the judge ruled that its ownership of B&M and Parrot stocks was illegal. Earlier in the year, Kohrs's smoke bills would have, in essence, given farmers in the Deer Lodge Valley the legal means to do the same thing: declare Amalgamated an outlaw.

About a week later, Judge Clancy issued two rulings. He awarded full ownership to Heinze's company of the Minnie Healy mine, which adjoined some B&M claims and had been tied up in apex litigation for several years. More significantly, Clancy issued an injunction prohibiting Amalgamated from owning stock in, and thereby deriving dividends from, the B&M and the Parrot companies.[52]

Amalgamated responded immediately. The same day as Judge Clancy's rulings, the trust's officials issued an order that operations of all Amalgamated companies in Montana be shuttered immediately. At a stroke, fifteen thousand men were thrown out of work in Butte, Anaconda, Great Falls, and in timber, coal-mining, foundry, railroad, and other operations that supported the mines and smelters. The three thousand men in Butte and Anaconda who had been off the job for nearly three months while the smelter furnaces were connected to the new flue system had been back on the ACM payroll only about a month. With winter approaching, Butte families were desperate, and Amalgamated was blaming Heinze for the disappearance of the community's payrolls. Three days after Amalgamated announced the shutdown, Heinze stood on the steps of the Silver Bow County Courthouse to address a crowd estimated at ten thousand. In a masterful speech, he explained that he and the minority stockholders were the only protection Butte and the entire state had against complete domination by the Rockefellers and the Standard Oil trust. Heinze explained his terms for a settlement that could put the thousands of Amalgamated workers back on the job; the minority stockholders would sell their shares to the Butte Miners' Union if Amalgamated promised to transfer to him the minority shares it owned in one of his mines that was tied up in apex litigation. Following those transactions, an arbitration committee agreeable to both parties would negotiate the remaining disputes between Heinze and Amalgamated. Finally, to assure that he had the crowd's support, he demanded that Amalgamated make a three-year commitment to maintain the $3.50 daily wage prevailing in the Butte mines.[53]

Amalgamated's officials refused Heinze's proposal and made their own demand: Governor Toole should call a special session of the legislature, which would pass a new law intended to meet the approval of the Montana Supreme Court. The law voided by the court earlier in the year had allowed for a change of trial venue, if a party believed it could not get a fair trial in the district where the case was filed. Amalgamated now proposed a law that would allow a party to petition for a replacement judge if the litigant believed the initial judge was biased. Once the bill became law, Amalgamated would bring back its workers. Toole at first was reluctant to bow to Amalgamated's demands, instead trying to assemble an arbitration committee of distinguished individuals, including both of Montana's U.S. senators and James J. Hill of the Great Northern Railroad. Amalgamated officials and Heinze would not meet with the committee. Still refusing to accede to Amalgamated's blackmail, the governor was soon flooded with petitions from unions and others urging him to call the special session, which he finally did on November 10. Amalgamated immediately began reopening its mines and smelters. The legislature convened on December 1 and passed the Amalgamated legislation on December 10.[54]

The year 1903 had a profound impact on all of Montana. Citizens of the state watched in awe as a giant corporation imposed its will on their elected representatives in government, and it did so by flexing its muscles as an employer. Nearly all of Butte and Anaconda relied on the Amalgamated Copper Company for their livelihoods, and much of the rest of the state did as well, directly or indirectly. People saw firsthand what life without all those jobs in the mines, mills, smelters, and ancillary works of the copper industry was like, and they did not like it. And people in Butte and Anaconda had already experienced a foretaste of life without those jobs during the July-to-September suspension intended to finish construction of the flue-and-stack system. People experienced directly the association that could arise between jobs and the mitigation of environmental impacts on farmers' lands, crops, and livestock. Amalgamated would exploit this association in the next few years, as the farmers continued to press their case.

Managing Tailings at Anaconda, Great Falls, and Butte

While the citizens of Montana were convulsed in 1903 with the growing realization that their state government appeared to be under the thumb of a giant trust so closely associated with Standard Oil that it seemed to be the same beast, farmers and ranchers in the Deer Lodge Valley were still trying to find ways to earn their livings in the face of threats far more palpable than the stark but still abstract threat of political domination. Residents of the valley could smell the smoke and could see their poisoned livestock dying from the arsenic that the smelter smoke spread on the range. Irrigators saw the damage that water, laden with heavy metals from tailings, caused to their thirsty crops. Residents believed their very livelihoods were being destroyed.

After the special session of the Montana legislature passed Amalgamated's legislation in December 1903, many Montana citizens remained alarmed at the prospect of having their state government controlled by a giant trust associated with Standard Oil. Even as the legislature was meeting, representatives from around the state gathered in Butte in hopes of forming a new political party intent on preserving Montanans' rights to self-government. Democrats and Republicans were invited to join, and they were assured they would not be asked to take positions on national politics; Montana politics would be the focus. Observers noted that Amalgamated controlled all the Republican newspapers in the state and controlled the leading Democratic papers as well. Meanwhile, Democrats, alarmed that several prominent figures in their party were serving Amalgamated's interests, started in early 1904 to form anti-Amalgamated clubs in Cascade, Fergus, Gallatin, Broadwater, and Meagher counties. Even though Amalgamated seemed to be able to control leading Democratic officials in the state, the clubs were determined that the trust not control the party's rank and file.[55] The hegemony that Montana's copper-mining giant enjoyed over the state's politics and press has attracted the attention of numerous historians, but it is not the focus of this book. Our attention returns to the legal action farmers and ranchers in the Deer Lodge Valley took to protect their property rights.

Because of the design of the Upper Works and Lower Works and their locations relative to the hills north of Anaconda, as already described, smoke from the old Anaconda stacks evidently had not bothered many farmers and ranchers in the Deer Lodge Valley until the Washoe smelter began operating. Tailings, on the other hand, had already been damaging farmlands along Silver Bow Creek downstream from Butte and in the Deer Lodge Valley along the Deer Lodge River. Tailings posed a complex set of problems for mining companies. Millmen knew, from assaying their tailings before discharge, that coarse tailings retained relatively little copper, but the very fine tailings, the slimes, had quite high assay values, approaching the grade of the ore being milled. Yet further treatment of fine tailings was more costly than the values that could be recovered, so the tailings had to be discharged to make way for fresh ore being run through the mill, in order to keep revenues flowing. Millmen also knew they could store tailings, in hopes that improved technologies of concentration would enable the companies to recover more values from the tailings at some time in the future, but engineered storage impoundments were expensive, and the concentrators at Butte and Anaconda were discharging thousands of tons of tailings daily. It was too costly for the mining companies to keep all those tailings out of nearby streams, so a considerable volume of copper and other heavy metals washed downstream with the water that was subsequently used by farmers for irrigation.

The smelters at Butte and Anaconda faced a problem of tailings disposal not initially experienced at the Boston and Montana's Great Falls smelter. With such a tremendous volume of water flowing past its Great Falls works, the B&M simply dumped its tailings and slag into the Missouri River, out of site and presumably out of mind. When the concentrator opened in 1892, it had a capacity of about 900 tons per day. By the early twentieth century, its daily capacity had been expanded to 3,000 tons. Of this amount, about 1,950 tons was discharged daily into the river as tailings. Even with such a convenient tailings disposal system as the Missouri River, however, the B&M started hearing complaints from downstream in 1905, when Fort Benton, a city forty miles below Great Falls, filed suit against the B&M, claiming that tailings were contaminating the city's water supply. The B&M quickly settled with Fort Benton, offering to install a system of wells for the city. In 1907, William Witt told the company that tailings in the river were clogging his irrigation system. He was followed in 1908 by complaints from another rancher, Carl Shilling, and from the city of Fort Benton, which again claimed that tailings and slag were fouling its municipal water system, this time by seeping into the new wells.[56]

The B&M paid relatively little heed to Fort Benton's complaints, offering various rationalizations in addition to offering bits of help for Fort Benton's water system. The company argued that there were no sanitary problems with its discharges into the river, an argument that ignored the inorganic toxins in the smelter's wastes. As the lawyer representing Fort Benton continued his complaints, he also began to hear from interests downstream of Fort Benton who were also seeking relief. John D. Ryan, president of Amalgamated, instructed the smelter manager in early 1908 not to expend any effort in finding ways to keep tailings and slag out of the river, a command that the manager did not question until after May 1908, when torrential rains in Montana caused significant flooding in rivers on both sides of the Continental Divide, including along the Missouri River. Heretofore, slag had not been carried so far downstream, but the flood moved much of the slag downstream, past Fort Benton, and into the stretch of the river that was under the jurisdiction of the federal government as a navigable river. Ryan authorized the smelter manager to initiate a study to explore options for disposing solid wastes into the river, but the smelter never implemented such a technical solution. Meanwhile, company lawyers negotiated a settlement with the downstream rancher who was registering the most strenuous complaints, thus keeping the issue largely out of public view. These complaints concerning the Great Falls smelter occupied some of Amalgamated's legal attention, but far more significant challenges to the trust's tailings disposal practices were arising in the Butte-Anaconda area.[57]

The B&M's Great Falls smelter practice of dumping tailings into the Missouri River, out of sight and out of mind, attracted the attention not only of people downstream but of critics in the mining industry as well. Noted metallurgical engineer Hiram Hixon published a 1909 article in the *Engineering & Mining Journal*, the mining industry's leading trade journal, titled "Hunting Metals for Their Hides." Hixon compared the practice of many western mining enterprises to that of the nineteenth-century buffalo hunters, who unconscionably wasted vast resources by killing innumerable bison only for their hides, leaving the rest of the carcasses to rot. About the Great Falls practice he wrote: "The treatment of the large accumulation of tailings at Butte and Anaconda by leaching will in a few years be profitable, owing to the partial oxidation of the sulphides [*sic*], and this illustrates the hide-hunting principle practiced at Great Falls, Mont., where all tailings are swept away by the river. When they go into the river they contain as much copper as many of the ores mined in the Lake Superior region, and if properly stored would in a few years yield a large portion of their copper contents at a small cost."[58] It should be remembered, of course, that many of those Butte and Anaconda tailings dumps survived only because the tributaries of the Clark Fork River did not have sufficient flow, as the Missouri River did, to carry the tailings away.

In the early twentieth century, the Washoe smelter was the largest source of tailings in the upper Clark Fork basin, but it was by no means the only source. There were still four copper smelters operating at Butte, each with its own concentrator discharging hundreds of tons of tailings daily. Except for the Butte Reduction Works, which by then had developed an extensive impoundment system to keep tailings from flowing into Silver Bow Creek, the Butte smelters had done little to prevent tailings from washing downstream. The main reason that relatively little tailings did wash downstream was because of the systems of impoundments the smelters had developed to settle solids, so that water could be reused in the concentrators. These impoundments, although quite highly developed in some cases, did little to keep tailings at the margins from washing away during spring runoff or other periods of high stream flow. The lengths to which concentrators in Butte went to manage their tailings show that waste disposal posed serious challenges, requiring engineered solutions, long before downstream property owners began to complain. A brief description of tailings management at each of the major copper concentrators in Butte, and of the lengths to which entrepreneurs went to recover copper from those tailings, will serve to illustrate how laden with metals Butte's tailings were, whether they remained impounded at Butte or washed downstream to the lower reaches of Silver Bow Creek and the Clark Fork.

In the 1890s, after the B&M had moved its smelting operation to Great Falls, the MOP smelter was upstream-most of the Butte smelters. Located on the east side of the Butte hill just upslope from Silver Bow Creek, the MOP built a trestle to launder its tailings across the creek to the flat on the east side. The company built timber and earthen dams to impound the tailings so solids could settle. Clarified water was then pumped back to the concentrator to be reused. By the early twentieth century, the large tailings dump east of the creek had grown to more than forty feet in depth. The MOP also intermittently trammed some tailings in cars to a dumping ground west of the creek, forming a pile of tailings with a steeper slope than the dump formed by launders east of the stream. The west-side pile also grew to a height of about forty feet. The MOP built a flume three hundred to four hundred feet long so Silver Bow Creek could flow into the east-side impoundment without eroding material from the west-side dump.[59]

The B&B disposed of its tailings along Silver Bow Creek on top of an area the Silver Bow mill had used to dispose of its silver-mill tailings. A county road served as an embankment to impound the tailings. In about 1894, the B&B began to launder tailings south of the county road toward the Northern Pacific tracks, and by 1899, the tailings dump had expanded beyond the N.P. tracks to a streetcar line running from the Butte hill and across Silver Bow Creek to the southeast. The company built a tailings elevator and a trestle in 1899 to sluice tailings to higher levels. By the early twentieth century, the B&B tailings dump was about forty feet high and covered nearly twenty acres.[60]

The Parrot smelter had closed in 1899, but the tailings that remained on Parrot flat held enough mineral values to make recovery economical using later technologies. Almost immediately after the smelter closed, Amalgamated remodeled the Parrot concentrator so that it could rework old Parrot tailings on the dump south of the plant. By May 1900, the new equipment in the concentrator was treating 300 tons of tailings daily, shipping the concentrates to the Anaconda smelters. This re-treating program lasted only a short time. In the years after the Parrot smelter closed, precipitation plants also operated on Parrot tailings to recover copper in solution.[61]

Around the turn of the century, there were also numerous precipitation plants located near the major mines to treat water pumped from the underground workings. As sulfide minerals along the walls of the underground workings oxidized by being exposed to the atmosphere, they became water-soluble. Groundwater moving through the underground working dissolved the minerals, hence the metals content of water pumped from the mines. (The ongoing process of oxidized sulfide minerals dissolving in underground water is the source, generally, of acid mine drainage, or ADM, that plagues abandoned mines all over the world and is the source today of the contaminated water in Butte's Berkeley Pit that makes it, arguably, the largest body of contaminated water in the world.) In the precipitation plants, water pumped from the mines was passed over piles of scrap iron. Copper sulfate, dissolved in the water, would react chemically with the iron to yield ferric (iron) sulfate in solution and a copper precipitate, called cement copper, that settled at the bottom of a precipitation plant and was then smelted. Enterprising entrepreneurs would also lease tailings piles and capture water percolating through the piles, water that dissolved oxidized copper minerals in the tailings. In the early 1900s, there were precipitation plants at nearly all copper tailings dumps along Silver Bow Creek.[62]

In 1891–1892, the Colorado smelter, downstream-most of the reduction works in Butte, had built a new ore-dressing works adjacent to its smelter with a capacity of 200 tons/day. The Colorado's new concentrator employed a system of launders to convey tailings in a northwesterly direction, across a spur of the smelter rail system, and toward Silver Bow Creek. The roadbed of the Northern Pacific Railroad formed the south side of the tailings dump. Tailings the Colorado dumped along the bank of Silver Bow Creek forced the course of the creek northward about five hundred feet. The creek constantly eroded the margins of the tailings dump, but the dump expanded northward because the volume of tailings disposed by the Colorado was greater than the creek's ability to carry the tailings away. By the early 1900s, the dump had reached a depth of fifteen to twenty feet.[63]

As the above descriptions demonstrate, each of the Butte smelters and concentrators at the turn of the twentieth century faced serious challenges in managing its tailings. Problems facing the Anaconda smelters also presented the ACM with engineering challenges, but the challenges do not begin to describe the damage that tailings were causing to downstream property owners. The combined outflow from the Butte and Anaconda tailings dumps eventually proved so devastating to riparian farmland downstream

FIG. 6.6. Tailings on the Miles Ranch along Silver Bow Creek. This photo was entered into evidence at the Magone trial in 1905. Even though taken a few years after the Miles brothers registered their complaint with the Butte Reduction Works, it illustrates the conditions about which they complained. All the light material is tailings that settled as fine solids onto the Miles' land during high water. Photo courtesy of the National Archives at Denver, cat. # NAID 298154.

that landowners began to complain formally to the smelter companies. The first known formal complaint by a downstream farmer to one of the smelter companies was that filed by Thomas Miles, who owned a ranch along Silver Bow Creek at the west end of Ramsay Flats near the point where the creek flows into Durant Canyon (also known as Silver Bow Canyon). In 1898, Miles took his complaint to A. H. Wethey, manager of W. A. Clark's Butte Reduction Works. Wethey sent two of his men, Fred Pratt, who managed the concentrator, and James Doull, superintendent of construction at the Butte Reduction Works, to investigate conditions at the Miles Ranch.[64]

Doull and Pratt interviewed Miles at his ranch, where he told them that fine solids had been spreading over his streamside lands during spring high water for at least ten years. After construction of the Butte, Anaconda and Pacific line along the creek in the mid-1890s, he began to notice that the tailings deposits had an adverse effect on his crops. Doull and Pratt assayed samples of deposits at several locations on Miles's land and found that the materials carried as little as 0.4 percent copper and 1.0 oz. of silver per ton and as much as 0.9 percent copper and 2.6 oz. of silver per ton. They reported to Wethey that the tailings they found exhibited the red color characteristic of roasted silver-mill tailings rather than the gray color of tailings from a copper concentrator. Nevertheless, Doull and Pratt concluded that there was no question that some Butte Reduction Works slimes were settling on Miles's land.[65]

In the early 1890s, the Butte Reduction Works, just upstream of the Colorado smelter, had merely dumped its tailings along the south bank of Silver Bow Creek, forcing the bed of the creek northward. In about 1892, the company began annually to build a retaining wall of gunnysacks filled with tailings along the south side of the creek. The purpose of the wall was not so much to keep tailings from washing downstream, but rather to keep the tailings dump, which grew ever larger, from migrating upstream into the supply of clean water for the concentrator. In fact, each spring during high water, workers removed the gunnysacks so the increased flow would carry away as much of the accumulated tailings as possible. When the spring runoff subsided, workers put a new set of tailings-filled gunnysacks in place.[66]

Perhaps having learned informally that its tailings were causing problems downstream, the Butte Reduction Works began making a special effort to prevent tailings and slag from flowing into Silver Bow Creek in 1897. The company switched from gunnysacks to an open culvert or flume, built of timber pilings and plank sidewalls, which allowed the creek to pass through the tailings dump without eroding tailings. Soon the company had to build a second culvert north of the original bed of Silver Bow Creek. Engineers designed the second culvert wider than the original creek bed had been so it could carry large volumes of water during periods of high flow. Elevated tailings discharged through launders to the tailings dump caused it to grow, both in extent and in depth. As the tailings dump expanded downstream, to the west, the company extended both culverts, trying to stay ahead of the toe of the dump. Originally about eight feet high, the plank sides of the north culvert had reached a height of about twenty-four feet by 1905. The north culvert, with a greater capacity, had a length of 2,350

FIG. 6.7. The Open Piling-and-Plank Culvert That Allowed Silver Bow Creek to Pass through Butte Reduction Works' Tailings Dump without Eroding the Tailings. The view is to the east, with the Butte Reduction Works in the background. Tailings are on either side of the culvert. The photo was entered into evidence at the Magone trial in 1905. Workers would periodically change the spillway's location to enhance the volume of solids settling before water spilled into the creek. Photo courtesy of the National Archives at Denver, cat. # NAID 298160.

feet by then. The south culvert, in the original bed of Silver Bow Creek, was partially covered by 1905. The eastern half, still built of pilings and planks, was about eight feet high and was covered with more than twenty feet of tailings. The western segment was still open, with plank sidewalls about thirty feet high. It was about 1,280 feet long in 1905.[67]

The next step for the Butte Reduction Works was to impound its ever-growing tailings dump by means of walls built of slag. The walls surrounded an area more than 2,000 feet long east to west and about 1,000 feet wide. To build the slag walls, the company's workers built a temporary trestle along

FIG. 6.8. Building the Slag Impounding Wall at the Butte Reduction Works. This photo appeared in a 1909 article describing improvements at the Butte Reduction Works, including the slag wall to impound tailings. The photo shows the steel-plate forms in place and ready to receive molten slag, carried to the site in ladles on the vehicle that runs along the tracks. After the slag at this level had solidified, crews would reinstall the steel-plate forms atop the wall and build a trestle for the tracks next to the wall, so molten slag could be poured into the forms for the next lift. *Engineering and Mining Journal* 88 (11 December 1909): 1153.

the walls' alignment. Cars carrying molten slag from the furnaces were run along the trestle to points where the slag was poured between forms sitting atop the walls of already-solidified slag. The Butte Reduction Works used steel plates as formwork for erecting the slag walls. After a section of slag had cooled, the steel plates could be removed and reused as forms to build the next lift of the wall, much like slip-form construction with concrete. The slag-wall tailings impoundment was bisected from east to west by the open culverts that conveyed Silver Bow Creek through the tailings dump. Later in the decade, the Butte Reduction Works used a similar technique of slag construction to rebuild a culvert running through the dump. In 1907, the first section of covered culvert through the tailings dump, 1,200 feet long, was built of reinforced concrete. Two years later, the company built another section of culvert using reinforced slag. The covered culvert allowed tailings to be placed atop the culvert as well as on either side of it.[68]

One of the problems the Butte Reduction Works faced in its effort to control the discharge of tailings down Silver Bow Creek was that Missoula Gulch emptied into the smelter's tailings disposal area. Cloudbursts in the Missoula Gulch watershed would cause enormous volumes of water to rush down the gulch, scouring silver-mill tailings from the slopes and bed of the gulch and depositing them in the Butte Reduction Works' tailings disposal area and filling the culvert that conveyed Silver Bow Creek through the disposal area. To address the problem, the Butte Reduction Works built a slag-wall diversion for Missoula Gulch in about 1903 or 1904. It discharged Missoula Gulch to the west of the Butte Reduction Works' tailings disposal area toward the Colorado smelter.[69]

Despite efforts by the Butte Reduction Works to impound its tailings, Thomas Miles and his brother filed suits in 1902 against the owner of the smelter, W. A. Clark's Colusa-Parrot Mining Company, as well as the owners of three other Butte smelters, the B&B, Colorado, and the Parrot (despite the fact that the Parrot smelter had ceased operating by then). Others filing separate suits against the Butte smelters were the Western Loan and Savings Company, which had acquired an interest in some land downstream of Butte, and E. J. Erickson. The plaintiffs' attorney was John A. Shelton. All the lands in question were upstream of where tailings from the ACM's concentrators at Anaconda flowed into the Deer Lodge River. The judge ruled against the plaintiffs because, while each of the smelters may have discharged pollutants into Silver Bow Creek, the plaintiffs could not determine what portion of the total each of the smelters had caused to the agricultural lands.[70]

As with the Butte smelters, management of tailings and concentrates at the Washoe smelter required considerable attention as well. And, as at the Butte smelters, the Washoe discharged copper and other heavy metals into the Deer Lodge River.[71] When it first opened, the concentrator at the Washoe produced four classes of product: 1) coarse concentrates, which went directly to the blast furnaces; 2) fine concentrates, which went to the roasting furnaces and then the reverberatories; 3) slimes, containing about 2.6–2.8 percent copper; and 4) tailings, containing about 0.8 percent copper, which went to the tailings dump. No tailings particles were larger than 1.5 mm. The slimes contained very fine solids; 90–95 percent of the particles would pass through 200 mesh (0.0067 mm). For a brief initial period, the slimes were discharged from the concentrator with the rest of the tailings. Because the slimes still retained considerable value in copper, the ACM built ponds northwest of the works to store slimes for further treatment. The first two ponds for processing slimes were built in January 1903, two were finished by November 1903, and two more ponds were added soon thereafter, bringing the total to six. Each pond, of about four acres, allowed the slimes to sit relatively motionless so solids could settle to the bottom and the water could then evaporate or flow away. Periodically, the ponds were drained, and the solids were scooped from the ponds and set in piles to allow more water to drain out. Water discharging from the ponds, along with water draining from the slag pile, flowed east in a ditch to join the rest of the tailings stream. There, solids still present in the water could settle out behind the impoundments for coarse tailings. In December 1903, a ditch was built to carry water from the slime ponds and the slag pile directly to the Deer Lodge River, bypassing the tailings impoundment. Sometimes, when the slime ponds were being emptied of their accumulated solids and there was no place to discharge slimes from the concentrator, slimes were simply allowed to flow down the ditch toward the tailings disposal area.[72]

To recover copper still in the slimes, they were used in two of the Washoe's departments, the briquetting plant and the converter department. Unwatered slimes hauled to the briquetting plant were mixed with equal parts of finely screened first-class ore and table concentrates from the settling tanks at the concentrator. A small amount of reverberatory coke (recovered from the ash from the reverberatory firegrates) was added to the mixture. A briquetting machine cast the mixture into briquettes, which were then charged to the blast furnaces, thus recovering the copper in the slimes. Unwatered slimes hauled to the converter department were mixed as a binder with siliceous ore to be used to form the converter linings. Again, copper in the slimes was recovered, in this case as the converter linings were eaten away during the bessemerization process. After several firings, discarded converter linings were charged to the blast furnaces, again to recover the copper content.[73] Despite these steps to recover the copper present in slimes, the ACM did not exercise special care to keep slimes out of the river.

At the time the Washoe smelter went into operation, the Washoe and

Anaconda companies owned several thousand acres of land north and east of the works used for the disposal of tailings. These tailings disposal areas had already been used for the disposal of tailings from the Old Works. Sand tailings from the Washoe concentrator were discharged through launders to the flat extending to the east and situated between Warm Springs Creek on the north and Mill Creek on the south. Coarser and heavier material settled almost immediately and tended to pile up under the ends of the launders. Lighter and finer materials in the tailings were transported by the water farther downgradient to the north and east.[74]

From the beginning of the Washoe's operations, the ACM employed a series of embankments to impound tailings and slow their flow so that water, when it left the impoundments, was as clear as possible. Within a year of the Washoe's opening in 1902, three such dams had been built. The first impoundment, very near the launders, was filled almost immediately. The second impoundment was actually the old Montana Union roadbed, with improvements so that it would function as a dam. An old bridge along the roadbed was rebuilt to serve as an overflow and sluiceway for dam no. 2. By 1905, the no. 2 impoundment was also almost filled with tailings. The third dam, built at the end of 1902, was northeast of the old roadbed. By 1905, a fourth dam was built farther to the east. From the spillway of dam no. 4, water flowed along a natural channel through a low swampy area and into the Deer Lodge River near what was called the French Crossing Ranch. John Gillie, general superintendent of Amalgamated's mines at Butte, claimed that the ACM's tailings impoundment system was far superior to any other in North America. He acknowledged, nevertheless, that sometimes the water discharging into river was cloudy. He also acknowledged that some of the slimes were oxidized, making soluble copper sulfate, which could then be dissolved and carried downstream.[75]

Silver Bow Creek also carried tailings and slimes from Butte, through Durant Canyon, and into the Deer Lodge Valley. With all those tailings flowing into the Deer Lodge Valley, farmers started seeing problems with their crops toward the end of the 1890s. James Saville had a ranch in the upper Deer Lodge Valley at Stuart, near where Silver Bow Creek flowed out of the canyon. He had used the creek for irrigation since 1880 and first noticed damage to his crops in 1898. Peter Valiton farmed along the Deer Lodge River about three miles south of Deer Lodge. He took his irrigation water from the river, below the mouth of Warm Springs Creek, which was also below the point where outflow from the Anaconda company's tailings impoundments joined the Deer Lodge River. After noticing that slimes settling on his irrigated fields damaged crops, he inadvertently discovered that, by conveying irrigation water from the river through a low-lying slough, he could cause most of the slimes to settle out of the water before he channeled it to his fields. Tailings from Butte and Anaconda smelters even settled on farmland north and downstream of the city of Deer Lodge. Charles Williams owned a farm six miles north of Deer Lodge and took water for irrigating from the Deer Lodge River. He had noticed how muddy his irrigating water was for years, but in 1898 he began to believe that it was damaging his crops. By the early twentieth century, he had many areas in his fields where nothing grew.[76]

The Magone Tailings Suit

Shortly after the turn of the century, some of the farmers and ranchers who believed their lands were being damaged by tailings filed legal complaints. The first known tailings suit in the upper Clark Fork basin was filed in Montana District Court in Butte by William Norton, Mary Ann Norton, Rachel Erhart, and Morgan Thomas in June 1901. Their land was located just north of Stuart. Because their property was situated along the Deer Lodge River upstream of the point where outflow from the ACM's tailings dumps entered the creek, the Nortons and their associates filed their complaint against the Butte companies, naming the Colorado, Colusa-Parrot (Clark's company that owned the Butte Reduction Works), the Parrot, and the MOP. In subsequent years, other property owners along Silver Bow Creek, the Deer Lodge River, and the Clark Fork also filed suit against the Butte companies and the ACM, but the largest of the tailings suits was brought by Hugh Magone against both the Butte and the Anaconda smelters in 1903.[77]

Filed in federal court in early September, more than a week before the Washoe resumed operations, following the eleven-week shutdown to connect furnaces and their flues to the new flue-and-stack system, Magone's suit sought $20,000 in damages and an injunction to prevent the smelters from continuing to discharge tailings into streams used for irrigation. In addition to suing the Washoe Copper Company, Magone named W. A. Clark's Colusa-Parrot Smelting and Mining Company (Butte Reduction Works), Heinze's Montana Ore Purchasing Company, and the three Amalgamated companies that operated smelters and concentrators in Butte: the Colorado Smelting and Mining Company, the Butte and Boston Consolidated Mining Company, and the Parrot Silver and Copper Company (even though its smelter was closed). Although Magone was seeking damages for harm done to his own property, he was filing the suit in cooperation with others in the valley, who would be helping to pay the costs of litigation against the collective might of all Butte's big mining companies. Magone's suit would grow to record size, but it attracted little attention in 1903, as most Montanans were more immediately concerned with the ramifications of Amalgamated's now evident political power.[78]

Hugh Magone was a rancher who had moved to the Deer Lodge Valley in the late 1880s. In 1891, he bought a half-section of land along the Deer Lodge River in the Race Track area (about ten miles south of Deer Lodge) from the Northern Pacific Railroad. The land included both benchland and bottomland. Starting with two cows and a calf, he had built his herd to about 150 head in 1901, his best year. When he bought the land, it was all in pasture, and the bottomland supported grasses, sage brush, and willow. Each year, Magone plowed more ground and put more acres into cultivation, starting with a few acres of grain, potatoes, and hay. He increased his acreage in hay especially, planting clover and timothy to support his growing cattle herd. As was typical of cropland in the Deer Lodge Valley, Magone's ranch relied on irrigation. An irrigation ditch had been put through his land in the late 1880s, and shortly after he bought the place he also bought an interest in the ditch so he could use the water. The source of the ditch was the Deer Lodge River.[79]

Magone had noticed when he began irrigating with water from the Clark Fork that it was muddy, but he claimed not to have noticed whether there were any deposits of tailings on his fields, either existing or caused by irrigation, because he did not know at the time that tailings could be harmful to crops. His operation grew through the 1890s, so that by the end of decade he was ready to expand his holdings. In 1899, Magone bought an adjoining quarter-section of bottom land, known as the Roseborough place, at sheriff's auction. He had not noticed whether there were tailings on the Roseborough place either. After his peak year of production in 1901, Magone's crops began to diminish, even though he applied manure to his land in an effort to maintain fertility. Declining production caused Magone to begin to suspect that tailings were the cause of his problem. He noticed that by the early 1900s tailings had settled over all the low-lying areas of his bottom land, exhibiting different colors in different areas. In some places, what he took to be tailings were white, some deposits were green, and some areas exhibited what he described as a slate gray appearance. Many such areas no longer supported vegetation. In September 1903, Magone filed his suit in federal court.[80]

In April 1904, Magone, Nick Bielenberg, and Mrs. Mary Norton, all of the Deer Lodge Valley, and Clowes Miles, who had a ranch along Silver Bow Creek just east of Durant Canyon (and who with his brother had filed suit against the Butte Reduction Works and other Butte smelters in 1902, complaining of tailings damage), called a meeting, to be held at the Powell County Courthouse, of property owners interested in gaining relief for damages caused by the smelters' pollution of the Deer Lodge River. Deciding that the preliminary name of their organization would be the Anti-Debris Association, they ratified articles by which to govern themselves. The initial name they chose for their organization was that used in the late 1870s by farmers in California's Sacramento Valley, who had successfully fought the hydraulic-mining industry's practice of sluicing debris from gigantic placer mining operations into the tributaries of the Sacramento River. Anti-debris associations continued to function in California well into the twentieth century.[81]

During 1904, the mining companies filed their responses with the court, each one denying that it was responsible for the injuries to crops and land from which Magone claimed to be suffering. The judge ruled that he wanted to hear evidence, so preparation for the trial proceeded. The Anaconda company began surveying tailings damage to farms in the Deer Valley, distinguishing between lands damaged by direct overflows from the stream and lands that were under irrigation ditches and subject to damage by that means. The farmers arranged to have three professors, two from Iowa State University in Ames and one from the University of Nebraska, analyze samples of water the farmers had taken from Silver Bow Creek. In early 1905, as the trial date approached, attorneys for the parties argued in court about how the farmers' experts could present their testimony. Meanwhile, attorneys for the mining companies argued that the city of Butte should also be named as a defendant in the case.[82]

The trial began in March 1905 in the court of U.S. District Judge William H. Hunt. Born in New Orleans in 1857, Hunt attended Yale University before transferring to the University of Louisiana (which later became Tulane University) to study law. He moved to Dakota Territory in the late 1870s and then to Montana, where he served as attorney general of the territory from 1885 to 1887. He was an associate justice of the Montana Supreme Court from 1897 to 1900. After serving as governor of Puerto Rico from

1901 to 1904, he was appointed by President Theodore Roosevelt to be federal district judge for the District of Montana, where he served until 1911, when President William Howard Taft appointed him to the short-lived U.S. Commerce Court. When Congress abolished the Commerce Court in 1913, Hunt was appointed to the Ninth Circuit of the U.S. Court of Appeals. For the Magone trial in 1905, Hunt appointed Oliver T. Crane as master in chancery to hear the testimony of the numerous witnesses that both sides planned to present. Crane was born in New York in 1855. He, too, attended Yale, graduating in 1879. After being admitted to the bar in New Jersey in 1882, he visited Dakota Territory and then moved to Montana in 1883. He tried his hand at sheep ranching for four years and then traveled abroad for several years. He moved to Helena in 1890 and took a job as reporter for the Montana Supreme Court. He was hired as librarian for the State Law Library in 1902 before Hunt appointed him master of chancery for the U.S. District Court.[83]

John A. Shelton of Butte and Thomas J. Walsh of Helena represented Magone. Shelton had represented the Miles brothers and the other plaintiffs in the 1902 suits against individual Butte smelters. Thirty-nine years of age at the time of the Magone trial, Shelton was a native of Iowa who had studied law at the University of Iowa and the University of Michigan and worked for a law firm in Des Moines before moving to Butte in 1892. He was admitted to the Montana bar in 1893. Walsh was born in Wisconsin in 1859. After earning a law degree at the University of Wisconsin in 1884, he practiced law in South Dakota before moving to Helena in 1890. Walsh would later become a U.S. senator from Montana.[84] Cornelius F. "Con" Kelley and L. O. Evans of the ACM's legal staff represented the four Amalgamated facilities named in the suit (ACM/Washoe, B&B, Colorado, and Parrot). W. A. Clark and F. A. Heinze each had his own attorneys to represent their respective smelters, the Butte Reduction Works and the MOP. One interesting sidelight to the trial was the way attorneys for the companies, which otherwise competed in apex litigation, cooperated closely in defending their respective clients in the Magone case. The year 1905 was near the peak of Heinze's legal battles with Amalgamated in other courtrooms, but during introductory motions on the first day of the trial, L. O. Evans reported to the master that, on that particular day, he was authorized to represent the MOP, because Heinze's attorney, James Denny, was unable to be in attendance.[85]

Magone's attorneys first presented Robert Ellis, a photographer who had taken photos of the tailings deposits at each of the Butte smelters. Ellis knew little about smelting, little about his camera, and little about how to present himself as a knowledgeable expert in court. By the end of his cross-examination by the companies' attorneys, Ellis was unwilling to attest definitively to almost anything that Magone's attorneys had hoped the photos would represent visually.[86]

Magone's attorneys also presented several farmers and ranchers, who described the impacts they had observed tailings had on their lands and crops. The first farmer to testify on Magone's behalf was Charles H. Williams, who lived on his ranch about six miles north of Deer Lodge. He had moved to Montana in 1882 and acquired his first parcel of land, the 160 acres along the Deer Lodge River on which he lived, in 1883. He continued acquiring adjacent land, and in 1897 he joined with Henry B. Davis of Deer Lodge to incorporate the Davis and Williams Company. By the time of the Magone trial, Davis and Williams owned about twenty thousand acres of land in the vicinity, Davis had retired from the firm, and Williams owned the company along with his son, Lee Williams, and Peter Pauley. Williams mainly grazed sheep, along with a few workhorses and very few cattle. He irrigated natural hay meadow along with cultivated timothy, clover, wheat, oats, and a little alfalfa. He testified that when he first moved to the Deer Lodge Valley, water in the river was clear, and he could drink from it. He also testified that fish lived in the stream, a statement to which the defendant mining companies objected, because the issue in the case was whether the water in the Deer Lodge River, when used for irrigation, was deleterious to vegetation, not whether fish could live in the stream. Magistrate Crane sustained that objection, but he overruled most of the companies' other objections. The transcript of the Williams testimony is very lengthy, because of the companies' frequent and lengthy objections to questions and Williams's answers in direct examination.[87]

Williams had first noticed the irrigation water he took from the Deer Lodge River becoming a bit muddy in the early 1890s, but he thought little of it until the mid-1890s, when farmers to the south (upstream) began commenting on it, thinking the substance in the water might be damaging to crops. Williams first began noticing adverse effects on his crops around 1898. A diligent farmer, he first thought his diminished crops might be due to tired soil, so he tried summer fallowing (letting the soil rest once every three years) and planting clover, knowing that clover could replenish nitrogen in the soil. Those efforts to return his crop output to levels he had previously achieved failed, and he concluded that the declining productivity of his land was due to tailings (particularly slimes) in water from the Deer Lodge River. He characterized the water as "roily," and often "slatey" in color, although sometimes the color of the fine solids varied. Water from the river deposited a layer of fine solids over most of his irrigated ground, three or four inches thick in some areas and very thin but noticeable in others. By the time of his testimony at the Magone trial, his irrigated land would no longer produce grains and produced little if any hay. He had also concluded that dust from the slime layer, when it dried, killed his sheep when he turned them into the irrigated land to graze. Williams did not believe that his sheep had died from smelter smoke because, he said, his place was so situated in the valley that it received little of the smoke. He had cut open some of his dead sheep and saw what their condition had done to their internal organs. When he first had had Dr. Knowles examine his sheep, the state veterinarian attributed their demise to a germ, but by the time of the Magone trial, Knowles had changed his diagnosis and agreed that the sheep had died from ingesting dust while grazing on Williams's irrigated land.[88]

Peter Valiton also testified on Magone's behalf. Since the mid-1870s, Valiton had owned and operated a ranch along the river about three miles south of Deer Lodge and also downstream of the mouth of Warm Springs Creek (and hence below where ACM tailings entered the river). He had his own suit against the ACM pending at the local court in Deer Lodge. Valiton described his diminished crop yields, and he also stated that the layer of slimes on his hay meadow was so thick that he had to clean the slimes from his mower's sickle bar after each pass through the meadow.[89] James F. Saville owned and operated a ranch at Stuart, which he bought in 1880. It was upstream of the Magone place, and also upstream of where ACM tailings entered the river. Having lived in Butte for a few years in the 1890s, while he rented his ranch to others, he was familiar with the smelters there. Saville described the stuff in the water that he took from the Deer Lodge River for irrigation as "dark gray matter, with the apparent of fineness [*sic*] of flour boiling up from the water all the time, and it gives a slimy, muddy appearance." He said that when the substance was deposited on his ranch and exposed to the atmosphere, it took on a "greenish cast," which he compared to copper.[90]

After forty days of testimony, Magone's attorneys finished presenting their witnesses in mid-April. Unfortunately for Magone, none of his side's five expert witnesses appeared personally before Crane, so they could not be cross-examined; Magone's attorneys only submitted depositions for the five experts.[91]

The mining companies, on the other hand, presented an impressive array of witnesses, both expert and lay. They presented old-time miners, who recalled the first days of mining in Butte and southwest Montana, including how muddy placer mining had made Silver Bow Creek and the Clark Fork. They presented superintendents and foremen from the various concentrators, who testified to the techniques used for concentrating ores and for disposing of tailings. The companies presented agricultural specialists, who testified that Magone and the others were having problems because they practiced poor farming methods, especially by not compensating for the alkaline soil conditions in the Deer Lodge Valley. And they presented soil chemists, who testified that, even though soluble copper sulfate leached from tailings and flowed down Silver Bow Creek and the Clark Fork, the alkali soils of the Deer Lodge Valley converted the copper mineral to an insoluble form, thus rendering it not harmful to plant life.[92]

Crane had decided in May that he would not visit the Magone place, explaining that he was going to receive only that evidence he could convey to Judge Hunt in the record. Crane finished hearing oral testimony, and his court recessed in mid-August 1905, by which time court reporters had filled 6,300 pages with typed transcript. In doing so, new technology was making its way into the trial. The hearing lasted about four hours each day, during which time the court reporters took notes on about thirty thousand words. After each hearing, the reporters adjourned to an upper room in the courthouse, where they read from their notes into a phonograph. The reporters then listened to the recordings of their own voices to type the transcript. Using such a process, a transcript was ready to be read and signed by each witness a day after giving testimony. The trial resumed in mid-October and continued for another month, as depositions were read into testimony. With the depositions, testimony in the case filled eight thousand typed pages, said to exceed by about three thousand pages the previous record for testimony in Montana litigation. The trial had been expensive for both sides, and observers considered it but a preview of the litigation that was sure to ensue if Magone prevailed.[93]

By the time testimony in the Magone trial had ended, the farmers had joined forces to file an even more monumental suit, described in the next section of this chapter. The new suit was only against the Washoe smelter, and in that complaint the farmers sought to enjoin the Washoe Copper Company and the Anaconda Copper Mining Company from operating the smelter in such a way that its smoke killed their livestock, damaged their crops, and, they alleged, rendered their farms and ranches in the Deer Lodge Valley worthless. The two cases were distinct, with Fred Bliss serving as plaintiff in the smoke suit, and none of the Butte smelters was named as a defendant, but the mining companies in the Magone case sought to link the two in the eyes of the magistrate. In July, when Magone's attorneys put three character witnesses on the stand to impeach the integrity of a defendants' witness, Ben Crosswhite, who had testified earlier about the poor character of Magone's ranch, the defense attorneys cross-examined each of the character witnesses by asking if he was a member of the association of farmers that had formed to sponsor the Bliss suit. On cross-examination, each of the three witnesses acknowledged that he had filed a damage claim against the ACM, that he was a member of the smoke association, and that he had paid one hundred dollars to help fund the organization's work. In fact, the three men had attended a meeting of the smoke association the day before they testified about Crosswhite's character. The intent of the questions seemed clearly intended to show that the witnesses were not disinterested when denigrating Crosswhite's character.[94]

In March 1906, Oliver Crane issued his findings of fact. He found that the several concentrators named as defendants by Magone had indeed been discharging tailings into the streams and that those tailings had found their way to Magone's land, but he did not find sufficient reason to grant an injunction against the companies. Employing the balancing doctrine, he found that it would not be in the best interest of the greatest number of people to halt operations of the mills and smelters. Crane did, however, recommend that Judge Hunt order the defendant companies to pay Magone damages totaling about $1,700.[95] The companies received little more than a slap on their wrists.

Crane issued his findings a month after the Ninth Circuit Court of Appeals had reversed a U.S. District Court ruling in California. The United States had filed suit against the Mountain Copper Company, whose smelter smoke was damaging timber resources on National Forest lands in northern California. The company appealed to the Ninth Circuit after the district court enjoined the company from operating its smelter in a way that damaged the adjacent forests. Circuit Judge William B. Gilbert wrote the opinion overturning the district court's ruling, with Circuit Judge Erskine

M. Ross consenting. (Thomas Hawley, judge for the U.S. District Court for Nevada, was the third judge who heard the appeal, and he dissented in the opinion.) The Ninth Circuit opinion stated that the United States' role in the litigation was that of a property owner, not the sovereign, so the United States' status was no different from any other property owner who might complain of damages. The judges found in favor of Mountain Copper because the economic harm caused by closing the smelter would be far greater than the economic harm smelter smoke was causing to the government's forests. The Ninth Circuit also rejected the government's argument that Mountain Copper should be required to install a flue-and-stack system as the Washoe smelter recently had done, because the Washoe's system could recover arsenic, which was of value to the ACM. Mountain Copper's ores had no arsenic, however, so such a system would be of no financial value to the company. Gilbert and Ross both believed in the development of natural resources in the American West, and the precedent they established in the Mountain Copper case helped to shape the outcome of the farmer's continuing complaints against the Washoe smelter's smoke. The dissenting opinion in the Mountain Copper case argued that profitable corporations did not have the right to destroy the property of neighboring landowners, a belief that would continue to animate the legal pursuits of the farmers in the Deer Lodge Valley.[96]

The Bliss Smoke Suit

Back in 1903, as their state senator, Conrad Kohrs, had tried to pass legislation aimed at enhancing their legal tools for addressing the smoke problem, and as they had held meetings aimed at organizing a collective effort at recovering damages from the smelter company, many farmers and ranchers in the Deer Lodge Valley initially resisted the ACM's efforts to settle claims for damages individually, especially in light of the release that the company demanded from each farmer in return for payment of damages. American farmers had historically joined agricultural organizations in large numbers to promote their collective welfare, and the Deer Lodge Valley farmers were engaging in local collective action to pursue their mutual ends.[97] The farmers had sustained serious losses in 1903, however, and their situation was dire, so they began settling individually. By the end of the year, all but three ranchers—Nick Bielenberg, the Quinlan brothers, and August De Rossier—had settled with the company for damages.[98]

Fig. 6.9. Washoe Smelter with Its New 300-Foot Stack. This view to the south shows the smelter and the flue extending up the hill to the base of the new stack. Note that on this day, the smelter smoke is not being carried aloft to the upper atmosphere but is drifting out over the Deer Lodge Valley to the northeast, where farmers and ranchers grazed their livestock. Photo courtesy of the World Museum of Mining, cat. # 01516.

In 1904, as the parties in the Magone case were preparing for the tailings trial to begin in federal court the next year, smoke continued to vex farmers and ranchers in the Deer Lodge Valley. In the spring of 1904, Sorn Beck, who raised sheep along the Little Blackfoot River north of Deer Lodge, moved some of his animals south to the Race Track area to graze on new grass and stubble in grain fields. Within a few weeks, his sheep began dying; when he passed through Deer Lodge in mid-April, he reported that they were dying at a rate of fifty per day. By the time he could move them back to safe grazing, he expected to have lost several hundred sheep. Within a week, Dr. Knowles, the state veterinarian, had been summoned to examine Beck's sheep. After inspecting their stomachs and other organs, he concluded that they had died of arsenical poisoning. Knowles took several of the organs to a pharmacy in Deer Lodge and placed them in jars with alcohol, so that they could be tested later by a chemist. Beck then traveled to Anaconda with his attorney to file a claim for damages at the smelter, but the company refused to pay him, asserting that the smelter could not be the cause of Beck's problems because the new stack-and-flue system was recovering all the arsenic from the smelter smoke. Beck reported that the company agent suggested that the problem may have been caused by smoke from the Butte smelters drifting more than twenty-five miles to the Race Track.[99]

A month later, Knowles invited W. D. Harkins, a chemistry teacher at the University of Montana in Missoula, to investigate conditions on the ranges in the valley. After graduating from Stanford University in 1900 with a bachelor's degree in chemistry, Harkins had taken the job in Missoula. He had first visited the Deer Lodge Valley in November 1902, when some residents asked him to investigate the extent to which arsenic from the Washoe smelter's smoke had settled on vegetation in the valley. His sampling had shown considerable arsenic on the vegetation and also smaller concentrations of arsenic in the organs of animals he tested.[100] He would play an ongoing role in providing scientific expertise in support of the farmers' complaints.

Farmers began negotiating with ACM officials in autumn 1904, but company officials soon asked that negotiations be suspended until after the November election. In December 1904, frustrated with the inaction, farmers and ranchers met at the Willow Glen School and agreed on a course of action. They formed the Deer Lodge Valley Farmers' Association, raised more than $3,000 among themselves, and agreed to hire a chemist and a veterinarian to begin gathering evidence in support of their cause. Meanwhile, property owners to the north held a similar meeting on the last day of the year, at the Powell County Courthouse in Deer Lodge, to organize support from the lower part of the valley. They read the minutes of the Willow Glen meeting, raised $2,000, and authorized Harry Quinlan and C. H. Williams to visit ranchers in the valley and encourage them to join. They discussed the possibility of joining forces with the Deer Lodge Farmers' Association, and they agreed to hire an attorney.[101]

In December 1904, the *Silver State* also began featuring regular coverage of the smelter smoke controversy underway in the Salt Lake Valley, where the Highland Boy smelter had commenced in 1899, and where smoke was not only damaging agricultural lands but also causing disturbance in Salt Lake City and Ogden. Farmers there presented evidence to the smelters of dead and ailing livestock, and residents filed complaints with the county board of health and state board of health. The Utah Consolidated Mining Company, which owned the smelter, suggested forming a board of experts to study the problem, which the farmers construed as a delaying tactic. Meanwhile, another smelter in the valley was under construction, adding urgency to the farmers' situation.[102]

The Deer Lodge Valley Farmers' Association experienced similar delays and intransigence at the hands of the ACM. The association held another meeting at Willow Glen School, at which nearly all the 107 members were present. They appointed a nine-member committee to receive claims for damages. The committee had the authority to evaluate claims and to exclude any claims considered to be excessive. After evaluating claims totaling $2,300,000, the committee concluded that claims worth $1,750,000 had merit. Negotiations with the ACM resumed in February, but then new ACM president, John D. Ryan, asked for another delay until April 1905. The committee sent the ACM a proposed settlement: members of the association would accept payment of about two-thirds of the damages claimed ($1,120,000), and they would also sell their land, water rights, and improvements to the company for about $920,000. If the company accepted the settlement by April 7 and made payment by May 1, 1905, the farmers would vacate the lands and turn them over to the company by November 1. If the company did not accept the settlement by April 7, the association would reserve the right to initiate legal action. At that point, the company broke off negotiations entirely, leaving the association no recourse but to sue. On May 4, 1905, Fred Bliss, an Idaho resident who owned a farm about six miles northeast of the Anaconda stack, filed suit on behalf of his neighbors in federal court, asking for an injunction to close the Washoe smelter if the Anaconda company did not cease spewing poisonous smoke across the Deer Lodge Valley.[103]

Attorneys R. L. Clinton of Butte and C. M. Sawyer of Anaconda represented Bliss, and ACM lawyers A. J. Shores, D. Guy Stivers, and C. F. Kelley represented their company. When the ACM lawyers filed their response to the Bliss complaint in June, they cited the $300,000 in damage claims the company had already paid for harm done by smelter smoke and the millions of dollars the company had recently spent at the smelter to prevent its smoke from further injuring the farmers' personal and real property. The company denied that the Deer Lodge Valley was ever a fertile agricultural area or that lands there were worth as much as the farmers claimed. And the company offered as its main argument that the city of Anaconda owed its existence to the smelter, that Amalgamated's Butte mines employed eight thousand men who depended on the Washoe smelter, that the Deer Lodge Valley farmers marketed much of their produce in Anaconda and Butte, and that if the Washoe smelter were forced to close, the Butte mines and the coal, timber, freight, and other operations that supported them would have to close. Butte and the entire state were very much dependent on the Washoe and Anaconda companies.[104]

Judge Hunt ruled that the trial would proceed and that Oliver Crane would again serve as master of chancery to hear and receive testimony. Then in late summer, the parties organized a tour of the Deer Lodge Valley for Hunt and Crane. Taking place over two days during the third week in August, the tour was conducted in two automobiles, with Hunt and representatives of the company in one car and Crane and representatives of the farmers, along with Con Kelley, in the other. On Wednesday, the ACM chose the sites to visit. After visiting the Bliss place and a number of ranches in the foothills on the west side of the valley, the party received a tour of the Washoe smelter and stayed at the Montana Hotel, Anaconda's fanciest, and which was owned by the ACM. The next day, the plaintiff's attorneys chose the sites to visit, focusing on farms and ranches in the south end of the valley, where damage from smelter smoke was the worst.[105]

The ensuing trial entailed fourteen months of proceedings, a total of 237 witnesses, and testimony filling some twenty-five thousand pages of typed transcript. It was said to have been the longest and costliest injunction suit ever brought before an equity court in the U.S.[106] In addition to Bliss's own testimony, his lawyers were able to marshal a broad assortment of farmers and ranchers from the area to describe for the court the extent of damages the Washoe's smoke had caused. The farmers' central complaint was that their animals were being poisoned by smelter smoke, but witnesses also described extensive damage to crops, timber, and other vegetation. One of the most common symptoms attributed to the smoke was sores or ulcers that appeared in the nostrils of horses. Cattle would scour (suffer from diarrhea), not shed their winter hair, develop watery eyes, and often deliver dead calves or slink their calves (deliver them prematurely). Calves born alive were unusually weak. Dairy cows quit giving milk. Animals' breath gave off what their owners described as the odor of garlic.[107] John Collins had been a state stock inspector since 1893. He remembered that the Deer Lodge Valley used to provide excellent range for stock, but in the years since the Washoe smelter started operating, stock populations had dropped precipitously. At the time of the trial in 1906, he claimed there were hardly any animals in the valley for him to inspect, and there were virtually no horses left.[108]

Farmers who raised small grains generally reported tremendous decreases in yield. They claimed they could tell that smelter smoke caused the heads of the grains to turn red. Farmers and ranchers did not necessarily report diminished hay production, but they were convinced that smoke caused their hay and pasture to be harmful to animals. Ephraim Staffanson's place was about four miles from the Washoe and less than two miles due north of the Anaconda Lower Works. He had noticed some damage to his hay while the Old Works operated, but he testified the damage increased remarkably in the fall of 1902, shortly after the Washoe smelter began operating. He claimed smoke from the smelter turned his hay yellow and brown and that it burned holes in the vegetation. Staffanson was one farmer who did claim that his hay crops diminished.[109] John Bohn was a truck farmer on Dempsey Creek in Powell County, about twelve miles north of the smelter. Even at that distance, he testified, the smoke was sometimes so thick he could not see his neighbor's house, less than a quarter mile away. Bohn had successfully raised produce for many years before the Washoe smelter opened. Since the smelter opened, he had seen his crops diminish: cabbages, peas, and tomatoes wilted, developing yellow and white spots on their leaves. His potato yield was reduced to 20 percent of what it had previously been. Once-productive berry bushes had grown sickly and quit producing, and he had ripped them out. Bohn also described the same kinds of damage to his herds and grain crops that other farmers were reporting.[110]

The Anaconda company also presented farmers as witnesses. One group of witnesses was dairy farmers from the Butte vicinity, who had bought hay from the Deer Lodge Valley after the Washoe smelter had opened. They reported their cows sustained no harm from eating the hay. Some of the dairy farmers were close enough to the Colorado smelter, Butte's westernmost smelter, to occasionally experience smelter smoke. They testified that it had done their farms and their cattle little or no harm. Charles Harrington operated a dairy just to the northwest of Anaconda, and he also testified that smoke, even from the Old Works, had not interfered with his business.[111] Another group of witnesses were experienced agriculturalists from the Deer Lodge Valley who testified that many of the individuals appearing on Bliss's behalf were not very good farmers or ranchers, that the livestock industry had been in decline in the Deer Lodge Valley for more than twenty years, and that diminution of crops was due to other causes, like quality of the soils.[112]

Peter Jessen, who had farmed along the North Branch of Mill Creek about four miles east of the Washoe smelter since 1869, testified that he believed his diminished grain crops were due to tired soil after thirty years of continuous cultivation. Even though he had lost several animals to smoke immediately after the smelter opened in 1902 and had been paid damages by the ACM, Jessen did not believe the smoke plume came in contact with his place much. He also believed worsening range conditions in the valley were from overgrazing.[113] Artemus M. Day, who farmed along lower Gray's Gulch about three miles west of Anaconda and raised vegetables for market, testified that smoke from the Old Works used to bother his garden crops, but since the Washoe smelter went into operation in 1902, smoke did little damage to his crops.[114]

Both sides also assembled teams of experts including academic scientists.[115] Two chemistry professors, Robert Swain of Stanford University and W. D. Harkins of the University of Montana, working under contract to the government, had completed some of the first field research in the case during the summer of 1905. By installing testing devices at the smelter, they tried to determine the quantity of arsenic leaving the smelter's stack. Swain and Harkins testified for the farmers that the Washoe stack discharged forty-four thousand pounds of arsenic across the environment daily. Daniel E. Salmon, the noted veterinarian who had been the first graduate of Cornell University's new program in veterinary medicine and founding chief of the Bureau of Animal Industry at the U.S. Department of Agriculture (USDA), testified about the symptoms that Deer Lodge Valley livestock exhibited of arsenical poisoning. The farmers were also successful enlisting the federal government to supply experts from other agencies in the USDA to assess the damage being caused by the smelter. W. G. Weigle of the U.S. Forest Service examined the effects of smelter smoke on forestlands in the vicinity of the smelter. J. K. Haywood of the USDA's Bureau of Chemistry assessed the chemical effects of emissions from the smelter, including sulfur dioxide on plant life and arsenic on animal life. He also examined the effects tailings had on crops when applied to fields by means of irrigation with streams impregnated with tailings. The USDA's Bureau of Animal Industry sent Robert J. Formad to the Anaconda area to investigate the impacts of smelter emissions on animal life.[116]

Fig. 6.10. Daniel E. Salmon, in fur coat, standing at a barn in the Deer Lodge Valley viewing a horse. Photo Courtesy of Cornell University Library, Rare and Manuscript Collections, V. A. Moore Collection.

E. P. Mathewson took charge of assembling the ACM's team of technical and scientific experts. A native of Montreal, Quebec, Mathewson had graduated from McGill University with a bachelor of applied science degree in 1885 and then worked briefly for the Dominion (Canadian) Geological

Survey. In 1886, he went to work as an assayer for the Pueblo Smelting and Refining Company's smelter in Colorado. Working his way up to chemist and assistant superintendent, he was made superintendent of the Pueblo works in 1889. Toward the end of the century, the Pueblo Smelting and Refining Company was acquired by the Guggenheims, but prior to that, in 1897, Mathewson had gone to work for the Guggenheims' Philadelphia smelter, also located at Pueblo. Within a few months, Mathewson was placed in charge of the Guggenheims' technical staff, managing smelters at Perth Amboy, New Jersey; Monterrey, Mexico; and Antofagasta, Chile. In 1901, the Guggenheims merged their smelting interests with the American Smelting and Refining Company (ASARCO), another large trust, similar to the Amalgamated Copper Company, that had been created in 1899. Henry Rogers and Leonard Lewisohn, who were instrumental in forming Amalgamated Copper to consolidate the mining industry at Butte, were also at the center of ASARCO's formation in 1899, to consolidate smelting and refining interests throughout the U.S. ASARCO, which now owned most of the lead-smelting capacity in the Americas, fired Mathewson, leaving him with few options for employment in the lead industry. He returned with his family to Montreal and was unemployed for more than six months before Frank Klepetko asked him to take over the blast furnace department at the Washoe smelter in 1902. Mathewson immediately began making operations of the department more efficient, so impressing his superiors that, when Klepetko resigned later that year, Mathewson was appointed superintendent of the entire Washoe smelter.[117]

One of the first places Mathewson turned for help in preparing the ACM's technical case was to his hometown. He asked Duncan McEachran, a prominent Canadian veterinarian from Montreal, to visit the Deer Lodge Valley during the summer of 1905 to inspect livestock conditions. Mathewson then asked McEachran to assemble a team of veterinary experts of national stature. Foremost among them were Leonard Pearson, Pennsylvania state veterinarian and member of the faculty at the University of Pennsylvania, Theobald Smith of Harvard University, and Veranus A. Moore of Cornell University. All three were graduates of Cornell, and Smith and Moore had earlier worked for D. E. Salmon at the Bureau of Animal Industry. McEachran and the ACM had also asked Salmon to be a part of the company's veterinary team, but Salmon declined, saying he would soon be leaving on a prolonged trip to Uruguay. The ACM hired Henry C. Gardiner, a veterinarian in private practice at Bozeman, Montana, to serve as

FIG. 6.11. Anaconda's Expert Veterinarians Conducting a Postmortem of a Horse in the Deer Lodge Valley. Duncan MacEachren is standing with a tablet of paper, V. A. Moore is squatting with his hands on the animal, and H. C. Gardiner is squatting next to Moore. Identities of the other individuals are unknown. Photo Courtesy of Cornell University Library, Rare and Manuscript Collections, V. A. Moore Collection.

its staff veterinarian throughout the fieldwork and testimony. The veterinarians performed hundreds of autopsies on horses and cows during the summer of 1906 and presented evidence intended to show that arsenic from the Washoe stack was not the cause of ill-health in the livestock. The ACM also presented such experts as F. W. Traphagen of the Colorado School of Mines and Harry Snyder of the University of Minnesota to offer soil analyses suggesting that agricultural problems in the Deer Lodge Valley stemmed from causes other than smelter smoke.[118]

Seeking a plant pathologist to determine the effects of smelter smoke on vegetation, Mathewson turned to another fellow Canadian, Samuel Fortier. Mathewson and Fortier had been classmates at McGill University, receiving their bachelor of applied science degrees in 1885. From 1899 to 1903, Fortier had worked as irrigation engineer for the Montana Experiment Station at Bozeman, where Traphagen and Gardiner were also on the staff.

He then moved to Berkeley, California, to conduct irrigation surveys and lecture at the University of California. In early 1906, Mathewson sought Fortier's recommendation for an expert plant pathologist to conduct fieldwork in the Deer Lodge Valley and to testify on the company's behalf in the Bliss case. At Fortier's recommendation, Mathewson secured the services of Prof. Ralph W. Smith of the University of California, who joined the ranks of the ACM's distinguished group of experts.[119]

The ACM's experts conducted their fieldwork during 1906 and testified on behalf of the company later that year. They agreed in their findings: the high mortality rate among livestock in the Deer Lodge Valley was caused by a parasite, and there was no evidence of arsenic poisoning. Because this testimony directly contradicted that of Salmon and the farmers' other experts, the Bliss case offers an interesting study in conflicting expert testimony, because the case was tried at a time when some of the scientific specialties involved were taking shape as recognized disciplines. It was also a time when the veterinary profession was going through important changes.[120] The Bliss case is interesting because, at least in the case of the veterinary experts, the individuals on both sides remained professionals held in high regard by their peers after the trial concluded. There is no evidence that the veterinarians deliberately entered false testimony, despite the fact that it appeared to be conflicting.

The uncertainty of the science of arsenical poisoning and the environmental impacts of smelter smoke in 1906 is illustrated by an examination of the participation of D. E. Salmon and V. A. Moore in the Bliss case. Moore was on the faculty of one of the nation's most prestigious schools of veterinary medicine, Cornell, during a time when the field of veterinary medicine was going through a significant transition paralleling the industrialization of the U.S. economy. In the nineteenth century, the profession served the needs of agriculturalists, but it depended to a remarkable extent on the market created by expanding urbanization and the corresponding boom in horse populations to power urban transportation. At the turn of the century, some veterinarians were alarmed at growth of electric street railway systems and the sudden popularity of bicycles. Seeing that motorcycles and horseless carriages might become just as popular, some veterinarians wondered what they would soon be doing for a living. Perhaps in part due to the shifting market for the veterinary profession, the faculty at Cornell diversified its activities early in the twentieth century. For example, Moore's work for the Anaconda company was the first large undertaking by a member of the Cornell faculty for an industrial corporation. In the context of an industrializing economy, performing services for one of the nation's great industrial corporations may have suggested promising new openings for the employment of veterinarians.[121]

It is also noteworthy that Moore claimed to have been able to find no evidence of arsenical poisoning, while he devoted a great deal of attention to bacteriological investigations of conditions in the Deer Lodge Valley. Indeed, all the initial letters he received from the Anaconda company's chief metallurgist requesting scientific services in the case were on ACM letter-head beneath an underlined heading: "bacteriological investigations." Bacteriology had taken its position as a distinct discipline late in the nineteenth century, and Moore had played an important role in its establishment while conducting research at the Bureau of Animal Industry. Bacteriology continued to be a major focus of Moore's work as a member of Cornell's veterinary faculty.[122] Disciplinary boundaries in these new sciences helped limit understandings of the effects of industrial pollution on human health, with some specialists recognizing only biological agents as causes of ill-health while others recognized inorganic toxins.[123] Similarly, disciplinary boundaries may have contributed to Moore's seemingly limited view of causes of ill-health and death in the livestock of the Deer Lodge Valley.

Further insights into Moore's participation in the Bliss case are possible through examination of his personal finances. It is at least interesting and perhaps significant to know that prior to his work for the Anaconda company, Moore never earned more than $4,000 in a single year. His Cornell salary was $3,000 per year, and each year he also took in several hundred dollars from outside consulting. In 1906, however, in addition to his university salary and several small outside contracts, Moore received more than $5,000 in gross revenue from the ACM, and his annual income tripled. Throughout the next decade and more, he continued to receive fees not only from Anaconda but also from an ACM subsidiary, the International Smelting and Refining Company, and ASARCO as well.[124] Moreover, during his trips to Montana to conduct field research for Anaconda Company, Moore joined with some of the other specialists in veterinary medicine dabbling in mining stocks. Letters among the veterinarians and related to the Bliss litigation always included information pertinent to the case itself, but occasionally they also included reflections on how the participants' mining stocks were doing.[125] Clearly, Moore had an economic interest in the prosperity of the mining industry.

Scientists, however, do not gain national prestige by crassly skewing scientific findings in support of financial gain. Scientists gain standing in their fields by making important findings concerning how the natural world works and by subjecting those findings to the scrutiny of their peers through publication in the scientific press. This Moore and his colleagues in the Bliss case were anxious to do. When the ACM's chief metallurgist recruited Moore and other specialists to assist the ACM in its defense against the farmers, he assured them that the company wanted facts, that the company wanted to know if it was indeed causing damage to livestock in the valley. He further suggested that Moore and the others could use the information they collected as a contribution to science once the litigation was concluded. As Moore and the others worked together in the field, they also discussed with ACM officials how results of their research might be published. Early in the case, meanwhile, Salmon, the farmers' expert, gave a talk before the Keystone Veterinary Medicine Association in Philadelphia on the damage smelter smoke had caused in the Deer Lodge Valley. Moore was anxious to present a counter view, and he was invited to do so before the New York City Veterinarians Society in April 1907. He received permission from the ACM legal department to address the New York City veterinarians, but he was cautioned to present only that which was already a matter of record and not to say anything that might open new questions as the two sides prepared for cross-examination and final arguments before the court.[126]

Moore and his colleagues appeared absolutely convinced of the veracity of their scientific findings. Meanwhile, they were quite skeptical not only of Salmon's findings but also of his capabilities as a researcher and of his integrity. For example, Moore had recommended Dr. Charles Flocken, another Cornellian working for the Bureau of Animal Industry, to the ACM's Gardiner to work on the case in the ACM's lab. Flocken had declined Gardiner's offer of employment, because he did not want to give up his permanent position at the USDA. He was also prescient in predicting that, as interesting as the scientific work in the case might be, the outcome of the case would probably be determined more by the skills of the lawyers than by the scientific facts of the matter. Flocken remained intellectually curious about the case, though, and wrote Moore several times expressing interest in how the research was going. Meanwhile, as a bureau employee, he also had opportunity to keep in touch with Salmon whenever the latter returned to his home in D.C. Flocken reported to Moore that Salmon was keeping extremely busy with the case, but he feared Salmon was not as skillful a researcher as he had been in younger days. Moreover, Salmon did not have access to the extensive laboratory and other facilities the ACM provided its experts.[127]

After the experts for both sides had testified, McEachran was even more blunt in his criticism of Salmon. In a January 1907 letter to Moore, he offered two explanations for Salmon's testimony: either Salmon had been given livestock that had been fed large quantities of arsenic before he examined them, or he was "swearing as he was paid to swear." Continuing about Salmon, McEachran wrote, "I feel sorry that any member of a liberal profession would so demean himself—for I cannot believe that he is not departing from the truth."[128]

In January 1909, Judge Hunt, who had also tried the Magone case, ruled in favor of the ACM and against Bliss. He ruled that there *did* appear to be arsenical poisoning, but his decision was based on what he considered the more important argument concerning the economic damage that would be done to the Butte-Anaconda area if the smelter were closed. The smelter paid more than half of the taxes that Deer Lodge County collected. Forcing the ACM to close the smelter would probably force most of the mines in Butte to close as well. Butte had grown to a population of more than 70,000 in three decades, thanks to mining, and most of that population depended on mining, directly or indirectly, for their livelihoods. Moreover, Hunt reasoned, the rest of Montana was quite dependent on Butte's mining industry and the Anaconda smelter, thanks to the coal, limestone, and timber produced and shipped there, and the freight hauled in support of the mining industry constituted a significant portion of the railroad traffic in the state.[129]

Even though the judge's opinion was founded more on the economics at issue than on an understanding of the injury the smelter was inflicting on the environment, the Bliss case is nevertheless important for the window it provides on the science of environmental impact as it existed in 1906, and on the legal principles in vogue at the time.[130]

After Judge Hunt's ruling in the Bliss case, Moore was anxious to publish his findings. In the meanwhile, however, the U.S. government was preparing to file suit against the Anaconda company over damage caused by smelter smoke to national forest resources near the company's reduction works, as will be described in the next chapter. The farmers also appealed Judge Hunt's ruling. The ACM, therefore, would not authorize Moore or its other experts to publish any of their findings. Although this frustrated Moore, he also saw a possible positive aspect to the farmers' appeal. He had been discouraged that Judge Hunt's ruling suggested there had been damage to livestock caused by smelter smoke. Moore hoped that the appeals court would leave the facts absolutely clear. When the company won in appeals court, Moore was still thinking about articles he would like to write on what he had learned during his research for the ACM, but the farmers took their appeal to the U.S. Supreme Court.[131] The Supreme Court refused to hear the farmers' appeal on a technicality, but the federal case against the company dragged on through the second decade of the century. Apparently, Moore never published anything on his scientific findings in the Bliss case.

And what of Salmon? What motivations might he have had to work for the farmers? He had served in the USDA for more than two decades with distinction. He was the founding chief of the USDA's Bureau of Animal Industry. He and his staff at the bureau had made important contributions to the establishment of the discipline of bacteriology, and they had identified and developed treatments for many of the diseases plaguing cattle, hogs, and other animals important to America's agricultural industry. He was among the veterinarians most highly regarded by his own profession and as well as by the American public at the turn of the century. But his career had not been without controversy. During the 1880s and 1890s, veterinarians and bacteriologists had struggled to understand and learn how to treat the disease now known as hog cholera. The struggle produced a bitter conflict between Salmon and a respected member of the faculty at the University of Nebraska, Frank Billings. Although somewhat scarred, Salmon prevailed in that controversy, and he was also able to preserve the reputation of the bureau.[132]

The Bureau of Animal Industry came under fire again early in the twentieth century with the onset of the beef-trust and meatpacking controversies, which were made sensational by Upton Sinclair's *The Jungle*, published as a book in 1906 but serialized in the socialist journal *Appeal to Reason* beginning in February 1905.[133] Shortly after serialization of *The Jungle* began, Salmon was implicated in an unrelated scandal. He was charged with having a financial interest in the publishing company that supplied the USDA with meat-inspection labels. Following an investigation ordered by the secretary of agriculture, Salmon was found innocent of any wrongdoing, although he had had an "unfortunate" prior relationship with the publishing company. Salmon had been in partnership with the owner of the publishing company but had severed the partnership in 1901. He made a loan to the owner of the company at the time the partnership was dissolved, however, and the owner used later profits from a 1902 USDA label-printing contract to repay Salmon, giving rise to the charges of impropriety. Unfortunately for Salmon, President Roosevelt was not satisfied with the results of the inquiry and asked for Salmon's resignation. Salmon resigned from government service in September 1905.[134] This turn of events disturbed Salmon's colleagues at the bureau and his peers throughout the veterinary profession, many of whom believed Roosevelt was using him as a scapegoat for the larger meat-inspection scandal.[135]

Despite his advancing age—he turned fifty-six during the summer of 1906—and the possibility that he had lost some of his research skills after serving so many years as the head of a federal agency, Salmon toiled tirelessly on the farmers' case. Descriptions of his work on behalf of the farmers suggest the efforts of a man trying to resurrect a tarnished reputation. McEachran had tried to recruit Salmon to the Anaconda company's cause, but it is understandable that a man whose distinguished career with the federal government had ended in a scandal associated with the beef-trust controversy would not want to go to work for the copper trust. Instead of enlisting with the ACM, Salmon went to work for the farmers, putting in numerous days of field research, presenting extensive testimony, and helping the farmers' lawyers develop questions for cross-examining the ACM's witnesses.[136] In the end, however, he found himself on the losing side. When his work for the farmers ended, he accepted a job with the government of Uruguay, establishing a new college of veterinary medicine there. After five years in that country, he returned to his home in Washington, D.C. In 1913, with no other means of employment, he accepted a position as manager of a small hog cholera serum manufacturing plant located, ironically, at Butte. He died there a year later of pneumonia, while his wife was still at home in Washington. One historian of American veterinary medicine says Salmon's dismissal from the Bureau of Animal Industry left him "broken in spirit" and that he never fully recovered from the episode.[137]

The editor of the *American Veterinary Review* (*AVR*) had corresponded with Moore during 1906, saying he had received letters from others concerning the Bliss case, all of which sounded biased. He determined that his journal would not say anything about the case, indeed would not even report any news about it, until a decision had been rendered. Nevertheless, he was anxious to hear Moore's findings at a meeting of the New York City

Veterinarians Society.[138] When the *AVR* did start reporting news of the case in late 1907, it quoted findings of the master of chancery in the federal court that the smelter was a significant component of the economic base of the area, on the one hand, and that arsenic from the smelter smoke appeared to be contaminating animal fodder, on the other. A subsequent notice in mid-1908 abstracted evidence compiled by Salmon and some of the other experts who testified on behalf of the farmers, calling special attention to the bulletin by Haywood just published by the federal government and titled "Injury to Vegetation and Animal Life by Smelter Wastes."[139] When the *AVR* finally published original scientific findings from the Bliss case, they came in the form of a lengthy series of articles not by Moore but by Salmon titled "Arsenical Poisoning from Smelter Smoke in Deer Lodge Valley, Montana."[140]

In the introductory installment to his *AVR* series, Salmon addressed the conflicting expert testimony given in the Bliss case. He suggested that the scientific investigations conducted in relation to the case had made significant contributions to understandings of veterinary toxicology, especially of chronic, as opposed to acute, poisoning.[141] Significantly, such an awareness of the dangers of chronic poisoning was just beginning to develop in the field of industrial hygiene at this time as well.[142] Salmon also advocated a change in the practice for experts engaged in legal battles: rather than experts for the litigants presenting conflicting testimony, they should compare their results, and if they differed the experts should conduct further work to resolve the differences. Of course, the problems that could arise in American courts because of the advocate system, in which each side in litigation had expert witnesses testifying on its behalf, were not new to the twentieth century, and a cynical public had learned to question the integrity of experts decades earlier.[143]

Faced with conflicting scientific testimony, Judge Hunt based his ruling in the Bliss case in favor of the ACM because of its more sophisticated legal arguments. In a two-part strategy formulated by smelter manager E. P. Mathewson and ACM attorney Cornelius Kelley, the company sought to cast doubt on the farmers' testimony by presenting its own conflicting scientific evidence and by calling the expertise or integrity of the farmers' witnesses into doubt. For example, ACM attorney A. J. Shores began his cross-examination of D. E. Salmon by asking if President Roosevelt had asked Salmon to resign his post at the USDA.[144] The second and more important prong of the company's strategy was to convince the court that the ACM could do no more to abate its smoke pollution, so an injunction would force the company to close the Anaconda smelter, and that closing the smelter, and thereby the Butte mines, would cause far more economic harm than had resulted from possible damage to the farmers' livestock.[145] In Montana during the first decade of the twentieth century, the threat of closing the Butte mines and the Anaconda was not an abstraction; workers and their families had had a partial experience of that reality during the late summer of 1903, when the ACM closed the smelter to install the flues and three-hundred-foot stack, and then much of Montana had experienced that reality in the fall of 1903, when Amalgamated closed all of its operations in the state in order to strong-arm the Montana legislature into passing the legislation the copper trust wanted.

Judge Hunt also based his ruling on recent court precedent, citing the Mountain Copper case, in which the Ninth Circuit had said that a court should exercise discretion in each case when deciding whether to issue an injunction. Hunt also cited a more recent decision of the Ninth Circuit in upholding a district court ruling against the plaintiff farmers in Idaho and in favor of the defendant mining companies (*McCarthy v. Bunker Hill & Sullivan Mining & Concentrating Co.*), in which Judge Ross wrote that "no one has an absolute and unqualified right" to an injunction. Each case has "its own particular facts and circumstances." Ross continued (as quoted by Hunt): "Nor should an injunction be granted in any case where it will necessarily operate contrary to the real justice of the case. Furthermore, where, as in the present case [*McCarthy*], it is sought to enjoin a lawful business, the court should give due consideration to the comparative injury which will result from the granting or refusal of the injunction sought."[146] Hunt did not note that he had joined Circuit Judges Ross and Gilbert in that Ninth Circuit Court of Appeals ruling that affirmed the lower-court decision.[147]

The Mathewson/Kelley strategy had succeeded.[148] Not pleased with Judge Hunt's measure of the costs and benefits of the Washoe smelter, the farmers appealed to the U.S. Court of Appeals and the U.S. Supreme Court, losing in both courts. Such an outcome was not surprising considering the way federal judges of the period used common law and the balancing doctrine to settle nuisance suits arising from industrial pollution. This was especially true in the territory under the jurisdiction of the Ninth Circuit, led by circuit judges Ross and Gilbert. Writing the appeals court 1911 opinion affirming Hunt's ruling in the Bliss case, Ross disingenuously wrote that "the court is always ready and willing" to protect farmers from giant mining companies, noting that he was himself a farmer (he owned one thousand acres in southern California). Indeed, he had had a strong record of protecting property rights in other sorts of cases, but when it came farmers seeking protection from polluting mining companies, he sided with the companies.[149]

Although the Deer Lodge Valley farmers lost the Bliss case, the experts they employed were the only ones to make significant contributions to the scientific literature. The fieldwork completed by government scientists, experts retained by the government, and experts hired by the farmers yielded an impressive array of publications in professional and technical journals and by the Government Printing Office. Swain and Harkins published their findings as three "Papers on Smelter Smoke" in 1907 and 1908 issues of *The Journal of the American Chemical Society*.[150] Salmon published his findings in the nine-part series of articles called "Arsenical Poisoning from Smelter Smoke in Deer Lodge Valley, Montana," in *American Veterinary Review* during 1911 and 1912.[151] J. K. Haywood, chief of the Miscellaneous Laboratory of the U.S. Department of Agriculture's Bureau of Chemistry, published several pieces based on his fieldwork. He had a short article, "Injury to Vegetation and Animal Life by Smelter Fumes," in a 1907 issue of *The Journal of the American Chemical Society*. In 1908, the Bureau of Chemistry published a longer version of the same article under a similar title, "Injury to Vegetation and Animal Life by Smelter Wastes," as Bulletin No. 113. It included an expanded description of damage caused by tailings, in addition to damage caused by smelter fumes. Bulletin No. 113 was reprinted in revised form in 1910, following additional fieldwork conducted during 1908–1909.[152] Robert J. Formad, pathologist for the USDA's Bureau of Animal Industry, wrote "The Effect of Smelter Fumes upon the Livestock Industry in the Northwest," also published by the USDA in 1910.[153]

None of the results or conclusions of the ACM's experts (whether employees or consultants), produced from fieldwork during or immediately after the Bliss case, is known to have been published in the technical or professional literature.

In the final installment of the *AVR* series, after taking pains to present the evidence of arsenical poisoning in the Deer Lodge Valley, Salmon then summarized the ruling of Judge Hunt, which was based on the anticipated economic impact of closing the smelter, not on whether smelter smoke was killing livestock. Salmon concluded his series by stating: "And thus ended what was, perhaps, the greatest contest in veterinary toxicology that has ever been heard, and the farmers lost their case, but not because of failure to prove the fact of poisoning."[154] Farmers in the Deer Lodge Valley were clearly suffering from harm being done by smelter smoke. It was their misfortune to be living at a time and place in which the law, many elected officials, and much of American society valued the copper and the economic stimulus that the mining industry offered more than they valued the fair treatment to which the farmers believed they were entitled. Such weighing of values would continue to be undertaken entering the second decade of the twentieth century, but now the contestants would be the Anaconda Copper Mining Company and the United States government, and the issue of smelter smoke in the Deer Lodge Valley would assume national proportions.

CHAPTER 7

Integration into a Global Corporation

> The years from 1895–1901, inclusive, were in certain respects among the most spectacular that the American copper industry has even known. In this period the same tendency was manifest in that industry as in other American industries—the tendency of large corporations and groups of capitalists to come into control of more and more units of the industry.
>
> —Ernest Richter, "Copper Industry in the U.S., 1845–1925" [1]

Entering the new century, the market for copper and other base metals was continuing to grow at an amazing rate, and there were no signs that growth would abate. Meanwhile, the price of copper had partially recovered. After dipping below ten cents per pound during the financial collapse of the early 1890s, the price of copper was back above fifteen cents by the turn of the century. Butte and Michigan's Copper Country were still by far the nation's leading copper producers. Miners had yet to find the depth at which Butte's spectacular mineral wealth reached its limit, but the shafts on the Butte hill were reaching the limit of how quickly they could produce ore. The Keweenaw Peninsula's production had plateaued in the late 1890s, but then the Copper Range Mining Company opened a new district of rich native-copper mines south of Portage Lake, bringing renewed growth in output for the Copper Country. Meanwhile, to supply the insatiable demand for copper of an industrializing nation and global market, investors were opening new mines in Utah; Shasta County, California; and in several districts in Arizona. Some of these developments featured major investments by capitalists affiliated with the Amalgamated Copper Company.[2]

In Montana, while Amalgamated was cementing its hegemony over the state's levers of government and fighting the heroic efforts by the farmers of the Deer Lodge Valley to challenge the trust's ability to discharge contaminants into the environment, the copper giant had other items on its corporate agenda as well, foremost among them the consolidation of the Butte hill under its legal and operational control. This would occupy the first decade of the twentieth century, when the main resistance to the mining industry's polluting methods was mounted locally, by the farmers and ranchers in the Deer Lodge Valley. During the second decade of the century, resistance to Anaconda/Amalgamated's pollution grew national in scope, with the federal government taking its own legal action against the company, described in the next chapter. Meanwhile the leaders of the enterprise worked out how they could manage operations not only in southwest Montana but across the continent as well, and the technological and managerial system they developed laid the foundation for being a vertically integrated global corporation, indeed one of the largest such corporations in the world. This chapter describes that birth of a global metals giant, the post-Amalgamated version of the Anaconda Copper Mining Company, in order to form the backdrop the next chapter, which describes the battle between the federal government and a much larger corporate giant than the Deer Lodge Valley farmers had faced. It also provides the context for the last chapters of the book, which analyze the winding down of environmental controversies, leading to a relative hiatus in concern over pollution, even though the pollution was ongoing, until the rise of the regulatory state in the last third of the twentieth century. This history also sets the stage for recognizing what would be the largest Superfund complex in the United States.

Consolidation of Operations on the Butte Hill

When Amalgamated gained controlling interest in four large Butte mining companies (Anaconda, Colorado, Parrot, and Washoe) in 1899, they remained as distinct corporate entities and operating companies. When Amalgamated added the Boston and Montana and the Butte and Boston to its stable of companies in 1901, they, too, retained their corporate and operational identities. Each had its own set of managers, who oversaw operations of its own group of mines, which sent ores to each company's smelter (although the Parrot smelter had closed, and the Parrot mine was sending its ore to Anaconda). This relative independence was evident in some of the episodes recounted in the previous chapter, such as the summer 1903 shutdown of the Washoe smelter to connect the furnace flues to the new flue-and-stack system aimed at recovering arsenic before it was discharged with the smelter smoke. While the smelter was closed from July 1 to mid-September, the

Anaconda, Washoe, and Parrot mines also closed. The Colorado's Gagnon mine closed during the summer as well, but not because of the Washoe closure—it was because the Colorado smelter closed briefly to implement repairs. Meanwhile, the B&B and B&M mines continued producing, sending their ores to the B&B and B&M smelters in Butte and Great Falls, respectively. Gradually, over the next two decades, all those operations would be consolidated into a single technological system, with mines in Butte, smelter in Anaconda, more smelting and a refinery in Great Falls, all owned by a single entity and managed by a central organization.[3]

Leading the enterprise into that future was a new group of top managers, foremost among them John D. Ryan and Cornelius "Con" F. Kelley, who took the helm as many of the founding generation of Amalgamated officials died or quit within a decade of the trust's formation. Marcus Daly had already died in 1900. H. H. Rogers died in 1909. He was succeeded as Amalgamated's president by John D. Ryan, the son of a prominent mining captain in Michigan's Copper Country. Born in 1864, the young Ryan had been interested in neither mining nor college so, after clerking at his uncle's store in Hancock, Michigan, for a few years, he moved at age twenty-five to Denver, where he found work as a traveling salesman of lubricating oil. His territory included Montana. On one of his trips to Butte, he met Marcus Daly, who took a liking to him. In 1900, after about ten years selling oil, Ryan moved to Butte, where he bought an interest in the Daly Bank and Trust Company, which Daly had founded. Shortly thereafter, when Daly died, his heirs named Ryan president of the bank. Amalgamated ran its local finances through the Daly Bank, so Ryan's business skills soon came to Rogers's attention. Rogers gave Ryan a seat on the Amalgamated board of directors in 1904, and the following year, when William Scallon resigned, Ryan was elected president of the ACM.[4]

The other top manager of the enterprise for the first half of the twentieth century was Con Kelley, who had grown up in Butte. He had moved there as a child in 1883, when his father, Jeremiah, assumed the superintendence of the Bell mine. Jeremiah Kelley and Marcus Daly had met in Nevada. After the Kelley family moved to Butte, Daly, already developing his Anaconda property, was a frequent guest at the Kelley home. After young Cornelius graduated from Butte High School in 1892, he asked his father's friend for a job, and Daly placed him with a surveying crew in the Anaconda mine. After a couple years working in the company's engineering department, Con Kelley decided to pursue a career in the law rather than in mining engineering, so he headed to the University of Michigan. During the summers of his college years, he continued to work in the ACM's engineering department. After graduating from the Michigan Law School in 1898, he returned to Butte, started to practice law, decided to enter politics, and that fall was elected to the Montana legislature. Meanwhile, after a year in his private practice, in which Kelley took cases involving the apex law, he accepted an appointment as chief deputy county attorney. Daly died in 1900, and William Scallon became ACM president. A year later, he hired Kelley to work in the company's legal department, headed by A. J. Shores. With Kelley's experience working underground, he was able immediately to move to the front lines of the ACM's legal battles with Heinze involving apex law. He also played prominent roles in defending the company in the Magone and Bliss cases.[5]

After Amalgamated had forced the Montana legislature to pass its change-of-venue legislation in 1903, F. Augustus Heinze was less able to harass Amalgamated's subsidiary companies through apex litigation. In 1906, Heinze and Amalgamated officials finalized an agreement that brought the conflict to a close. Heinze would sell all his Butte copper properties to Thomas F. Cole, an iron-mining magnate in Duluth, Minnesota, who, like Ryan, had grown up in Michigan's Copper Country. After his father had been killed in a mine accident, Cole went to work, at the age of eight, at a mine sorting ore. Despite the demands of work, he was able to continue educating himself, enabling him to find jobs with increasing responsibility, as a brakeman for a railroad and then as a clerk in the office of the Calumet and Hecla Mining Company. From there, he moved to become the cashier for an iron-mining company on one of Michigan's iron ranges, and by 1881, at age twenty-one, he was chief clerk. Continuing to work his way into positions of greater responsibility, he was named vice president of the Oliver Iron Mining Company when it became a subsidiary of U.S. Steel in 1901. While overseeing the Oliver company's iron mines on the Mesabi Range in Minnesota, he began launching other successful ventures as well, entering the Butte scene in 1905, when he and others formed the North Butte Mining Company, which bought the Speculator mine and adjacent claims. The North Butte would become an important supplier of ore to the Washoe smelter.[6]

In the agreement negotiated between Heinze, Ryan, and the ACM attorneys, Heinze also agreed to convey all his legal actions against the

Amalgamated companies to Cole. Cole would then convey the former Heinze copper properties to the newly formed Red Metal Mining Company, of which Cole was president, and he would drop the legal actions against the Amalgamated subsidiaries. (Heinze would retain his Lexington group of silver/zinc properties in Butte.) Amalgamated in turn agreed to drop its litigation against Heinze. Red Metal was established for the expressed purpose of owning and operating the former Heinze properties. The Red Metal stock was in turn owned by the Butte Coalition Copper Company, a holding company established in 1906 to own stock. Cole, Ryan, and other Amalgamated officials occupied seats on the Butte Coalition board, which elected Cole president and Ryan vice president. A. C. Carson, manager of the North Butte operations, took charge of Heinze's copper mines in Butte. In the short term, Red Metal continued treating its ores as Heinze had, but on July 15, 1906, Red Metal closed the MOP smelter and began shipping its ores to the Washoe in Anaconda. By this time, the Colorado and B&B smelters had also closed, leaving W. A. Clark's Butte Reduction Works and the new Pittsmont smelter as the only smelters still operating in Butte.[7]

When Henry H. Rogers died in 1909, John D. Ryan succeeded him as president of Amalgamated. B. B. Thayer took over for Ryan as president of the ACM. Educated in mining engineering at Harvard University's Lawrence Scientific School, Thayer had moved to Butte when he graduated and began working for the ACM as a miner. He worked his way up the hierarchy to the position of vice president, prior to his appointment as president in 1909. Soon thereafter, in March 1910, Amalgamated took its next step in consolidating the Butte holdings. Under Ryan's and Thayer's consolidation plan, each of Amalgamated's Butte operating companies transferred its property and operations to the ACM. That included the Boston and Montana Consolidated Copper and Silver Mining Company, which transferred not only its Butte mines but its Great Falls smelter and refinery as well. At the same time, the ACM also absorbed the properties and operations of Red Metal Mining Company (Heinze's old mines). Within a short time, Red Metal, Butte Coalition, and most of Amalgamated's subsidiary corporations in Butte dissolved, leaving the ACM as Amalgamated's sole operating company in Butte. Two months later, Ryan negotiated a deal with W. A. Clark whereby the latter would sell all his Butte copper-mining and -smelting properties to Ryan, who would in turn transfer them to Amalgamated, which then turned the former Clark properties and operations over to the ACM. Now the ACM would operate Clark's Original and Steward mines, just north of Butte's central business district. Their ores would be shipped to the Washoe, as the ACM closed the Butte Reduction Works.[8]

With that series of transactions, the ACM came to own and operate nearly all the copper mines and reduction works in Montana, with only a handful of small Butte companies remaining to operate some peripheral mines. The ACM also acquired extensive coal and timber lands, including those of Amalgamated subsidiaries, the Diamond Coal and Coke Company in Wyoming and the Big Blackfoot Milling Company near Missoula. Observers in the mining industry considered the ACM's consolidation of the Butte hill the most significant development in the copper industry in 1910.[9] The ACM's consolidation of all Amalgamated's Montana operations also made the company the fifth-largest industrial corporation in the world at the time.[10]

Prior to 1910, Amalgamated's subsidiary mining companies in Butte had been distinct operating entities, and Amalgamated's corporate officials merely coordinated the operations of its several subsidiaries to a limited extent. One way they did so was through the office of John Gillie, a mining engineer who had worked in and around Butte since 1880. A native of Ottawa, Ontario, he had graduated from the University of Ottawa with degrees in mining engineering and civil engineering. After working briefly at a silver mill in Philipsburg, Montana, he became a consulting engineer in Butte. In 1900, after two decades of independent work, often for Butte's major mining companies, he was made superintendent of operations of the B&B, including its mines and smelter. In 1901, after Amalgamated acquired controlling interest in both the B&M and the B&B, the trust named Gillie general superintendent of mines for the six Amalgamated companies operating mines in Butte.[11]

With consolidation of all Amalgamated's Montana operating companies into the Anaconda company in 1910, it became possible for the ACM's engineers and corporate managers to transform the whole Butte hill, indeed the company's entire array of properties, into a single, efficient, smoothly running technological system. Prior to that time, each of a company's mines operated as a nearly independent unit. Each mine had its own machine shop, blacksmith shop, and framing shop for mine timbers. Mineshafts typically had a station at every one hundred feet of depth, but the levels of adjoining mines were rarely aligned horizontally. As mines grew deeper, companies often joined the underground workings of their mines to enhance natural ventilation. The mineshaft at a mine that was lower on the hill would function as a downcast shaft, and the shaft at a mine that was higher on the hill served as an upcast shaft. Each mine had to pump its own water. When Amalgamated acquired its several companies, there was little standardization among their operations.[12]

Once the ACM owned all of the mines in 1910, one of the first things it did to standardize and centralize operations, thereby reducing cost and increasing efficiency, was to integrate energy delivery and conversion into a single integrated system. The clearest example of this transformation was the ACM's integration of the hydroelectric developments at Great Falls with improved hoisting methods at the Butte mines. The ACM began implementing that integration of hydroelectric power and mine hoisting almost immediately.[13]

Until 1910, Butte's copper-mining operations had largely been powered by human and animal muscle and, in ever-increasing ways, coal-fired steam. Steam engines hoisted ore from the mines, and steam engines drove air compressors, which in turn supplied the underground miners' rock drills by means of piping and hoses. Steam engines drove the pumps that lifted water out of the mines. Each mineyard was its own energy conversion center, with railroads delivering coal to the coal bins for supplying the mineyard's steam boilers. Coal-fired steam locomotives both delivered the coal and hauled ore from the mines to the smelters. In the nineteenth century, several Butte mines had installed their own small steam-powered electric generating plants, and a Butte utility company had organized to serve Butte and a few mines with electricity generated with coal-fired steam in Butte. Uses of electricity began to expand with improvements in long-distance electrical transmission technology around the turn of the century. As demand for electricity in Butte grew, entrepreneurs developed hydroelectric dams and generating stations dozens of miles distant in order to supply the Butte market. The first such dam was located on the Big Hole River, south of Butte and on the other side of the Continental Divide. It began transmitting electricity to Butte in 1898, and it was followed by a dam on the Madison River, north of Ennis, in 1901, and a dam on the Missouri River at Canyon Ferry, south of Helena, in 1902. The application of electricity to mining at the turn of the century remained largely limited to lighting and surface tramming.[14]

After 1900, mining companies began using electricity to drive air compressors for rock drills (the largest single source of demand), to drive underground water pumps, and for underground tramming in addition to surface tramming. The companies also began to ventilate the mines, which grew hotter as the shafts were driven deeper (below 2,000 feet by 1900), and which became dustier and more unhealthy as natural ventilation grew less effective at depth. Even though electricity was available in the mines for lighting in the late nineteenth century, it did not completely replace the open flame for lighting until the 1930s. Electric lights were used underground only to light shafts and drifts. Miners relied on candles for light in the stopes until about 1912, when the carbide lamp was introduced into the Butte mines (battery-powered miner's lamps replaced the carbide lamp in the 1930s).[15]

In 1907, the Missouri River Power Company's second plant, Hauser Dam, between Helena and Great Falls, began serving Butte. For several years, until 1908, the Missouri River Power Company had a contract to supply electricity to Amalgamated and thus enjoyed a dominant position in the Butte market. That position changed suddenly when the steel Hauser Dam failed in 1908, interrupting the supply of electricity to the Butte mines. Other corporate interests moved to try to gain control of the Butte market. Conditions quieted greatly in 1912 when most of the companies producing electricity in Montana were consolidated into one company, the Montana Power Company, under the control of John D. Ryan. As president of Amalgamated, Ryan's effort to consolidate Montana's electric utility industry was closely coordinated with Amalgamated's venture in corporate consolidation and in systematizing the copper-mining industry.[16]

To provide Amalgamated with a reliable source of electrical power, Ryan had begun negotiating in 1908 for the purchase of the Great Falls Water Power and Townsite Company from James J. Hill of the Great Northern Railway. At the same time, Ryan set his engineers to work designing a new dam and power plant at Rainbow Falls, a few miles downstream of Black Eagle Falls and the Great Falls smelter. To supervise the work, Ryan retained the services of Henry Herrick, who was employed by Charles T. Main of Boston and who had been involved in the Black Eagle project in the early 1890s. The new facility at Rainbow was designed to take advantage of three sets of falls, with a timber-crib, rock-filled dam set atop the uppermost falls and the powerhouse built below the lower falls. A 2,350-foot steel-pipe system connected the dam to the powerhouse. The completed dam was 29 feet high and, with the falls, yielded an operating head of 105 feet. Six pairs of Francis-type, horizontal-shaft turbines drove six

4,200-kilowatt generators. Power from two of the generators was transmitted to the B&M smelter at Great Falls. The power from the other four generators was stepped up to 102,000 volts and transmitted 130 miles to Butte and Anaconda over lines carried by steel transmission towers (the route of the transmission lines through the mountains was more direct and therefore shorter than the Great Northern Railroad route between Great Falls and Butte). This transmission system was one of the earliest in the U.S. to operate at more than 100 kilovolts. The Rainbow plant and transmission system began operating in 1910.[17]

Beginning in 1908, the ACM had begun to investigate the possibility of converting its mine-hoist engines from steam to electrical power. Hydroelectric power at Great Falls was much less expensive than coal-fired steam. The ACM conducted a thorough evaluation of the application of electricity to hoisting. Because of the great depth and number of the ACM mines, the volume of ore hoisted, and the irregularity of the actual hoisting at each mine, however, the ACM decided not to use electric motors to directly drive the hoists, because of the likelihood of two or more hoists on the Butte hill starting simultaneously and thus causing large peaks in electrical demand. Control technologies did not yet exist to allow distant hydroelectric generating stations to respond to such violent load fluctuations. The ACM decided instead to use electric motors to drive air compressors at Butte on a continuous basis, to store compressed air in a series of giant receiver tanks, and to connect the mine hoists to the centralized compressed air system. This meant that the company merely had to modify, but not replace, its existing steam-powered hoist engines so they could be driven by compressed air. The compressor plant was located on the Butte hill near the High Ore mine.[18]

The centralized compressed-air system went into operation in 1912. With the electrification of the ACM's centralized compressed-air system, the Butte hill transitioned from a scattering of independent coal-burning nodes, distributed across the hill, to a single technological system powered by a giant central compressor plant, which was itself powered by electricity generated from waterpower more than one hundred miles away. At the time, the demand for electricity by the ACM mines was divided as follows: compressed air for hoisting, 8,000 electric horsepower; compressed air for rock drills, 13,455 h.p.; pumping, 4,780 h.p.; ventilating, 2,500 h.p.; tramming, 1,860 h.p.; and lighting, 1,300 h.p. Total electrical consumption for mining in Butte was about 31,895 electric horsepower, or about 23,800 kilowatts. About two-thirds of this power was used to drive air compressors for hoisting and drilling. The other major demand by the copper industry at this time was for milling, smelting, and refining at Anaconda and Great Falls. These operations consumed about 24,177 electric horsepower, or about 18,000 kilowatts. Thus the total consumption of electricity by the ACM in Montana in 1912 was about 42,000 kilowatts.[19] With the conversion to hydroelectric generation of power, the transmission wires linking the hydroelectric generating stations to the mines, mills, smelters, and refinery embraced a number of large facilities, separated by as much as 150 miles, into a single industrial landscape of energy generation and use.

Shortly after completion of the Rainbow plant in 1912, Ryan was able to consolidate his Great Falls facilities with the four other largest utilities in the state to form the Montana Power Company (MPC). With this control of electrical generation and transmission in Montana, Ryan continued to coordinate the electrification of the copper industry with the development of new hydroelectric facilities.[20]

The ACM also converted another aspect of its operations to electrical power during this period. In 1911, the company decided to electrify its subsidiary railroad, the Butte, Anaconda and Pacific Railway. When its new electric locomotives began operating in 1913, the BA&P became the first railroad in the United States to convert its entire operation from steam to electric motive power. Moreover, it was the first to convert any of its operations solely for economic reasons. Electrification allowed the railroad to haul more ore at a lower cost. The success of the BA&P electrification prompted the Chicago, Milwaukee, St. Paul and Pacific Railway (Milwaukee Road), a transcontinental railroad that had built its line through Butte a few years earlier, to electrify 440 miles of its Rocky Mountain Division between Harlowton, Montana, and Avery, Idaho. John D. Ryan, who was president of both the Montana Power Company and Amalgamated at the time, also sat on the board of the Milwaukee Road. His power company supplied the electricity, and his copper company supplied the wire for the Milwaukee Road's electrification.[21]

To provide power for the expanding electrification of the ACM's mines, mills, and smelters, for the enlarged electrolytic copper refinery at Great Falls, for the electrolytic zinc refinery built at Great Falls in 1916, and for the electrified railroads, Ryan initiated development of hydroelectric generating stations at the Great Falls (downstream of Rainbow Falls and the largest in the series of falls along the Missouri River near the city of Great

FIG. 7.1. Electrified Butte, Anaconda and Pacific Railway. Electrified locomotives, pulling ore cars between Butte and Anaconda, are crossing Silver Bow Creek and about to enter Durant Canyon. The dense railroad traffic around Butte is evident in this photo as the BA&P also crosses two other railroad lines, that of the Milwaukee Road on the far side of the creek and against the rock face, and that of the Montana Union, crossing the creek just before it passes under the BA&P line. Photo courtesy of the Montana Historical Society, cat. # PAc 82–62.0412.

Falls) and at Thompson Falls, on the Clark Fork in western Montana, downstream of Missoula. The sixty-five-foot-high concrete arch-gravity dam at the Great Falls, now called Ryan Dam, was built atop the falls to yield an operating head of about 145 feet. The plant went online in 1915. The following year, the Montana Power Company completed Hebgen Dam, a storage-only facility located near Yellowstone Park on the upper Madison River, one of the three principal tributaries of the Missouri. At the time, it was the second-largest storage reservoir in the U.S., and it allowed Montana Power to increase the minimum average flow to its plants downstream along the Madison and Missouri Rivers. Completion of Hebgen Dam allowed the company to expand capacity of the Rainbow plant in 1917, adding four turbines and two generators and bringing the total capacity to 35,000 kilowatts. In 1918, Montana Power completed construction of Holter Dam, between Helena and Great Falls near Hauser Dam. Holter Dam had a capacity of 50,000 kilowatts.[22]

Along with electrification of the Butte hill, the ACM made numerous other operational changes to integrate the mines into a single technological system. It reduced the number of operating shafts from thirty in 1900 to about twenty in the mid-1910s, and it placed ventilating fans atop several of the shafts taken out of service, in order to make them upcast shafts, leaving the remaining operating shafts, because of forced ventilation, as downcast shafts regardless of their relative locations on the hill. The ACM organized its underground workings so that all water drained by gravity to two centralized pumping locations underground. The company developed centralized machine and blacksmith shops on the Butte hill, and it developed a central timber framing shop at Rocker, three miles west of Butte, for shaping all standardized mine timbers. Each mineyard retained small blacksmith and machine shops for emergency repairs and a small framing shop for cutting miscellaneous timbers for that mineyard's particular needs.[23]

One of the most profound changes initiated by the ACM, which consolidation of the Butte hill fostered, was conceptual; it concerned geology and mine engineering, and it thereby had a physical impact on mining and mine development, making them more efficient. A mine is a human-made artifact. The veins of ore may be natural and not created by humans, but the network of underground workings that make it possible for miners to exploit the ore is of human design. A network of adits, shafts, drifts, crosscuts, and stopes is more than the consequence of miners following the vein (although it may have been only that in the discovery phase of a mine). The underground workings that comprise a mine serve multiple purposes: they give miners access to the working face of the ore body; they give miners a space within which to break the ore loose from the Earth; they provide a conduit for extracting ore, water, and waste (including waste rock and bad air) from the underground; and they provide a conduit for delivering necessary supplies (including tools, explosives, and fresh air) and services throughout the underground network. As such, the design of the underground workings can make the difference between a mine that operates efficiently and at a profit or a mine that does not.[24]

In the early years of mining in Butte, before mine owners knew what minerals were beneath the surface, and before they had begun hiring geologists to help understand the nature of the mineral deposits, companies sank mineshafts on outcroppings of promising minerals, following the formation to create inclined shafts. In time, miners on the Butte hill came to learn that veins usually had a strike toward the south, and that veins sometimes branched, but miners did not yet know this in the early years of mining. At depths of about one hundred feet, miners began to recognize the abundance of mineral wealth beneath the Butte hill. Still without assistance from geologists or engineers, miners abandoned their inclined shafts in favor of vertical shafts, which would be more practical for extracting large tonnages of ore. The uncertainty in what lay beneath the surface can be seen by comparing development of the Parrot and Anaconda claims, both located along the Anaconda-Parrot vein system, the Parrot claim to the west and the Anaconda to the east.[25]

Early developers of the Parrot claim sunk the Parrot shaft just south of the outcrop, while developers of the Anaconda claim sunk the Anaconda shaft just north of the outcrop. Both developments drove crosscuts to the north or south, from stations established in the shaft roughly every hundred feet of depth, toward where they hoped a vein of ore would be. Thus, the Parrot shaft passed through a branch of the vein at a depth of about one hundred feet, while miners had to drive a crosscut about fifty feet north of the shaft to reach the main vein. As miners continued to sink the shaft deeper into the hill, it passed through the main vein at a depth of about five hundred feet. At greater depth, miners had to drive crosscuts to the south in order to reach the main vein. Meanwhile, as the Anaconda mine was sunk deeper, each new level entailed a longer crosscut south from the shaft to reach the vein, because of the vein striking to the south.[26]

Until the advent of the apex litigation in the 1890s, individual mining companies knew little more about the structures of mineral formations in the Butte hill than what they could see in their mine workings. Parties in the litigation, among them the ACM, hired professional geologists and mining engineers to investigate the underground workings of various mines in order to present expert testimony at trial about the broader geological formation of the hill. Marcus Daly brought David Brunton to Butte in 1897 to serve as consulting engineer to the ACM. Brunton in turn brought H. V. Winchell, a professional geologist, to Butte to take charge of geological mapping of the mines in question. After Amalgamated took control of the ACM, and soon the B&M, the trust made Winchell its chief geologist, and in 1900, Amalgamated asked Winchell to organize a new Geological

Department. By mapping numerous sets of mine workings, and the veins and faults they intersected, Winchell came to understand the complex geological structure of Butte's subsurface in three dimensions.[27]

Winchell's successor as head of Amalgamated's Geological Department was Reno H. Sales, a native of Iowa, who as a boy had moved with his family to Montana's Gallatin Valley. In 1898, he graduated from nearby Montana State College in Bozeman with a B.S. in mining engineering, after which he went to Columbia University, earning an engineer of mines degree in 1900. Moving to Butte, he first took a job in the B&M's engineering department. In 1901, after a year at the B&M, he accepted a position as assistant in Amalgamated's Geological Department, where he continued to focus on the B&M and B&B mines. In 1906, after Winchell resigned, Amalgamated appointed Sales chief of the department. Between then and 1910, Amalgamated crews surveyed the mines and developed maps showing how the underground workings corresponded to surface features (already well surveyed). To that point, the focus of surveying and mapping was to accurately depict features, like shafts and chutes, relative to the horizontal drifts and crosscuts, and to depict geological features visible from the mine workings.[28]

With the consolidation of operations under the ACM's management in 1910 and completion of mapping the underground geology a few years later, Sales was able to further consolidate certain management functions under his purview, and by 1914, the company's mining engineering work was also placed under his direction. From 1914 onward, the Geological Department was responsible for all development work (i.e., deciding when and where shafts would be sunk deeper, designing stations at new levels, designing where to extend crosscuts toward anticipated ore bodies, where to extend drifts along the veins, and where to install chutes and underground ore-loading facilities). Sales's Geological Department made such development decisions with a view to the multiple functions the underground workings would serve, including breaking and extracting ore, removing waste rock, delivering supplies and services, and exploring for new ore. As the importance of effective development work grew at the ACM, so too did the size of Geological Department's staff, and Sales, its chief geologist, populated his staff of geologists, engineers, and samplers (the entry-level position for geologists) mainly with men trained in mining engineering, and not specialists in geology. Based on the rationalized development of mine workings within the Butte hill and conceptual understanding of its mineral structure, Sales claimed in later years, the ACM's Geological Department discovered millions of dollars of additional ore reserves.[29]

Having consolidated Amalgamated's Montana operations under the ACM's corporate umbrella, the enterprise then moved to rationalize its metallurgical operations. By the end of the 1910s, the ACM conducted all concentrating and smelting at the Washoe works in Anaconda, and it conducted all its refining at the Great Falls works. A few details of those metallurgical transitions will be found in the next chapter.

Expansion beyond Montana

Entering the 1910s, Montana was no longer the nation's greatest copper-producing state; Arizona's copper output had surpassed Montana's in 1907. All of Montana's copper came from the Butte hill, while Arizona had several distinct mining districts that were becoming prominent in the copper industry. Nevertheless, Butte's output in 1910, about 282,500,000 pounds of copper, represented more than a quarter of the nation's total and remained twice that of Arizona's most productive district, Bisbee. In 1910, copper mines, smelters, and refineries in the United States produced just over a billion pounds of the red metal, representing more than 56 percent of global copper production (about 1.9 billion pounds). The U.S. was a net exporter of copper and, while the U.S. was producing more than a billion pounds of copper per year, no other nation was producing more than 200 million. Spain, Mexico, and Japan were the only nations producing more than 100 million tons.[30] With growth in the domestic and global markets for copper and other metals showing no signs of slowing, the principal capitalists promoting Amalgamated and the ACM had been looking for new investment opportunities, and they continued to do so in the 1910s. The ACM would emerge from those developments a much larger and geographically expansive enterprise.

During the five years of 1910 to 1915, Anaconda moved quickly to restructure itself into a corporate giant of international proportions. To appreciate that restructuring, a few developments outside, but closely associated with, the Amalgamated Copper Company must be summarized. Leonard and Adolph Lewisohn were brokers of copper on the international market who had been principal investors in the Boston and Montana and the Butte and Boston companies. In 1900, the Lewisohn brothers, along with H. H. Rogers and William Rockefeller, incorporated the United Metals Selling

Company to take over the business of selling metals from the Lewisohn brothers' agency. United Metals also acquired a large electrolytic copper refinery on the New York Harbor, the Raritan Copper Works at Perth Amboy, New Jersey. The refinery treated much of the copper matte produced by Amalgamated's subsidiaries, and United Metals sold, as a broker, all the copper produced by Amalgamated.[31]

The Utah Consolidated Mining Company incorporated in 1903 to take ownership of the Highland Boy smelter, which was the source of the smoke that was bedeviling farmers in the Salt Lake Valley. Utah Consolidated's president was Urban H. Broughton, H. H. Rogers's son-in-law. The company was forced to close its smelter in 1907, when the judge hearing the farmers' smoke complaint issued an injunction. A year later, the company entered a contract with W. D. Thornton of Butte, in which Thornton agreed to build a smelter at Tooele, Utah, to smelt Utah Consolidated's ore. Thornton was the son of Col. John C. C. Thornton, a prominent early Butte miner who had launched the Gagnon mine, principal supplier of ore to the Colorado smelter. A graduate of Columbia University's School of Mines, Thornton was a close associate of John D. Ryan. In December 1908, Ryan and Cole incorporated the International Smelting and Refining Company, with United Metals Selling and Utah Consolidated owning considerable portions of the new company's stock. International's president was John D. Ryan, and the secretary and treasurer was Joseph W. Allen, who served in that position for many of the companies affiliated with Amalgamated. Thornton, Broughton, Cole, and others in the Amalgamated circle served on International's board of directors, and the directors named E. P. Mathewson, who was manager of the Washoe smelter, as the new company's general manager. Construction of the Tooele smelter commenced in early 1909, and it began operating in July 1910. At the same time, United Metals sold its Raritan refinery to International.[32]

Another important development took place in the Cananea Mountains of Mexico, just south of the border and near the Bisbee district in Arizona. Prior to the ACM's entry into the Cananea district, the American mine promoter William C. Greene had begun to assemble mining properties there. Observers in the early twentieth century believed that the Cananea copper deposit would prove to be one of the richest in the world, but Greene was in financial difficulty. In 1906, John D. Ryan and Thomas Cole visited the area and took options from Greene on some mines. The two also formed the Greene Cananea Copper Company, a U.S. company incorporated in

Minnesota, to acquire the stock in Greene's companies. Cole was president of the new company, and J. W. Allen was its first assistant secretary. Ryan and Cole, as well as H. H. Rogers and William Rockefeller, participated in purchasing major blocks of shares in the Mexican properties. Cole soon also became the president of the Mexican operating company as well, and W. D. Thornton and J. W. Allen took over as president and secretary/treasurer of Greene's old holding company, which moved its offices to Amalgamated's address in New York. Amalgamated owned 100,000 shares of Greene Cananea stock. In the process of all these transactions, Greene Cananea voided an old contract for marketing Cananea copper through another broker and entered a new brokerage contract with United Metals. Through the transactions, Greene was eased out of the picture, but one key person who remained with the Cananea enterprise was Louis D. Ricketts, a gifted, Princeton-trained metallurgist whom Greene had brought to the organization earlier in the century.[33]

By 1910, the mining world believed that most of the conventional sulfide copper deposits had been discovered, and engineers turned their attention to the so-called porphyry coppers, which were massive deposits of relatively low-grade copper ore, some of it in sulfide form and some of it having been oxidized across geological time. With the backing of the Guggenheim interests, D. C. Jackling demonstrated in the early twentieth, at Bingham Canyon, Utah, that low-grade copper ore, disseminated in porphyry rock, could yield a profit by utilizing highly mechanized methods of mass mining and mineral processing. Jackling's Bingham Canyon mine launched a series of new porphyry-copper mine developments, mostly in Arizona and Chile, with ores typically assaying less than 2 percent copper. Ryan and Cole decided to enter that action as well, and in 1910 they purchased a major stake in the Live Oak Copper Company, which was developing a porphyry-copper property in Arizona's Globe-Miami district. In 1912, they formed the Inspiration Consolidated Copper Company in order to consolidate the properties and operations of Live Oak and the nearby Inspiration Copper Company. Amalgamated owned 150,000 shares of Inspiration Consolidated stock, and Ryan, Cole, and Rockefeller were members of the new company's board of directors. William Boyce Thompson, who would go on to establish the Newmont Mining Corporation, was the principal promoter of the Inspiration Copper Company, and he was made the nominal president of Inspiration Consolidated. W. D. Thornton was vice president, and J. W. Allen was secretary and treasurer. Inspiration

Consolidated's corporate headquarters was 42 Broadway, the New York address that housed Amalgamated's and the ACM's corporate offices.[34]

At the same time that Ryan and Cole were organizing Inspiration Consolidated, International Smelting and Refining was expanding from its copper-smelting operations at Tooele and its copper-refining operations in New Jersey. In 1912, International added custom lead-smelting capacity to its Tooele smelter and it developed a huge new lead refinery in East Chicago. In 1913, International entered agreements with both Inspiration Consolidated and the Miami Copper Company (a neighboring company that was mining porphyry copper) to build a new smelter at Miami to smelt their concentrates. The following year, International's shareholders voted to sell all of its assets to the ACM and to dissolve the company. At the same time, the ACM formed the International Smelting Company, incorporated in Montana, and placed all the former International Smelting and Refining property in the hands of the new company.[35]

The transaction between International and Anaconda in 1914 paved the way for the final step in consolidating all Amalgamated's property in the hands of the ACM the next year. Anaconda paid Amalgamated $4,769,900, and in return, Anaconda acquired all Amalgamated's shares of Inspiration stock, Greene Cananea stock, and some blocks of stock in other entities. At the same time, United Metals Selling Company transferred all its assets and business to the ACM in exchange for ACM stock, conveyed those shares of stock to Amalgamated, and dissolved. Anaconda in turn created its own United Metals Selling Company of Delaware and turned the brokerage business over to the new company. Later in 1915, Amalgamated gave its shareholders the opportunity to exchange their shares in Amalgamated for shares in Anaconda. Having disposed of all its stock, Amalgamated dissolved. Its president, John D. Ryan, returned to the position as president of the ACM. Now Anaconda was a vertically integrated global company, with mines, smelters, and a refinery in Montana, with copper and lead smelters in Utah and Arizona, a lead refinery in Chicago and a copper refinery on New York Harbor, and significant interests in major copper-mining operations in Arizona and Mexico.[36]

First World War and Labor Unrest in Butte

The ACM's emergence as a global mining giant and its consolidation of most mines on the Butte hill under one corporate umbrella had implications beyond the integration of the hill into a single technological system; it had ramifications for the mine workers as well. Instead of their union being able to negotiate with several companies, all competing for scarce labor, the Butte Miners' Union now had only one major employer with which to contend. Within two years of the 1910 consolidation, the ACM began to flex its power as nearly the sole employer of miners in Butte, launching the first major period of serious labor unrest at Butte.[37]

For three decades, the "Gibraltar of Unionism," as Butte came to be known, had experienced little strife between miners and mining companies. The Butte Miners' Union, local No. 1 of the Western Federation of Miners, concerned itself mainly with caring for its members needs and supporting union-organizing efforts elsewhere in the mining industry. Meanwhile, Butte's several competing mining companies paid high wages and acceded to many of the Butte Miners' Union's demands concerning working conditions. In December 1912, however, two years after ACM emerged as the single dominant factor in Butte's mining industry, the company declared that no one could work in the mines without a company-issued "rustling card." This was a direct affront to the authority of the union in employment matters. The new situation revealed a deep division between the relatively conservative and primarily Irish leadership of the Butte Miners' Union, which was content to maintain harmony with ACM management (also largely of Irish stock) and a radical wing of the union, which wanted to push a more radical working-class agenda against the company and especially against the rustling card. The radicals believed that the union's leadership was in collusion with the ACM to keep the rank and file at bay. The split became a conflagration in 1914, during the second term of Butte's mayor, Louis Duncan, a Socialist. On June 13, disgruntled union members drove the union's leaders from the Miners' Union Hall and then dynamited it. No one was killed, but the incident led to the removal of Mayor Duncan from office, under pressure from the ACM for failing to defend private property.[38]

The labor unrest during the summer of 1914 was largely internecine, and the miners did not call a strike against the mines. The ACM and other Butte mining companies did fan the flames, however. With the onset of war in Europe and a temporary drop in demand for copper, the price of copper dropped as well, so the ACM reduced wages. Responding to the surplus of copper on the world market, the ACM reduced production at the beginning of August by closing several of its mines, throwing about two thousand miners and several hundred smeltermen out of work. Then in late August, the ACM and most other mining companies in Butte declared an open shop, saying they would no longer recognize the Butte Miners' Union or other labor organizations. With the Butte Miners' Union in organizational shambles, workers could take no united action. The Industrial Workers of the World (IWW) and other radicals formed the Butte Mine Workers Union, but the new organization did not gain enough members to be effective. Fights flared among factions of miners. On August 31, someone dynamited a building at the Parrot mine, doing minor damage. Montana's Governor Stewart, who had already been petitioned by Butte businessmen to call out the National Guard, was informed immediately of the blast at the Parrot, and on September 1 he declared martial law in Butte. Butte remained under military control for most of the rest of 1914.[39]

When the ACM closed those mines in August, ACM vice president Con Kelley announced that the action was solely in response to the depressed copper market and had nothing to with the local labor unrest. The Great Falls smelter was closed entirely, production at the Anaconda smelter was curtailed considerably, and from August through the end of 1914 the company's copper output was about half of normal. During the period of reduced operations, the company retimbered mineshafts and underground workings, made major improvements to the surface plant at some of its mines, and completed much of the overhaul at the Washoe and Great Falls smelters. The ACM did not get back to full production until nearly the middle of 1915. By then, wartime demand for copper had reached expected high levels, and the ACM was favorably positioned to meet its share of that wartime demand profitably. Trying to curry favor from its miners, the ACM pledged to raise wages to $3.75 per day, if the price of copper remained above fifteen cents per pound. The company said that it would abide by the spirit of the old contract it had signed with the Butte Miners' Union, even if the company did not consider the contract to be in force. When the previous three-year agreement with the Butte Miners' Union expired in June 1915, there was not a union that the miners recognized to represent them in trying to negotiate a new contract and regain the ACM's recognition as the representative of the miners.[40]

The 1910s were tumultuous years on the labor front, not only for Butte but also for the nation, especially because of World War I. Many in the American working classes, including significant numbers of Butte miners, were opposed to the U.S. entry into the war. Tensions were heightened in Butte because of the high proportion of the city's population that was of Irish heritage. Many of Butte's Irish were antagonistic toward England and so sympathized with Germany in the war, and that in turn antagonized patriotic Americans. As the United States' entry into the war grew more probable, radicals led by the IWW agitated against that eventuality as well. Conflict in Butte flared again in 1917, first in the face of military conscription and then in the wake of the Granite Mountain mine disaster.[41]

The Granite Mountain and adjacent Speculator mines were owned and operated by the Thomas Cole's North Butte Mining Company, an independent firm with close ties to the ACM. On June 8, workers let some electrical cable slip while lowering it down the Granite Mountain shaft, causing the casing to rupture and allowing oil used for insulation to leak from the cable. When the night shift reported to work, the new men did not realize the seriousness of the situation, and a spark, perhaps from a miner's carbide lamp, lit the oil-soaked shaft ablaze. Poisonous gases spread throughout the underground workings and billowed from the Speculator shaft, then being used for ventilation. Some miners were rescued, but many were trapped and 167 died, most suffocating or succumbing to the noxious gases.[42] The Granite Mountain fire is the worst hardrock mining disaster in U.S. history.[43]

Believing that the high death toll could be used to kindle support for his radical union, IWW organizer Frank Little moved to Butte in July and began organizing miners to participate in a strike that had begun three days after the Granite Mountain fire. A longtime IWW organizer, Little had been working in Bisbee, Arizona, when the Butte disaster occurred. Ignoring advice from IWW officials that it would be unsafe for him in Butte, he proceeded tirelessly to rally support for the union and to speak out against the war. On the night of August 1, masked gunmen broke into his rooming house in Uptown Butte, beat him, tied him to the bumper of a car, and dragged him to a railroad trestle near Silver Bow Creek, where they hanged him. His body was found the next morning with a note pinned to it saying, "3-7-77," the same mysterious code used by Montana vigilantes in the 1860s and 1870s when they executed thieves and rustlers. No one was ever arrested or tried for Frank Little's murder, although it was widely believed in Butte that the gunmen were agents of the Anaconda Copper Mining Company.[44]

Several radical labor leaders fled Butte or sent their families away, fearing that they might be the next targets of the masked "vigilantes." Nevertheless, the strike continued, and Butte copper production continued to suffer. In an effort to prevent further violence, and to protect nonstriking miners, the federal government sent troops to Butte on August 17. Smelterworkers at Anaconda went out on strike on August 23. Organizers called off the strikes in December 1917, but the troops remained in Butte until January 1918. Miners organized a few other strikes during the period, but with no success. The last strike ended in May 1920, three weeks after the "Anaconda Road Massacre," when ACM guards fired on picketing strikers. One miner was killed and sixteen wounded. All were shot in the back as the pickets had tried to flee the surprise attack by the company guards. Again, no one was arrested for the crime, and the ACM enjoyed an open shop in its mines until the 1930s.[45] The Anaconda Copper Mining Company's dominance in Butte and elsewhere in Montana would remain unrivaled for several decades.

Further Expansion and Diversification

The ACM was not only the world's greatest supplier of copper in the 1910s; it produced gold, silver, arsenic, coal, and timber—and also entered the zinc market. The ACM's move into zinc mining came after several other companies had demonstrated its profitability in Butte. The zonal structure of ores in the Butte hill is characterized by veins rich in copper minerals in the central zone, veins with both copper minerals and sphalerite (the sulfide of zinc) in the intermediate zone, and sphalerite and manganese mineral in the peripheral zone. Butte's big silver mines had been in the intermediate and peripheral zones, north of Butte, in the Walkerville area.[46] Because sphalerite is a sulfide mineral, entry of the ACM and others into zinc production added to the pollution the mining industry would generate.

Growth in the global demand for zinc was due to a tremendous increase in use of the metal for galvanizing iron and steel to prevent corrosion. Zinc's effectiveness to prevent corrosion of iron had been known for a couple hundred years, and then in the late 1900s innovators developed methods for dipping pieces or sheets of iron or steel in molten zinc, thus effectively applying a thin coating. The application was especially effective for preventing corrugated iron or steel siding from corroding. By 1905, when interest in Butte's zinc resources started to percolate, half of the zinc produced in the U.S. (100,000 tons out of the total 200,000 tons) was being used for galvanizing, with about a quarter used to make brass (the alloy of copper and zinc) and one-sixth used for sheets of zinc for various uses. Two years later, the use of zinc in the U.S. for galvanizing had increased by 50 percent, while the other uses remained about the same. A decade later, when the ACM would enter the zinc market, the use of zinc for galvanizing had increased tenfold, to more than a million tons, and zinc was now being used to galvanize pipe and barbed wire in addition to corrugated sheeting. Uses of the metal in making brass and zinc sheeting were increasing at about the same rate.[47]

In the early 1900s, investors had leased W. A. Clark's Black Rock mine, in the peripheral zone northeast of Walkerville, with an eye toward developing its zinc ore. After the first lessee failed, Clark sold the Black Rock mine to a new company, the Butte and Superior Mining Company, in exchange for shares in the company. He retained ownership of the adjacent Elm Orlu and Poser silver mines, which tapped the same ore body. The Butte and Superior's sinking of the Black Rock shaft reached a thousand feet of depth in mid-1908, and it became apparent that the vein was indeed rich in sphalerite, the sulfide of zinc. Clark's miners discovered the same thing during development of the Elm Orlu mine, and his metallurgists began to experiment with techniques for concentrating zinc ores at the Butte Reduction Works. This gave the Butte and Superior a place to ship its zinc ore, hoisted during development work. With demand for zinc expanding, the Butte and Superior decided to build its own concentrator near the mine. In the meantime, the company entered into an agreement with Basin Reduction Company to treat zinc ore from the Black Rock at Basin, some thirty miles northeast from Butte.[48]

When Clark sold his copper mines and the Butte Reduction Works to the ACM in 1910, he retained the right to use the concentrator at the Butte Reduction Works to mill his zinc ores. After a fire destroyed the concentrator, he temporarily closed the Elm Orlu mine until he could build a new concentrator, which he located about a mile south of the Butte Reduction Works, on the north-facing slope of Timber Butte. The Timber Butte mill went into operation on Elm Orlu ore in June 1914.[49]

While construction of the Timber Butte mill was underway, Clark's metallurgists began experimenting with the new technology called flotation, which the Butte and Superior had been successfully using at the Basin mill to concentrate zinc ores from the Black Rock mine. Gravity concentration, which had been the conventional process until then, exploits differences in the way particles of similar size, but of greater or lesser density, behave in moving water in order to separate more dense metal-bearing particles from less dense gangue. Flotation, on the other hand, is counterintuitive, in that the metal-bearing particles "float." It exploits the tendency of fine particles of sulfide minerals, being worked in appropriately designed equipment, to adhere to air bubbles and thereby float to the surface of a slurry of fine solids, water, and certain chemical reagents, while fine particles of gangue remain suspended in the pulp. Paddles then sweep the froth, bearing the particles of sulfide mineral, from the surface of the slurry. Clark obtained a license to use flotation from the Minerals Separation Company, which owned a patent on the process. The Butte and Superior had not obtained a license to use the process. Rather, James Hyde, who had worked for Minerals Separation, left that company and moved to Basin in early 1911 to install flotation equipment in the mill that the Butte and Superior was leasing. The process proved immensely effective, but by the end of the year, Minerals Separation had filed suit against the Butte and Superior for patent infringement. Hyde claimed that he was using a different process, because the concentrations of chemical reagents and frothing agents he used were different than specified in the patent. Details of the litigation are beyond the scope of this book; suffice it to say that, on one hand, Butte and Superior was significant for showing the U.S. mining industry the efficacy of flotation, and on the other hand, that Minerals Separation prevailed in the litigation, with appeals eventually reaching the U.S. Supreme Court, securing the company's patent rights to the process.[50]

In early 1915, the ACM began developing some of its mines, including the Poulin, Lexington, Emma, Gray Rock, and Alice, as zinc producers. At the same time, it began construction at the Washoe smelter of a zinc concentrator with a capacity to treat 2,000 tons of ore daily. The company's research department was also developing a sequence of processes to produce metallic zinc: 1) roasting the zinc concentrates at the Washoe to oxidize the sphalerite, 2) leaching the zinc calcines with sulfuric acid to dissolve the zinc, and 3) treating the zinc solution in electrolytic cells, in a process called electro-winning, to extract zinc metal from solution.[51] After developing a trial, 10-ton electrolytic plant at Anaconda in 1915, the company built its electrolytic zinc plant at the Great Falls smelter in 1916, in order to take advantage of the low-cost hydroelectricity there. The Great Falls plant initially had capacity to produce 100 tons of electrolytic zinc per day and was expanded to 150 tons in 1918. It quickly became the nation's leading supplier of electrolytic zinc, producing half the world's electrolytic zinc in 1920. The development of Butte as a major zinc-mining center is a chapter of Montana's mining history that will not be elaborated here. Suffice it to say that the Butte and Superior concentrator at the Black Rock mine and Clark's Timber Butte mill south of Butte became two new and major sources of tailings in the Summit Valley, the Anaconda company's zinc concentrator at Anaconda became a major source of tailings in the upper Deer Lodge Valley, and roasting zinc concentrates at Anaconda and Great Falls contributed considerably to the SO_2 pollution in the vicinity of Montana's metallurgical plants.[52]

As the ACM was consolidating its hegemony over the working conditions in Butte and diversifying its Butte output by expanding into the zinc production, it was also maintaining its focus on expansion and diversification of its enterprise outside Montana. One area of diversification was to cement its control of the Inspiration Consolidated Copper Company. The ACM restructured Inspiration's board of directors in 1918, when President W. B. Thompson resigned. He had been president of Inspiration Consolidated since its incorporation, but, according to L. D. Ricketts (who had joined the board in 1914), Thompson only held the position nominally. The real authority among the officers, according to Ricketts, was Ryan's close associate, W. D. Thornton. Thompson sold his Inspiration shares on the market in 1918, and the ACM's Cornelius F. Kelley replaced Thompson on the board and as corporate president. Also added to the Inspiration board that year were two other men from the ACM: its vice president, Benjamin B. Thayer, and William Wraith, a top ACM metallurgist. After graduating from the Michigan College of Mines in 1897, Wraith had taken a job at the Boston and Montana's Great Falls smelter. After Amalgamated took control of the B&M, Wraith transferred to the Washoe smelter, moving up the hierarchy, so that, in 1913, the ACM would move him to Tooele to manage operations at the new International smelter. In 1918, he moved to ACM headquarters in New York, where he had administrative charge of development of Andes Copper Company (described below). Another way that the ACM exercised control over the Inspiration operation was through its general manager, Charles E. Mills, a Ricketts protégé recruited from the Detroit Copper Company at Morenci. Mills served a joint appointment as general manager of Inspiration Consolidated and of the smelter at Miami, owned and operated by International Smelting and Refining, the ACM's wholly owned subsidiary.[53]

The ACM also acquired new copper-mining properties elsewhere, including the Walker mine in California and some promising copper properties at Potrerillos, Chile. The purpose of acquiring the Walker mine was to ensure a supply of ore and concentrates for International's smelter at

Tooele. In 1916, International took an option to acquire controlling interest in the Walker Mining Company. After sufficient exploration and development at the Walker mine, International exercised its option, purchased 51 percent of the Walker company's stock, and placed Wraith on the Walker company's board.[54]

The Potrerillos copper deposit was being explored in the 1910s by William Braden, a New York–based mining engineer who had an option on the property. In 1916, he offered his option to the ACM, which sent L. D. Ricketts, B. B. Thayer, and Reno Sales to Chile to inspect the property's potential. They were favorably impressed, and the ACM quickly incorporated Andes Copper, a subsidiary that would continue development of the Potrerillos property. The ACM made Ricketts president of Andes Copper, put Wraith in charge of operations, and sent L. R. Wallace to Chile to serve as general manager. Wallace, a metallurgist, had begun working for Inspiration in 1912, when Ricketts had him appointed as superintendent of construction for the International smelter at Miami. When the International smelter began operating in 1915, he continued as superintendent of operations, until he left for Chile in 1916.[55]

As the second decade of the twentieth century drew to a close, the Anaconda Copper Mining Company was a giant global corporation with properties and operations throughout the U.S. and the hemisphere, and with an ambitious and talented group of executives and managers at its helm at 42 Broadway in New York. John D. Ryan was chairman of the board, C. F. Kelley was president, and B. B. Thayer was vice president. In Montana, Reno Sales was head of the Geological Department in Butte, John Gillie was manager of mines in Butte, and metallurgist Frederick Laist was manager of the Washoe smelter in Anaconda. In addition to its Montana operations—copper and zinc mines in Butte, metallurgical works at Anaconda and Great Falls, sawmills near Missoula, coal mines near Great Falls and Red Lodge, and the Butte, Anaconda and Pacific Railway—the ACM owned the International Smelting Company (with its Tooele and Miami smelters, its Chicago and New Jersey refineries, and its share in the Walker mine), and the ACM owned almost 33 percent of the shares of Inspiration Consolidated stock, nearly 38 percent of the Greene Cananea stock, 78 percent of the shares in Andes Copper, and all of United Metals Selling Company. Each of the subsidiaries, whether majority-owned by the ACM or not, had its corporate offices at 42 Broadway. C. F. Kelley was president of both Inspiration Consolidated and International Smelting, and W. D. Thornton and

L. D. Ricketts were the vice presidents of both companies. Thornton was president of Greene Cananea, and Kelley was its vice president. Ricketts was president of Andes Copper, and Kelley and William Braden were vice presidents. Ryan was president of United Metals Selling Company, which was the world's largest copper broker.[56]

While the ACM was on the ascendency as a corporate giant of global proportions in the 1910s, pollution emanating from its Washoe smelter continued nearly unabated. It was against that corporate ascendency that the United States government made a partially successful effort to compel the ACM to abate its pollution in the 1910s, as described in the next chapter.

CHAPTER 8

The United States Files Suit

> If the Anaconda company should do all the good things its engineers recommended, it would never pay a dividend. They are good engineers, and most of the projects they urge are good, but if we carried out all of them our capital would be perpetually tied up.
>
> —John D. Ryan (1916)[1]

The Bliss case was not the end of legal challenges to the Anaconda Copper Mining Company and the smoke its Washoe smelter spewed into the air. The company's next opponent, the United States Department of Justice, had far greater legal resources, and it was backed by the political power of Theodore Roosevelt's White House, but the ACM's officials fought the government's efforts to compel the company to abate its polluting activities with the same tools that they had used to confront farmers' efforts to stop the destruction of agricultural resources in the Deer Lodge Valley. In resisting the challenge to its hegemony in Montana posed by the United States, the ACM stood on the principle articulated by John D. Ryan, his obligation to return dividends to the capital invested in the enterprise.

The farmers had brought their plight to Roosevelt's attention early in the century. In doing so, they brought him an issue that fell within two of his top priorities, the regulation of giant corporations, especially trusts, and the conservation of natural resources, especially forest resources. The farmers let him know that not only was the Washoe smelter damaging their livelihoods but also it was damaging trees. And the damage was being done by a giant trust, Amalgamated, that was closely associated with the Standard Oil Trust, which was already the focus of much of Roosevelt's attention in his effort to create a federal Bureau of Corporations. Roosevelt had had the opportunity to see the scene of the Montana conflict on his visit to Butte in May 1903, during his much-celebrated national tour. Said to have been the longest national tour, to date, by a sitting president, Roosevelt's trip began on April 1. Traveling by rail a total of fourteen thousand miles and passing through twenty-four states, the president's tour began in Maryland and continued westward through the Great Lakes states, the Dakotas, and into Montana.[2]

When he arrived at Butte the evening of May 27, Roosevelt entered a hornets' nest. He had been forewarned of the bitter conflict underway between Heinze and Amalgamated, but he also had enemies among Butte's union men. Shortly before the trip, Roosevelt had antagonized the Western Federation of Miners by sending federal troops to Arizona during a strike. Heinze, who controlled local politics, had first invited Roosevelt to dine in Butte as his guest. When Amalgamated officials learned of Heinze's invitation, they extended their own. Advisors to the president suggested that he dine as the guest of Butte's mayor, and that an equal number of Amalgamated and Heinze supporters be invited to the dinner and seated at the head table. Representatives of both conservative and radical factions of the Butte Miners' Union were also invited. After a tour of Butte the next day, during which radical members of the WFM jeered the president, Roosevelt attended the banquet in his honor at the Thornton Hotel. Mayor Mullins sat at the center of the head table with the president to his right. Further on the president's right sat seven men, including Heinze and John MacGuinniss, who often was the named plaintiff in Heinze's litigation against Amalgamated companies. To Mullins's left were eight men, including ACM president William Scallon, W. A. Clark, and, at the far end, two of Montana's notable pioneers, gold miner Thomas Cruise and rancher Conrad Kohrs, Powell County's state senator who had recently failed in his legislative effort to give farmers in the Deer Lodge Valley better legal tools to fight the mining industry's pollution.[3]

At about the time Roosevelt visited Butte, an agency in his administration began investigating the problem of smelter smoke. Not long after the president visited Butte, scientists of the U.S. Department of Agriculture (USDA) were in northern California studying the damage that smoke from the Mountain Copper Company's smelter near Keswick was doing to trees on the public domain. The U.S. Department of Justice (DOJ) had filed suit against the company and, to collect evidence in support of its case, the DOJ asked the USDA to send a scientist to those California forests to conduct fieldwork. J. K. Haywood, of the USDA's Bureau of Chemistry, spent part of the summer conducting experiments by subjecting several kinds of vegetation to various concentrations of sulfur dioxide and sulfur trioxide in their respective experimental atmospheres, and by collecting samples of vegetation, both dead and thriving, from the area around the smelter to ascertain the sulfur content of the foliage. He concluded that sulfur dioxide emitted by the Mountain Copper smelter was causing severe injury to vegetation within several miles of its stack. After hearing testimony in 1904 from Haywood and others, including witnesses for the company and farmers who complained of damage caused by smelter smoke, the federal judge in San Francisco ruled in favor of the government in 1905 and enjoined the smelter from continuing to pollute. Mountain Copper immediately appealed to the Ninth U.S. Circuit Court of Appeals.[4]

Haywood next made trips in 1905 and 1906 to the vicinity of the copper smelters at Ducktown, Tennessee, to conduct field research in the area's forests. Important litigation was underway there, as the neighboring State of Georgia had filed suit in the U.S. Supreme Court against the Ducktown smelters and on behalf of Georgia farmers, who were suffering damage from smelter smoke drifting from Tennessee into Georgia. Because Georgia's complaint was against operations in another state, Georgia could take its case directly to the nation's highest court. Haywood investigated conditions in forests of North Carolina, Tennessee, and Georgia that were near the smelters. He found that smelter smoke was killing vegetation as far as twelve miles from the smelters, and he concluded the damage likely extended even farther in areas he had not studied. The Supreme Court ruled in Georgia's favor in 1907, ordering the Ducktown smelters to either remove the harmful sulfur from their smoke or cease operating.[5]

Responding to complaints from the Deer Lodge Valley farmers, the USDA next sent Haywood to the Anaconda vicinity in 1906 and 1907 to investigate damage that the Washoe smelter was causing to farmers' lands, crops, and livestock. Accompanied by a forester, W. G. Weigle, Haywood studied not only conditions north and east of the smelter in the Deer Lodge Valley but conditions in the mountainous forests south and west of the smelter as well. Haywood expanded the chemical scope of his Anaconda investigations beyond the damage from sulfur he had investigated in California and Tennessee; he also analyzed the deposition onto vegetation of arsenic from smelter smoke, and he investigated the contamination of streams downstream of the smelter by tailings. Haywood drew several conclusions, including that damage to forests extended at least thirteen miles from the smelter and probably as much as twenty miles, that large amounts of arsenic were emanating from the smelter stack and settling on forage in concentrations great enough to be poisonous to cattle, that tailings discharged into the

Deer Lodge River made it unfit for irrigation, and that lands so irrigated had been damaged by dissolved copper. Weigle also wrote a report, noting that many hillsides in the area of the smelter had already been denuded by cutting fuel for early smelters and that forest fires had also ravaged some areas. Focusing his attention on forested areas not previously damaged by fire, he concluded that the cause of death to vast swaths of lodgepole pine and Douglas fir was sulfur in the smelter smoke. Weigle reported that, if the wind was blowing in the right direction, he could smell sulfur from the smelter smoke as far as twelve miles from the Washoe works.[6] He ended his report: "In conclusion it may suffice to say that unless the smelting of ores at Anaconda rich in sulphides [*sic*] and arsenic is suspended or that the plant be so constructed that the arsenic and sulphur [*sic*] may not escape, the plant and animal life within an area of about 300,000 acres of which about 1/3 lies within the bounds of the Hell Gate Forest Reserve will continue to be doomed to more or less destruction."[7]

In 1907, even before Judge Hunt had ruled in the Bliss case, President Theodore Roosevelt and his administration concluded that, with copper smelters causing so much damage to vegetation on national forests, the government had to take legal action, and the U.S. Department of Justice began preparing cases against the Anaconda Copper Mining Company and other western mining companies that were causing the damage. By 1907, the Mountain Copper Company, in Shasta County, California, had won its appeal to the Ninth Circuit and the U.S. was about to appeal its case to the U.S. Supreme Court, the nearby smelter of the Mammoth Copper Company had been operating for two years, and Balaklava Copper Company's new smelter was under construction.[8]

With so many cases to handle pertaining to smoke from copper smelters, the government retained the services of Ligon Johnson, an Atlanta attorney who, working with the Georgia attorney general, had successfully prosecuted the case against the Ducktown smelters before the Supreme Court. Born at Tuskegee, Alabama, in 1873, Johnson was educated at St. Mary's College in Belmont, North Carolina, Emory College in Oxford, Georgia, and the University of Georgia. He received his law degree from the University of Virginia in 1891 and was admitted to the Georgia bar in 1893, when he began his law practice in Atlanta. Besides his work on the Ducktown case, Johnson shared an important concern with Roosevelt, that of conserving forest resources and creating national forests. Johnson was a founder and first president of the Appalachian National Forest Association.[9]

Roosevelt's Department of Justice asked Johnson to take over the Mountain Copper case and also investigate the possibilities for legal action against other smelter companies in the West. Late in 1907, Johnson toured several western states from which the government had received complaints of smelter smoke damaging surrounding lands, including both government and private property. Johnson negotiated a settlement in the Mountain Copper case and prepared cases against several other smelter companies in California, but without doubt the largest smelter and the most extensive damage he faced was that at Anaconda. The legal and technical matters involved at Anaconda were also the most complex.[10]

Sympathizing with the farmers, Roosevelt determined to take legal action, but he wanted to wait until after the November 1908 election, so that the litigation initiated by the federal government would not become a political issue. Meanwhile, he had Attorney General Charles Bonaparte instruct Johnson to draft a legal complaint that would be filed with the federal court in Montana to initiate litigation. Johnson had his "bill of equities" drafted by August 1908. The DOJ would argue that the government, in its capacity as sovereign, could move to require that the Amalgamated smelter abate its air pollution, and the government could collect damages from Amalgamated for injury to federal property, for revenues of lost timber sales and grazing fees, and for the costs required for reforestation. Moving into autumn, other DOJ attorneys and the farmers' attorney R. L. Clinton refined the draft. A DOJ attorney took the draft to the White House in October for a meeting Roosevelt scheduled with Joseph Dixon, a budding Progressive Republican and Montana's newly elected U.S. senator. Dixon was not an ally of Amalgamated, unlike Montana's senior senator, the conservative Republican Thomas Carter. Nevertheless, Dixon opposed the president's plan to sue the company, because the town of Anaconda was wholly dependent on the Washoe smelter, and many sectors of the state's economy were dependent on Amalgamated. Despite Dixon's caution, Roosevelt and his attorney general were determined to proceed with the suit, because the damage being done to plant and animal life in southwest Montana was so dire. Besides, Roosevelt was at the end of his term.[11]

As the government began preparing its case against the Anaconda company, the DOJ collected affidavits from residents in the vicinity of the Washoe smelter who observed that damage to their environment was continuing. Henry Eccleston, who farmed about ten miles southeast of the smelter, noticed that the timber in the hills south of the smelter was sickly and dying. For example, an area known as the McCune Cutting, where contractors had cut timber for the ACM on the public domain during the nineteenth century, had been regenerating itself until the Washoe smelter went into operation in 1902. Eccleston believed that smelter smoke was now killing the young trees. Margret Quinlan, who lived about fifteen miles northeast of the smelter, said that sometimes the smoke from the smelter dropped down into the Deer Lodge Valley and was so thick she could smell it and hardly see through it. She occasionally had to send her daughter out of the valley to get away from the smoke. Mrs. Quinlan claimed that doctors believed "smelter poisoning" caused her daughter's poor health. Lizz Newsome had a small farm less than two miles southwest of the smelter and along the boundary of the U.S. Forest Reserve. From her place, she had noticed that fir trees were the first to be affected by smelter smoke. She had once raised chickens, but she claimed smelter smoke turned their combs black and yellow and made their feathers curl, preventing the birds from molting. Because these symptoms led to death, she could no longer make a living from her farm, and she had left it.[12]

John E. Jones raised stock on Lost Creek directly north of the smelter. He estimated that 90 percent of the timber in the vicinity of his place had been destroyed by smelter smoke. Sometimes, he said, smelter smoke would suddenly come upon his place, causing him to cough and his eyes and nostrils to burn. Both he and Lizz Newsome claimed that, after such events, they could notice an immediate effect the smoke had on plants. Many residents recalled that the Deer Lodge Valley had once had some of the best agricultural land in Montana, but smelter smoke had made it nearly impossible to raise livestock, and crop yields had diminished markedly. They also observed that damage to timberlands surrounding the valley was getting progressively worse each year. Walter Staton, who farmed and ranched a few miles southeast of the smelter and was one of the stalwarts of the Deer Lodge Valley Farmers' Association, believed that damaged timberlands extended at least fifteen miles in every direction from the smelter. South of the smelter, in a wide swath up Mill Creek to the Continental Divide, all the fir and nearly all the lodgepole had died by 1908, due, he was certain, to damage from smelter smoke.[13]

Before initiating the suit against Amalgamated, however, Roosevelt decided to try negotiation. He invited representatives of the Deer Lodge Valley Farmers' Association and Amalgamated to join Montana's two senators for a meeting at the White House. On a Saturday morning in early

December, Nick Bielenberg and W. C. Staton, of the association, and their lawyer, R. L. Clinton, joined Amalgamated's John D. Ryan and A. J. Shores at the White House for a meeting with the president, the attorney general, the USDA's chief forester, Gifford Pinchot, and Senators Carter and Dixon. Sen. Henry Cabot Lodge, Massachusetts, somehow also gained an invitation to the meeting, saying he wanted to represent his constituents who were Amalgamated stockholders. Roosevelt proposed three potential remedies, and he hoped that the participants could discuss them. One remedy was to move the smelting operations to Great Falls, where there were no nearby forests that could be damaged by smelter smoke. Another was to move smelting back to Butte, whose environment was already severely damaged. The third alternative would be to install devices necessary to allow the Washoe smelter to convert all the sulfur in its furnace emissions to sulfuric acid before the sulfur went up the stack. The sulfuric acid would in turn be used to make phosphate fertilizer, as was now being done successfully at the Ducktown smelters.[14]

The meeting did not go as Roosevelt had hoped, and the Amalgamated officials responded that none of the government's proposed solutions would be acceptable. The tenor of the White House meeting may be suggested in a description by a muckraker, Jerre Murphy, who claimed to have received an authoritative report of the meeting from a participant. Early in the meeting, each of the speakers, observing protocol, rose to stand when addressing the president. When it came Ryan's turn to respond to a question from Roosevelt, he remained seated and crossed his leg before replying, according to Murphy's informant.[15]

Roosevelt took Amalgamated's objections seriously and wanted government experts to investigate the trust's claims concerning the impracticality of the government's proposed solutions, especially the one that would convert sulfur in the smoke to sulfuric acid to be used in the manufacture of fertilizer. Roosevelt was deeply disturbed by the situation. On one hand, he recognized that the Washoe smelter was destroying extensive forests and grazing lands, as well as polluting streams; on the other hand, he recognized that, if the government forced closure of the smelter, even more property could be destroyed—not only that of Amalgamated but that of residents of Anaconda as well. He wanted to use the government's coercion to compel the trust to implement a solution to the problem, without destroying an industry and the livelihoods of thousands of people. Moreover, Roosevelt recognized that, if the government could not prevail on this issue in Montana, it would be hard-pressed to abate pollution from smelters elsewhere, including in California.[16]

At Roosevelt's request, and with the president's term in office nearing its end, Attorney General Bonaparte mobilized more government-sponsored research. Pinchot enlisted the services of chemist R. E. Swain, who had testified for the farmers in the Bliss case, and his Stanford University colleagues G. J. Peirce, a professor of botany, and J. P. Mitchell, an assistant professor of chemistry, to collect more scientific evidence of the damage smelter smoke was causing in the vicinity of the smelter. Among other findings, the Stanford scientists concluded "that the barrenness and desolation surrounding a smelter handling sulfurous ore are not mere coincidence but are the inevitable effects of which sulfur dioxide is the specific cause."[17] The U.S. Forest Service also had several consultants and its own experts in the field around Anaconda collecting evidence. For example, Forest Supervisor P. S. Lovejoy found in 1910 that damage in many areas around Anaconda had increased over what he had observed when he first surveyed the areas in 1907 and 1908. He estimated that the cost of damage to trees from smelter smoke was greater than $1,750,000.[18]

Meanwhile, the ACM asked outside experts to respond to the feasibility of the government's proposal that the company employ technologies, like those used at the Ducktown smelters, to recover sulfur and then use the acid to manufacture phosphate fertilizer. Among the objections the experts raised to the United States' proposal were: 1) the Washoe smelter treated much more ore than did the Ducktown smelters and discharged far more sulfur; 2) to treat Butte ores, the Washoe smelter used different smelting technologies from those employed at Ducktown; 3) Butte ores and Anaconda smelter smoke contained arsenic and the Ducktown ores and smoke did not, thereby posing a different set of technical hurdles to the recovery of sulfur at Anaconda; 4) there were no deposits of phosphate rock near Anaconda as there were near Ducktown (such deposits would soon be found, eliminating this objection); and 5) the region around Anaconda (the northern Rocky Mountains, or indeed the entire western portion of the United States) did not offer the markets for sulfuric acid or fertilizer that the southeastern region did. The Washoe was indeed discharging tremendous volumes of sulfur; one of the ACM's experts, Charles Doremus, estimated that the smelter was daily discharging over 800 tons of sulfur, more than half of it being given off by the MacDougall furnaces in the roasting department.[19]

FIG. 8.1. Typical Damaged Forest, August 1910. This photo appears in Lovejoy's 1910 report on damage to trees on the Deerlodge National Forest. The site is about twelve miles west of Anaconda, near today's Wraith Hill Ski Jump. The caption in the report reads, "Stand originally nearly pure, old, fir, with scattering lodgepole. Fir practically all dead; nearly all pine badly affected." Photo courtesy of the National Archives and Records Administration, College Park, MD.

One of the central concerns of both the ACM and of Ligon Johnson was how to eliminate waste, as it was then understood, in a profitable way.[20] Frederic J. Falding, a New York City consulting engineer, was hired by the ACM in 1908 as an outside expert to examine the possibility of recovering sulfur from the Washoe's smelter smoke. After writing, in a 1908 letter to Con Kelley, that it would not be technically possible to recover sulfur from the Anaconda smoke and that there would not be a market for the sulfur if the company could recover it, Falding commented, "You are, it is true, wasting out of your stack nearly 1000 tons of sulfur daily. . . . Such a waste naturally would not commend itself to your Directors."[21] J. Parke Channing, president of the Tennessee Copper Company at Ducktown, had the same understanding of waste when he wrote to Ligon Johnson in 1908 that his company's new sulfur-recovery process served "the two-fold purpose of relieving any injury which may be done by the sulfur fumes to vegetation in the surrounding country, and of conserving a very valuable constituent of the ores which has heretofore gone to waste." He went on to write, "I may supplement this by saying that this matter has been on my mind from the very inception of the Tennessee Copper Company in 1899, but that never until a couple of years ago were we enabled to so modify our smelting operations as to make the desideratum practicable."[22] This last remark suggests that Channing, like many others in the mineral industry, was imbued with the ethic that he should make an effort not to let resources go to waste in the byproducts discharged from his plant. Economics had prevented him from acting immediately on that ethic.

By mid-February 1909, it became apparent to the attorney general and his attorneys that Amalgamated would not agree to abate its smoke pollution in the short time that Roosevelt had remaining in office. Roosevelt recognized that the only way to deal with a recalcitrant trust as powerful as Amalgamated would be to file suit, and he was satisfied that the government's case stood on solid ground. He informed Bonaparte that, were his administration to continue, he would prosecute the case, but Roosevelt had less than a month remaining in office; he believed it would be best to leave the suit in the hands of his successor, William Howard Taft, because Taft's administration would have to see the case to completion.[23]

It took some time before President Taft and his administration decided to proceed with legal action against the ACM. During the first year of the Taft administration, Johnson continued working on the Anaconda and other copper-smelter smoke cases for the Justice Department, and he continued negotiating with Amalgamated, trying to reach an agreement that, if the government could find a commercially feasible method for removing the sulfur (or most of it) from the smelter's smoke, the company would implement it. The government's experts also continued to study the impacts of smelter smoke on the environment around Anaconda. Meanwhile, Taft's attorney general, George Wickersham, studied the case that Johnson and the DOJ had prepared and concluded it should have a high priority in the new Taft administration. Finally on March 16, 1910, with Amalgamated remaining intransigent, U.S. Attorney for the District of Montana James W. Freeman filed suit in the Montana District Court of the Ninth Circuit Court of the United States against the Washoe Copper Company, owner of the Washoe smelter, and the Anaconda Copper Mining Company, the smelter's operator. The suit sought to enjoin the ACM from operating the smelter in such a way that it discharged harmful smoke and gases over the property of the United States.[24] In the words of the complaint, the government sought to enjoin the ACM from "discharging any foul or dense smoke or any gas, vapor, odor, fume or substance damaging, destroying or injuring the property of your orator or its employees or tenants, or causing or producing any physical or bodily harm, injury, discomfort or inconvenience to the tenants and employees or denizens of said lands of your orator, or depriving your orator or its tenants, employees, and inhabitants, prospective or otherwise, of said region, of pure air and sunlight or pure water."[25]

Shortly after the government filed suit against the Washoe and Anaconda companies, an important corporate change took place. One week after the suit was filed, on March 23, 1910, the ACM's stockholders authorized the board of directors and officers of the company to acquire certain Amalgamated and related companies. All the companies controlled by Amalgamated transferred their assets and liabilities to the ACM, which then became Amalgamated's sole operating company. One of the related companies the ACM acquired was the closely allied Red Metal Mining Company, which since 1906 had been operating Heinze's former Butte copper mines. Shortly after the consolidation of the Amalgamated companies into the ACM, Amalgamated's president John D. Ryan also negotiated the deal with W. A. Clark wherein Clark deeded the properties of his copper companies to the ACM. These included the Original and Steward mines, owned by the Original Consolidated Mining Company, and the Butte Reduction Works, owned by the Colusa-Parrot Mining and Smelting Company.[26] With the sale of Clark's copper properties, the Butte Reduction Works closed, but Clark remained active in Butte mining and was soon to become a pioneer in the district's zinc-mining industry, as already described. These transactions in 1910 did not leave the ACM as the only copper-mining company in Butte, just the largest by far. By 1910, the Pittsmont smelter was operating steadily, and through the early 1900s, a number of other small mining companies arose, shipping their ores either to the Pittsmont smelter or to Anaconda for treatment.

The 1911 Agreement

As the government and the ACM prepared for trial, their attorneys continued trying to negotiate an agreement that would avert a trial. The government's position was that it was entitled to collect payment for damages to federal property, but it would be best to wait and determine what those damages had been until after the ACM ceased causing damage. Therefore, it was in the interests of all concerned for the U.S. to embark with the ACM to find ways to eliminate the causes of injury through technological improvement. A joint effort would be possible under the auspices of a "board of experts," convened to examine various technical means of abating the pollution problems at the smelter and to recommend their adoption by the ACM. During negotiations, the two parties had to come to agreement on a variety of issues, such as the composition of the board and what kinds of technical remedies the ACM would be expected to adopt. The ACM wanted to be certain that it would only be obligated to consider processes that were "practicable" (i.e., both technically and commercially feasible) at the works.[27] Negotiations proved fruitful, and on April 13, 1911, Benjamin B. Thayer, president of the Anaconda Copper Mining Company and representing the ACM, and Attorney General George W. Wickersham, representing the United States, signed an agreement under which the U.S. would suspend prosecution of the case as long as the ACM complied with the terms of the agreement.[28] The first paragraph of the agreement stated: "The defendant Anaconda Copper Mining Company agrees that it will at all times use its best efforts to prevent, minimize and ultimately to completely eliminate the emission and distribution from its smelting works at Anaconda, Mont., of all deleterious fumes, particularly those containing sulphur [*sic*] dioxide. But the said defendant, by entering into this stipulation does not concede that it has caused to be emitted from such works any fumes which are injurious to any of the interests of the complainant."[29] The agreement also stipulated that 1) a three-man board of experts would be established to conduct its own research and oversee ACM investigation of emissions from the smelter and of techniques and processes the ACM could implement to reduce its emissions in compliance with the agreement; 2) the ACM would implement methods and equipment that the board of experts recommended would move the smelter toward compliance with the agreement; 3) the ACM would pay for the costs of the board of experts and of other experts the board might retain; and 4) as long as the ACM complied with the terms of the agreement, the government would suspend its litigation against the ACM, but if the ACM ceased complying with the agreement the government could proceed with the suit. The agreement also named the first three members of the board and stipulated how each the three would be replaced, if he resigned or died.[30] The board of experts soon came to be known as the Anaconda Smelter Smoke Commission.

Belief in the unique abilities of engineers and other professionally trained experts to regulate society and its institutions, indeed belief in the desirability of having them do so, was a hallmark of the turn-of-the-century Progressive Era. Thus, the concept of creating public utility commissions to regulate electric and other utilities arose during the period. Progressives believed that such utilities were most efficient if they operated as natural monopolies, that such monopolies could be owned and operated by private corporations, and that the stockholders of those corporations were entitled to a reasonable profit on their investments, but the Progressives also believed that the overall conduct of those utilities was best regulated by impartial experts (i.e., public utility commissions). Such commissions would not be influenced by the irrationalities of politics or economics. Engineers and Progressive reformers agreed on some principles, like the supreme value of efficiency, but they disagreed on issues like the extent to which businesses needed to be regulated. Engineers, who identified closely with the business community, were much more sanguine about the possibility that corporations would act in the best interests of society.[31] This characteristic of engineers as experts was well exhibited by the engineers who served on the Anaconda Smelter Smoke Commission.

Under the terms of the 1911 agreement, John Hays Hammond was designated to chair the Smoke Commission, and should he ever leave the commission his successor would have to be acceptable to both parties to the agreement. Hammond was one of the most highly regarded mining engineers in the U.S., having already served on a commission of geologists and engineers appointed by Theodore Roosevelt to recommend changes in the controversial apex law, which had provided the basis for much litigation in the mining industry, including the aforementioned Heinze disputes with Amalgamated. Both the ACM and the government had confidence Hammond would chair the commission with impartiality. Yet in hindsight, his loyalty to the mining industry was evident. In a 1924 letter to the director of the Bureau of Mines expressing his desire to resign the commission, Hammond said he would be willing to remain as chairman if it was in the mining industry's interest. Indeed, in his autobiography Hammond recalled accepting the chair of the commission at its beginning "for the benefit of the industry."[32]

L. D. Ricketts occupied the seat on the commission filled at the ACM's recommendation. He was the respected metallurgical engineer who, since 1906, had overseen operations at Cananea, the large Mexican copper-mining property controlled by Amalgamated. Ricketts was also not averse to cooperating with government agencies, serving on a five-member General Advisory Committee appointed by the American Institute of Mining Engineers to advise the newly established U.S. Bureau of Mines. The third seat on the commission was filled at the pleasure of the attorney general, who designated J. A. Holmes, first director of the U.S. Bureau of Mines. Much of Holmes's prior experience was in government service with the U.S. Geological Survey.[33]

Despite the commissioners' collective experience in mining and metallurgy, much of their initial agenda appears to have been based on work previously done by Johnson, who as an attorney displayed an impressive technical grasp of the issues facing the ACM and the commission. In prosecuting the Anaconda case, as well as the Ducktown and other western copper-smelter cases, he had amassed considerable understanding of a complex of interrelated problems, ranging from the damage caused by smelter smoke to the difficulties of producing sulfuric acid from the byproducts of smelting different copper minerals in various mining districts, to the evolving agricultural and fertilizer markets in different regions of the country. Despite the creation of the Smoke Commission and its composition, including one member explicitly designated to represent the government, Johnson continued for a few years as an advisor to the attorney general on the progress the commission was making in various technical areas of investigation. In this regard, his role remained much as it had been since early 1908, when he was appointed special assistant to the attorney general and began working on the Anaconda case.[34] Although he was determined that the ACM should be made to discontinue the damage its smelter was causing, he also remained cognizant that solutions to the pollution problems had to be commercially feasible (which was the meaning of the phrase "practically applicable" in the agreement).[35] In a late-1911 report to the attorney general on the solutions being considered at Anaconda, Johnson wrote: "While all the harmful fumes cannot now be converted into sulphuric [*sic*] acid on a commercially feasible basis, or even a major portion of them, still, if the greatest good to the nation, the highest form of conservation, and in the end, the ultimate benefit of the company is to be considered, I believe that the conversion of the fumes into acid will prove the true solution of the fume problem."[36] Even the terms under which commercial feasibility were calculated were negotiable, however. A month after he wrote the above report, Johnson mused in a letter to the attorney general that an important influence on whether or not a process was considered commercially feasible was whether or not the calculations included savings attributable to reduced future damage claims, both those made by the government and those filed by private property owners. As the history of the Smoke Commission unfolded, it became clear that possible future damages were not to be part of the cost-benefit analysis.[37]

Research of the Anaconda Smelter Smoke Commission

The new board of experts first met June 10–20, 1911, in Washington, D.C., although Ricketts, who was in Mexico, was not able to attend. The commissioners met again the next month at Anaconda, where they inspected the operations of the smelter and the condition of the surrounding area. Only Holmes and Ricketts traveled to Anaconda. After viewing the damaged areas and considering some of the expert reports that had been prepared for the Bliss and U.S. cases, the commissioners decided to have a summary prepared and to have its own damage assessment conducted. The commissioners took this step recognizing that their task was not to determine the extent or character of the damage but rather to eliminate the causes of the damage. The commissioners nevertheless believed they needed to have a good understanding of the kinds of damage being done so they could set priorities in finding solutions. Toward this end, they contracted with renowned botanist J. W. Blankenship, of Berkeley, California, to conduct additional fieldwork in the area around Anaconda. Blankenship, who had earned a PhD in botany from Harvard University in 1898 and then took a job teaching at Montana State College in Bozeman, had also worked at the Montana Agricultural Experiment Station. He had been one of the ACM's experts in the Bliss case. His was the only smoke-damage investigation the Anaconda Smelter Smoke Commission conducted. Throughout the rest of the decade, however, the commissioners continued to receive reports on conditions in the surrounding area from the Forest Service and from H. C. Gardiner, the ACM's staff veterinarian who had been put in charge of such monitoring.[38]

After considering the smoke being discharged from the Washoe smelter and the damage it was causing the surrounding environment, the Anaconda Smelter Smoke Commission decided to focus its attention in two areas: 1) the gaseous, invisible portion of the smoke, of which sulfur dioxide required the most serious attention because of the damage it caused to plant life, and 2) the visible, or liquid and solid, portion of the smoke, which proved harmful mainly to animal life. The visible portion of the smoke consisted primarily of very fine dust particles comprised of the various mineral compounds found in Butte ores. Arsenic was of special importance, because it was in gaseous form at higher temperatures and sublimated to a fine powder at lower temperatures. (The process of a substance moving between the gaseous and solid phases, without going through the liquid phase, is called sublimation.) The first research undertaken by the commission, then, was to determine the quantity and character of the various constituents of the smoke, not only as they left the stack but also as they made their way through the entire array of smelter processes. The Smoke Commission analyzed the testimonies of the numerous experts in the Bliss case and began to conduct its own tests at the smelter almost immediately. These tests were performed by E. M. Dunn, an ACM chemist who had begun work in the testing department at the Washoe smelter in 1907 and was assigned to work with the Smoke Commission in 1911.[39] He presented results of his tests, some taken even before the Smoke Commission was created, to the American Institute of Mining Engineers, meeting at Butte in August 1913.[40]

To perform the on-site data-collection at the Washoe smelter, the Smoke Commission assembled a team of engineers and scientists to work with Dunn. Together with the ACM, the Bureau of Mines, and the attorney general, the Smoke Commission decided that the research would be under the supervision of Frederick G. Cottrell of the Bureau of Mines' San Francisco office. A native of Oakland, Cottrell had graduated from the University of California in 1896 and then taught high school chemistry for three years before traveling to Germany for graduate studies. After a year each at the University of Berlin and the University of Leipzig, he received a PhD from the latter institution. Returning to the University of California, he taught physical chemistry until 1911, when he went to work as a chemist and metallurgist with the newly created Bureau of Mines. While Cottrell remained in San Francisco, the Anaconda Smelter Smoke Commission hired two employees to work at the Washoe smelter: H. V. Welch, who began work in 1911, and James O. Elton, who started in 1912. Welch was a physical chemist with the bureau in San Francisco and was detailed to work for the Smoke Commission. Elton was an ACM metallurgical engineer who had already been working on Dunn's testing department staff, although he had worked briefly in Idaho between his ACM employment and when he joined the Smoke Commission staff. The ACM detailed Dunn to work on the team, but he remained an ACM employee, rather than a Smoke Commission employee, and was often referred to as the company's representative on the team. The ACM also assigned one of its staff chemists, W. E. Mitchell, to work with Dunn on the team. Other members of the ACM's testing department rendered assistance to the team as needed.[41]

After a short time, the Smoke Commission and its researchers came to realize how variable emissions from the smelter could be and how dependent those variations were on such factors as season of the year, type of ore being treated, and capacity at which the plant was operating, all of which influenced draft conditions in the flue system and therefore the extent to which materials condensed or sublimated and dust settled. Because of variable operating conditions, it was more than a year before the commissioners believed they were able to begin planning the remedial measures to be tried and the methods for testing them. During the first year, the commissioners visited other smelters to examine techniques being used to abate smoke problems, and they also began to assess markets for some of the materials to be recovered from the Washoe smelter's smoke, including arsenic, bismuth, and sulfur or sulfuric acid.[42]

After the initial study of emissions from the Washoe smelter, the Smoke Commission revised Doremus's estimate to conclude that the plant was daily discharging about 1,000 tons of sulfur dioxide gas (containing about 500 tons of elemental sulfur) in the smoke from the various furnaces. (During the years of peak production at the height of World War I, the Smoke Commission would estimate the smelter discharged more that 1,700 tons of sulfur dioxide each day.) The commission then embarked on its most difficult task, trying to develop a means of recovering all that sulfur from the smoke. Commission researchers, aided by members of the Bureau of Mines staff, conducted experiments using techniques that, at the outset, appeared to hold promise for recovering the contaminant either as elemental sulfur, liquid sulfur dioxide, or sulfuric acid. Among the processes tested were those developed by W. H. Kelly in Los Angeles, by William A. Hall in Brooklyn, and by a Professor Young at Stanford University (the so-called Thiogen process). Tests were conducted at the laboratories of the inventors as well as at the bureau's laboratory in San Francisco and at the Balaklava smelter in northern California. None of the new technologies proved applicable to the Anaconda situation.[43]

By the end of 1913, it had become apparent to the Smoke Commission that a technical solution to the sulfur problem at Anaconda was not readily apparent:

> The Board recognizes the fact that at the present time there appears to be fundamental chemical difficulties in the way of satisfactory measures for the elimination of the sulphur [*sic*] dioxide fumes, owing to the fact that no process has yet been discovered which renders practicable the elimination of these fumes in connection with smelter operations. And the smelter industry not only in Montana, but in other of the Western States, is one of such magnitude and large economic importance that the Board does not deem it wise to submit recommendations that would involve radical changes in the Washoe Smelter plant until it is certain that such changes can be made without destroying this industry.[44]

The only method for recovering sulfur that appeared technically feasible was the conventional one of converting the sulfur dioxide in the roaster and blast furnace smoke—because the SO_2 in the smoke from those two sets of furnaces was of sufficient concentration—into sulfuric acid. The problem that still lingered was how to market sulfuric acid once it was made from the smoke. Serious attention to this vexing problem would await the second half of the decade.[45]

Based on calculations of how much arsenic was passing through the smelter system and how much was being collected in flue dust and then recovered for commercial purposes in the arsenic plant, the commission estimated that in 1912 only about 25 percent of the arsenic in the smoke was settling out with the flue dust and only about 40 percent of the arsenic available in the flue dust was being recovered as a marketable product. Flue dust for arsenic-recovery purposes was taken from the six hundred feet of the main flue nearest the stack, that being the section where the flue-gas temperature had dropped to the point where arsenic sublimated and settled as dust. Nevertheless, an estimated 40 tons of As_2O_3 was going up the stack each day in 1912. Shortly thereafter, the commission completed tests on furnace configurations for the arsenic plant to improve the plant's ability to recover arsenic from flue dust. Based on those tests, the commission recommended, and in 1915 the ACM installed, a new arsenic plant equipped with a Howell-White roasting furnace in place of the Brunton furnaces, and with Cottrell electrostatic precipitators to remove flue dust from the

FIG. 8.2. Anaconda Company Workers in Flue Dust near the Base of the Stack, October 1913. The flue dust was rich in arsenic, copper, lead, and other metals. Note the minimal respiratory protection. Photo courtesy of the Montana Historical Society, cat. # PAc 82–62.2196.

volatilized arsenic prior to condensing the arsenic. At about that same time, however, significant changes were taking place in the overall smelter operations, causing flue-gas temperatures and velocities to increase greatly. As a consequence, virtually no arsenic was settling out with the flue dust, and an estimated 60 tons of arsenic trioxide (As_2O_3) daily went up the stack from 1916 to 1918. From 1915 until a new flue-and-stack system went into service in 1919, the Anaconda arsenic plant served little purpose other than to treat flue-dust material shipped there from the Great Falls smelter for arsenic recovery.[46]

The Great Falls smelter had greatly enhanced its ability to capture flue dust in 1909, when the Boston and Montana built a new flue-and-stack system. In technical articles about the new system, the company was quick to assert that the stack and flue were built "not so much to avoid possible litigation with so-called 'smoke farmers,' as to prevent the loss of the valuable dust."[47] Nevertheless, the company had received complaints of smoke, even if limited to workers at the plant or living in the nearby Black Eagle community. Furthermore, the capacity of the smelter (4,500 tons per day) had outgrown the capacity of the existing flues and stack, meaning that under certain conditions the company had to curtail operation of some furnaces. The insufficient draft supplied by the existing flues and stack to the furnaces had also meant that the work environment in the smelter buildings was unnecessarily smoky, leading to inefficiencies in the tasks performed by the smelterworkers.[48]

Shortly after the B&M had installed MacDougall roasting furnaces at Great Falls, engineers began monitoring the amount and value of flue dust being lost up the old stack, which sat atop a hill back of the smelter. Flue dust losses were considerable, in large part because there was not enough cross-sectional area in the existing flues to allow the velocity of flue gases to drop to the point where fine solids would settle. The existing flues traversed a distance of about 1,500 feet between the smelting furnaces and the stack, which was 185 feet tall and had an inside diameter of 20 feet. The B&M also noted that its existing flues were not designed to facilitate the recovery of dust. They remained quite hot, so workers could not empty them of dust unless the furnaces ceased operating. Because of the flues' limited cross section, after the dust got a few feet deep it began to restrict the draft the stack was intended to provide. Finally, the irregularly spaced access doors made it difficult for workers to extract the dust. The men complained of skin irritation when assigned to dust removal. The work was so disagreeable that smelter manager Frank Klepetko had imposed a rule in 1898 that any man who refused to take his turn at cleaning the flue would be fired.[49]

To improve the recovery of flue dust, the B&M designed a new flue, the main feature of which was a dust chamber more than four hundred feet long and nearly two hundred feet wide. With a height of 21 feet, the dust chamber had a cross-sectional area of nearly 3,700 square feet, almost ten times that of the three old flues combined. The greater cross section reduced the velocity of gases passing through the dust chamber so that fine solids could settle out. Along the bottom side of the dust chamber, more than one thousand hoppers were arranged in twenty-two rows to allow workers to readily empty the dust into trains of cars. Correspondingly, twenty-two sets of narrow-gauge tracks were arranged beneath the hoppers to guide the cars. Inside the dust chamber, more than one million wires were suspended from the ceiling to facilitate the settling of dust, a method developed in the 1880s by an individual in Silesia named Roesing. Some dust would adhere to the wires, so a mechanism was installed allowing workers to shake the dust loose periodically. The dust chamber was equipped with adjustable inlets for regulating the introduction of outside air for cooling purposes. If desired, the temperature within the dust chamber could be lowered to the point where arsenic trioxide would condense. When the B&M first turned smoke and fumes from the roasting and smelting furnaces into the dust chamber, management did not wish to collect arsenic, for lack of a market. In about 1913, by which time the ACM owned the Great Falls smelter, an arsenic plant was built there with a capacity to treat about 20 tons of flue dust per day.[50] Soon thereafter, as already mentioned, the ACM began shipping flue dust from Great Falls to the Anaconda smelter for arsenic recovery.

To provide a draft for the furnaces, new dust chamber, and flue, the B&M contracted with the Alphons Custodis Chimney Company of New York, which specialized in the design and construction of tall stacks. The new stack, built of brick, was 506 feet tall and had an inside diameter of 50 feet. When completed, it became the tallest stack in the world and the stack of the greatest capacity. Construction of the new flue and stack cost $1,100,000. Through increased recovery of copper, silver, and gold in the flue dust, the new stack and flue allowed the smelter to save about $130,000 per year.[51]

Spurred in part by the Great Falls example, researchers working for the Anaconda Smelter Smoke Commission believed that improving the recovery of flue dust at the Washoe would serve the overall intent of the 1911 agreement. Improved dust recovery would increase the ACM's recovery of copper, silver, and other metal values in the ore. Those improved recoveries, the commissioners hoped, would in turn help pay for the costs of recovering arsenic. Toward that end, the Smoke Commission initiated some experiments with briquetting flue dust, on the theory that the resulting briquettes could be smelted in the blast furnaces to effectively recover copper from the flue dust. It was also expected that the fumes exiting the blast furnace would be especially high in volatilized arsenic, which when condensed and collected would make a good feed for the arsenic plant. The cost of briquetting proved too high, however, and in any event, it became apparent that the blast furnaces were to be phased out at the Washoe smelter. Consequently, the briquetting experiments were dropped.[52]

To improve the recovery of flue dust and condensed arsenic, the Smoke Commission also turned its attention to an invention of F. G. Cottrell, the Bureau of Mines' scientist who headed the Smoke Commission staff. Cottrell had begun experimenting with apparatus for collecting flue dust by electrostatic means while he was still teaching at the University of California. As he was quick to say, ideas of using the electrostatic properties of small particles to remove them from smoke were nearly a century old, having been described in Europe by various scientific journals. Cottrell took it upon himself to engineer those scientific principles to commercial applications during the early 1900s, filing for a patent on his process in 1907. In his design, he passed high-voltage direct current through a system of electrodes of both positive and negative charge placed in a flue. The electrodes would attract particulates of the opposite charge. Cottrell had hoped to assign his patent rights for the process to the Smithsonian Institution, with the intention that proceeds from licensing the process would be used to support other scientific research. It proved too cumbersome for the government to administer Cottrell's intellectual property, so he assigned his patents to an entity called the Research Corporation. Through the Research Corporation, he began experimenting with commercial-scale installations of his electrostatic process, first at the Hercules powder works in Pinole, California, and then at a cement plant and several lead and copper smelters in that state. By the time the Anaconda Smelter Smoke Commission was ready to recommend experiments with Cottrell electrostatic precipitators at the Washoe smelter, the system was also being installed in metallurgical plants at Garfield and Tooele, Utah; Trail, British Columbia; Omaha, Nebraska; and Perth Amboy, New Jersey.[53]

Beginning in late 1913, the ACM installed small experimental electrostatic precipitators on the flue serving the arsenic plant and between the flues of the roasting department and the blast furnace department so researchers could test the precipitators' effectiveness on gases leaving those departments. During the fall of 1915, the Smoke Commission recommended that the ACM install two Cottrell units on the flue of the new no. 2 roaster, which was being built to accommodate the additional roasting needs of the Washoe smelter. Eventually, four units were installed. By the time the four units were placed in service in September 1916, the amount of material being treated in the no. 2 roaster and the velocity and temperature of the gases being sent through the Cottrell treaters were much greater than anticipated and the system had to be completely reconfigured in order to determine the effectiveness of the electrostatic precipitators.[54]

Simultaneously, it became clear, even as the Smoke Commission was arriving at conclusions in support of recommendations for smoke-abatement equipment and processes the ACM should install in compliance with the 1911 agreement, that the ACM was implementing momentous changes in the fundamental ways it extracted copper and other metals from Butte ores. The Bureau of Mines' Arthur Wells reported on this turn of events as follows: "The period between 1913 and 1916 was one of very rapid change and improvement in the whole copper smelting industry and this was no where more in evidence than at Anaconda. This has necessitated the Commission's planning not only for adapting its recommendations for the works as they were at the time of investigation but as they would probably be by the time the construction it wished to propose could be completed."[55]

In other words, the ACM was leading the copper industry in technological innovation for copper metallurgy quite independent of the pollution-abatement research and development being conducted by the Smoke Commission, and the ACM was doing so at a more rapid pace than the Smoke Commission could complete its work. Before moving to the recommendations that the Smoke Commission made during the second half of the decade, then, it is important to review the technological changes the ACM was making at the smelter. These involved almost every department.[56] Moreover, the changes were the direct result of a concerted program of research and development at the Washoe intended to improve operations in all facets of the works. An interesting aspect of the ACM's research and development program at its reduction works is the way the ACM incorporated the pollution-abatement research it was doing in cooperation with the Smoke Commission into the company's broader, system-wide program.

ACM Research and Development

Embarking on the new century and the opening of its new smelter, the ACM had decided upon a fairly complex system to extract mainly copper but also silver and gold from Butte ores. The company already had nearly twenty years' experience in changing not only individual components of the system, such as jigs or roasting furnaces, to improve its operating efficiencies; it also had experience in changing whole arrays of components, comprising subsystems, to achieve overall system efficiencies. As the case of the switch to steam stamps showed in the 1880s, the company was generally seeking to reduce the waste of resources in its byproduct stream, but the company was willing to increase waste if it meant improved overall profitability. Each time, having made a larger systemic shift, engineers again turned their attention within that system to finding ways to reduce waste.

Knowing that the entire smelter complex would continually be going through technological change, as engineers developed new means to reduce waste and increase efficiencies, the ACM had designed the Washoe smelter so that changes could be made in any given department without impinging on another. With a smelter designed to accommodate research, development, and change, the ACM's metallurgical engineering staff proceeded to determine what that change would be.

For about the first ten years of the Washoe's operation, changes made at the works could be characterized as expanding and fine-tuning the smelter. Beginning in the early 1910s, by which time the ACM was the owner and operator of nearly all the copper-mining and -smelting properties in Montana, and while the company was in the early period of operating within the 1911 agreement it had entered with the United States to investigate technologies that would abate the smelter's polluting smoke, Anaconda decided to take a systematic look at its entire ore-processing operation. The effort was headed by the ACM's chief metallurgist, Frederick Laist. After graduating with a degree in chemistry from the University of California in 1901, the Cincinnati native had taught that subject at the University of Utah before taking a position in the laboratory at the Washoe smelter in 1904. Laist quickly rose through the metallurgical departments to become superintendent of the smelter and then, in January 1915, superintendent of all ACM metallurgical operations in Montana.[57]

The ACM's systematic research and development began at about the same time as did the work of the Anaconda Smelter Smoke Commission. Two facets of this simultaneity are noteworthy. First, as already described, much of the research done under the auspices of the Smoke Commission was actually performed by the ACM's research staff. Several ACM publications suggest that the Anaconda company simply absorbed this Smoke Commission work into its larger research program. Second, technological progress leading to improved profitability for the ACM clearly took precedence over the pollution-abatement objectives to which the ACM had stipulated in the 1911 agreement that established the Smoke Commission.

The ACM's prioritization of objectives is evident in the fact that results of the system-wide research program pointed the way, as will be seen, to a radical reshaping of operations at the Anaconda Reduction Works, characterized by the abandonment of gravity concentration in favor of flotation, and the abandonment of blast furnaces in favor of reverberatory furnaces. The new structure of operations at Anaconda meant that even first-class ores, which previously had been considered direct-smelting ores, would be concentrated rather than being smelted directly, and it meant that all ores would be ground more finely. The net result was that a significantly greater volume of gases, fumes, and dust would be charged through the existing flue-and-stack system. The ACM made this profound change in its physical plant and operations without requesting approval of the Smoke Commission. Rather, the Smoke Commission was left, as already noted, to respond to a new set of operating conditions that were significantly different from when it had begun its work in 1911.

The Anaconda company's absorption of Smoke Commission research into the larger ACM research program is suggested by a series of publications the company produced through the middle years of the decade. In 1915, the ACM published a bound volume titled *Report of the Research, Concentration and Testing Departments for the Years 1912, 1913 and 1914.* In a forward to the volume, ACM chief metallurgist Frederick Laist wrote:

> It has been realized for a long time by all of us who have been engaged in the development or application of new metallurgical processes that

> much experimental work, costing thousands of dollars, is repeated from time to time, because of the absence of adequate records in readily accessible form.
>
> In order to avoid this in the future as much as possible, it has been decided to establish what we call permanent records. These records are compiled during the year by the men in charge of the experimental work, and at the end of the year they are printed in book form.
>
> We aim to keep out of these printed records anything which would obviously be useless to future investigators, but have attempted rather to err on the side of putting in too much, rather than too little; since a piece of work which may seem useless to us now may have considerable importance at some future time.
>
> This first report covers three years' work, as it was thought best to carry the record back to the time when the rather active investigation period, in the midst of which we now are, commenced. From now on each book will represent one year of work.
>
> The work which is recorded in these pages is the basis upon which the present reconstruction of the Reduction Works in Anaconda was undertaken. The results being obtained in the large installations are the best possible indication of the accuracy with which the experimental work was done.[58]

Over the next three years, the company published comparable reports for 1915, 1916, and 1917, respectively. Each volume contained sections for the same three departments: research, concentration, and testing. The research department explored leaching to recover both copper and zinc, looking not only at various leaching procedures but also at various roasting techniques to prepare pulp for leaching; explored techniques for producing phosphate fertilizer and phosphoric acid; explored the manufacture of building brick from tailings; and examined another Butte company's process for recovering copper from leachate by means of electrolysis, among other research.[59] The concentration department conducted experiments on various kinds of grinding mills, sizing and classifying equipment, dewatering appliances, and jigs, tables, and other machines for concentrating tailings and slimes.[60] The testing department conducted tests throughout the smelter to determine such things as the character of the dust collected in the blast furnace dust chamber, the performance of materials and equipment in the MacDougall roasting furnaces, and the effectiveness of the smelter flue system. These latter were topics of concern to the Smoke Commission, but the testing department also engaged such questions as how various kinds and sizes of coal performed in the reverberatory furnaces, how firebrick made of various materials performed in furnaces, and how much water was consumed by the various operations at the Anaconda Reduction Works.[61] Reports of all three sections included reprints of papers ACM staff presented at meetings of the American Institute of Mining Engineers and published in the institute's *Transactions*, but they also included numerous tables and figures representing results of the departments' research as well as narratives describing background to the research, methods employed, and conclusions drawn.

These ACM reports reflect an effort on the part of the company's engineers and management to rationalize more thoroughly 1) the processes by

FIG. 8.3. Anaconda Copper Concentrator, Table Section. As with sections for other steps in the process of concentrating ores (e.g., fig. 6.1), the table section was a vast room housing many tables used to enhance the overall recovery of copper in ore by further treating the materials in the fine-particle-size range (very fine sands and slimes). Photo courtesy of the Montana Historical Society, cat. # PAc 82–62.1601.

which they tested equipment and techniques, 2) their projection of how promising equipment and techniques in each department would perform within the operational and financial contexts of the larger Washoe works, and 3) how their understanding of proposed changes in equipment or technique would affect other aspects of the reduction works. ACM engineers concerned themselves with a wide variety of research questions, among them—but not limited to—how to conduct operations consistent with pressures to abate air pollution brought against the company by neighboring property owners, private and government alike. Foremost among the research questions, however, was whether proposed equipment or techniques would return a profit to the company's investors.

Significant changes in the method of concentration at the Washoe began in 1912, when the ACM began remodeling one of the eight sections at the concentrator to house a new system of concentration, which had been developed over the previous few years at the ACM's Great Falls smelter. It is not surprising that metallurgical engineers at Great Falls would have devised a more efficient system of gravity concentration. In the early years, the convenience at Great Falls of being able to dump tailings directly into the Missouri River came at the expense of being a site that did not afford the acreage necessary for the vast tailings storage system established at Anaconda. Whereas Anaconda engineers could—and did—discharge their tailings with the assurance that most of those tailings would be on the dump at some future date when a technological innovation might make it possible to re-treat them, Great Falls engineers did not enjoy that luxury. They had one chance during concentration to separate as much copper as possible, and then the tailings went down the river. The Great Falls method of concentration, introduced at Anaconda in 1912, achieved relatively higher recoveries of copper from ore.

The most significant change in the Great Falls system was that Hancock jigs replaced Evans jigs, which were prominent in the original array at the Washoe concentrator. Although the Hancock jigs required more water, they improved the recovery of copper. Both the Great Falls and Anaconda systems used Wilfley tables to treat very fine materials. Tests of the remodeled section at the Washoe showed that it recovered 85.5 percent of the copper in the ore, compared with an 82.6 percent recovery of copper using the system of Evans jigs still in place in the other seven sections of the concentrator. Both systems yielded the same loss in slime tailings (5.6 percent), following final concentration of slimes from the mill; the improvement came from the improved performance of the Hancock jigs. Wastes from the remodeled section contained only 8.9 percent of the copper originally in the ore, while wastes from the original Evans system contained 11.8 percent of the copper in the ore. Sand tailings leaving the concentrator when using the Great Falls system assayed at only 0.48 percent copper.[62]

At the same time the ACM was installing the Great Falls method of concentration in a section of the Washoe concentrator, it was also experimenting with the use of convex round tables, also developed at Great Falls, for treating slimes from the mill. Tests at Anaconda were conducted in an experimental plant with a capacity of 50 tons of slimes per day. The ACM simultaneously conducted tests on a variety of other settling tanks and centrifugal separators for recovering copper and other values from the slimes. Slimes from both the remodeled section of the concentrator and the sections using Evans jigs were treated in the experimental plant, and the ACM found that it could recover another 3 percent of the copper originally present in the ore. As a result of the tests, the ACM decided to build a large slime-treating plant, in which round tables, twenty feet in diameter, were stacked atop each other in configurations twenty decks high. In this way, the plant could effect economies of scale while saving considerable floor space. The slime-treating plant had twenty of these sets of round tables, giving the plant a capacity to treat 2,500 tons of slimes per day, the amount that was currently being sent to the slime ponds. The tables could recover more than 50 percent of the copper in the slimes. Instead of containing more than of 2 percent copper, the 1,900 tons of slimes discharged daily from the concentrator dropped to about 1.1 percent copper. It was estimated that, without treating more ore, the slime-treating plant would recover an additional 1,000,000 pounds of copper per month. The new slime plant went into operation in March 1914, and the experimental slime plant was demolished.[63]

The introduction of the slime treatment plant in 1914 was another example of the principle under which engineers often introduced methods to reduce the waste of natural resources, but only if those methods proved profitable. The round tables used in the new plant were a technology that had been around for some time; indeed, they were essentially the same technology the ACM had used at the Old Works to treat the excessive amount of slimes produced by the steam stamps. Those nineteenth-century convex round tables, called buddles, had been supplanted by the superior Wilfley table. The devices tested by the ACM in 1912 took advantage of new opportunities afforded by the industrial era such as steel frames, concentrating surfaces of modern, high-quality cement, and mechanical drive-trains allowing many tables to be stacked atop one another and operated from a single drive, saving considerable floor area in the mill. Not surprisingly, the early development of improved round tables, beginning in 1904, also took place at Great Falls, again undoubtedly due to the lack of land area there for the storage of slimes for future re-treatment. The ACM was able to produce copper from its slimes with the new twenty-deck round tables at a cost of only about $0.04/lb., including the cost of smelting. An observer called it "the cheapest copper the ACM produces."[64]

The third set of experiments initiated during this period to improve the concentration of ores involved the leaching of copper values from ores and tailings. As early as 1907, the ACM had recognized the potential for using leaching and precipitation, similar to that used on old tailings at inactive Butte concentrators, for the recovery of copper values from the vast volumes of tailings deposited by the Anaconda smelters. In 1913, the ACM built an experimental 80-ton leaching plant for treating ores, tailings, and slimes. This plant was distinct from an experimental plant begun in 1911 at the Washoe to test the Bradley process, which used a calcium-chloride solution to dissolve metals from copper ores. The experimental Bradley plant was built near the slime ponds and had a capacity of 250 tons per day. Although completed by the end of 1912, the Bradley plant was fraught with mechanical problems, which delayed its full-scale operation. The ACM finally closed the Bradley plant in April 1913.[65]

The ACM's experimental leaching plant, meanwhile, was based on the use of sulfuric acid to dissolve copper from finely crushed material. The ACM tested the experimental plant on ACM tailings, which were mixed with salt (as a chloridizing agent for silver sulfides) and roasted in a MacDougall furnace to oxidize the copper sulfides.[66] The leaching was then accomplished with sulfuric acid in large tanks. Copper dissolved by the acid solution was then precipitated out of solution by allowing the fluid to flow over scrap iron. The ACM also experimented with other means of recovering the copper from the leachate, such as by using electrolysis. Two small companies at Butte, the Bullwhacker and the Butte-Duluth, were experimenting simultaneously with leaching copper oxide minerals in the ores along the eastern edge of the Butte district. They, too, were trying to recover the copper in solution by electrolysis. The ACM decided the most economical way to recover the copper was by precipitation.[67]

ACM metallurgists contemplated using the leaching process as the final stage in the treatment of its ores as well as to treat tailings already discharged from the Washoe concentrator and from earlier operations. After successful development work in 1914, the ACM began construction of the first stage of its full-scale leaching plant with a capacity of 2,000 tons per day. This was enough only to treat slime tailings as they were produced by the concentrator. The leaching plant was located just east of the concentrator. The building housed twenty-eight MacDougall roasting furnaces for roasting tailings; fifteen redwood tanks, each 14 feet deep and 50 feet in diameter, for leaching roasted tailings; and three 180-foot precipitation launders for recovering copper in solution on scrap iron. The copper precipitate, called cement copper, was sent to the briquetting plant. Leached tailings were discharged to the normal tailings dump by means of a launder running east from the leaching department. The company expected it would soon expand the plant to 9,000 tons per day. The ACM estimated that when all its leaching and concentrating improvements were implemented, the company would be able to recover as much as 50,000,000 pounds of copper per year. In conjunction with its leaching plant, the ACM also built a 100-ton sulfuric acid plant, which could extract sulfur from the smelter smoke to supply the leaching plant with acid. The leaching and acid plants began operating in mid-1915.[68]

At the beginning of 1914, the ACM had decided to cease operating the concentrator at Great Falls, concentrating all its Butte ores at Anaconda instead, and then shipping some concentrates to Great Falls for smelting. In order that production of copper from both the Anaconda and Great Falls smelters could remain the same, the Anaconda concentrator would have to treat about 12,500 tons of ore per day, an increase of about 3,000 tons daily from the previous year.[69] This made it even more important to operate the Anaconda concentrator as efficiently and economically as possible.

Concentration by Flotation

During a December 1914 trip to Butte, Amalgamated president John D. Ryan and ACM president B. B. Thayer announced that the ACM would spend $6,000,000 on improvements to the Anaconda and Great Falls smelters in the coming year. The company officials promised that the changes being implemented at the two works would increase the percentage of metals the company could recover from the ores it was mining and treating, and the changes would yield reduced costs, the net result of which would enable the ACM to mine and treat ores of lower grade, and that in turn would extend the life of the Butte mines. Of the total capital investment that the ACM intended to make in its plant, two-thirds would be spent on improvements to the Washoe smelter. None of the improvements were intended to comply with the ACM's 1911 agreement with the United States or otherwise reduce the discharge of pollutants.[70] In fact, the modifications to the smelter *increased* the volume of pollutants being discharged from the smelter and gave the Anaconda Smelter Smoke Commission a new set of realities to which it had to respond in developing its recommendations for technical fixes.

As the ACM was implementing its slime treatment and leaching plans at the Washoe smelter, the company also entered negotiations with the Minerals Separation Company, which held patents on the flotation process, to see if flotation might be applied to the slime tailings leaving the slime treating plant. Flotation was the revolutionary, counterintuitive method for concentrating ore that was being used at Butte by the Butte and Superior Copper Company on zinc ores. Minerals Separation, however, had the Butte and Superior enmeshed in litigation for patent violation, so the ACM proceeded with its experiments legally, by purchasing a license to do so. Actually, the ACM entered the flotation arena through its association with Inspiration Consolidated Copper Company, which was perfecting its own methods for mining and milling low-grade porphyry copper in Arizona.

As Inspiration was conducting experiments to determine how best to design a new mill to treat its low-grade ores, it acquired a license from Minerals Separation in 1912. This was the opposite approach taken by its neighbor, the Miami Copper Company, which began performing tests using flotation in 1913, but without licensing the process from Minerals Separation. The latter filed suit against Miami Copper in July 1914. Meanwhile, Inspiration began its experimentation by sending ore samples to the Minerals Separation laboratory in San Francisco and, via Amalgamated's Raritan Copper Works at Perth Amboy, New Jersey, to Minerals Separation's headquarters in London. Based on favorable results in those tests, and tests in a 10-ton flotation machine placed near the mine, Inspiration decided to adopt flotation and secured a license to use the process in a 600-ton experimental mill, which began operating in January 1914. In the larger experimental mill, Inspiration tested several kinds of concentrating machines in conjunction with the Minerals Separation process, and it also tested several other approaches to flotation that competing innovators had developed. During those experiments, Inspiration metallurgists developed their own flotation machine for use with the Minerals Separation process. In the eighteen months the 600-ton mill operated prior to the opening of Inspiration's permanent 15,000-ton mill, the test mill treated 257,272 tons of ore. It recovered 88.56 percent of the sulfide copper and 79.95 percent of the total copper in the ore (oxidized copper minerals in the porphyry ore were harder to recover). The first unit of the new mill began operating in June 1915, all eighteen 800-ton units were operating by mid-1916, and Inspiration became the first operation in the U.S. to implement flotation on a large scale and fully authorized by the patent holder.[71]

In order to be authorized to use the flotation process for full-scale production, Inspiration prepared to obtain the necessary license from Minerals Separation. Because of its ties to the ACM, metallurgists at the Washoe smelter became aware of the performance of the flotation process on copper sulfide minerals. After some quick tests on materials at the Washoe concentrator in June 1914, the ACM joined Inspiration in signing a joint licensing agreement with Minerals Separation in February 1915. The joint agreement allowed the two companies to use their combined tonnage of material in calculating, on a sliding scale, the royalties due Minerals Separation for using the flotation process. The two companies would pay a royalty of twelve cents per ton if they treated a combined 4,000 tons daily. The royalty would drop to four cents per pound if they treated a combined 30,000 tons daily. Flotation was a tremendous success at Anaconda. Whereas the Washoe's gravity concentrator used to lose 12.4 pounds of copper out of every ton of 3 percent ore (i.e., ore with 60 pounds of copper per ton), the flotation concentrator lost only 3 pounds of copper.[72]

The ACM's tests showed that flotation was a superior method for concentrating all copper ores from Butte and that flotation equipment could replace all the equipment and steps in the gravity concentration at the lower end of the mill. Implementing the flotation process thus obviated the need for: 1) installing the Great Falls system of concentration in the other seven sections of the Washoe concentrator, 2) using the giant decks of round tables in the slimes treatment plant, and 3) leaching as the final treatment of tailings leaving the concentrator. With the installation of flotation equipment, the upper end of the concentrator still produced concentrates and coarse tailings. The latter were then reground and treated by flotation, which would yield a concentrate assaying 8 percent copper and tailings assaying about 0.1 percent copper. With flotation and other improvements, the ACM was now able to recover, through the entire sequence of processes, about 17 percent more copper from a given amount of ore. In the course of concentration, roasting, smelting, and refining, the ACM could now recover 91 percent of the copper in the ore rather than 78 percent (during concentration itself, the ACM was able to recover 95 percent of the copper by flotation, compared with 82 percent using gravity concentration).[73]

As a result of this radical change in the method of concentration, the ACM not only decided to demolish the recently built slime treatment plant and devote the leaching plant, for which plans for expansion were dropped, solely to existing tailings on the dumps. The ACM also decided to send all ores it mined to the concentrator and from there to the roasting department and the reverberatory furnaces, thereby discontinuing use of the blast furnaces. To treat the larger volume of concentrates that now had to be calcined, the ACM built a new roasting building, which housed forty new MacDougall roasting furnaces. When the concentrator was completely remodeled for flotation in early 1916, each of the eight sections had a capacity of 2,000 tons/day, giving the entire Washoe concentrator a capacity of 15,000 tons/day (allowing for normal shutdowns for repairs). Thus, by 1916, when both the Anaconda and the Inspiration concentrators were operating at full capacity, the two companies were able to qualify for paying the lowest royalty rate provided in the licensing agreement with Minerals Separation.[74]

Flotation figured significantly in a profound change, around 1916, in the quantity and quality of pollutants discharged by the Washoe smelter's treatment of copper ores. First, all the tailings discharged from the concentrator were very fine, and the copper content of the tailings was now so low that the ACM could dispose of them permanently, with no thought of ever trying to re-treat them. Because all the ore sent to the smelter was concentrated, instead of some portion of the first-class ore going directly to the blast furnaces, the volume of material being roasted was much greater. Moreover, that material was very fine, having been ground thoroughly enough to be treated by flotation. Consequently, the smelter's MacDougall furnaces sent a greater volume of dust into the flue system.[75]

With the introduction of flotation and the retention of the leaching plant, the ACM possessed two means of re-treating old tailings, and over the next decade the company used them both to do so. Between 1915 and 1921, inclusive, the ACM hauled about 3,000,000 tons of sand tailings from the Washoe's original tailings dump, just northeast of smelter, to the leaching

FIG. 8.4. Roaster #2 Building and Furnaces under Construction. This photo shows MacDougall roasting furnaces before they are enclosed in their building. Each cylindrical furnace features several hearths, or levels. Concentrates are charged into the top hearth, where revolving rakes move the charge toward holes where they can drop to the next hearth, and again until the roasted charge drops out of the bottom hearth. Heat generating by the oxidizing sulfide minerals sustains the roasting process without an additional heat source. Photo courtesy of the Montana Historical Society, cat. # PAc 82–62.2498.

plant to be reworked. During the period 1918–1926, inclusive, the ACM shipped another 3,000,000 tons of sand tailings from the old dump to the concentrator to be re-treated by flotation.[76]

John D. Ryan boasted that by means of the improved metallurgical performance that the ACM was able to achieve at the Washoe smelter through the introduction of flotation and other improvements in 1915, the company was able to increase its annual production by 55,000,000 pounds of copper, without increasing either the grade or volume of ore it was treating. Ryan added that 40,000,000 pounds of that increased output would be realized without any increase in costs.[77]

The ACM's major overhaul of the Washoe smelter was watched closely by mining industry. The *Engineering & Mining Journal*, the industry's leading trade journal, extolled the ACM's boldness in a March 1915 editorial, noting that the new technology will "make more money for the company." "That the change of plan was so completely, boldly, and swiftly made indicates conclusively that the Anaconda staff rides no hobbies and is not fettered by prejudices or preconceived ideas."[78] The ACM's research program and costly overhaul paid handsome dividends. The ACM's increased output of 55,000,000 pounds of copper per year was about equal to the total amount of copper Anaconda had produced in 1887 (57,000,000 pounds), the year Montana's copper production first exceeded that of Michigan's. The greater output would increase revenues by more than $7,000,000 per year. The ACM's chief metallurgist, Frederick Laist, estimated that the overhauled reduction works would also cost about $1,500,000 less to operate each year. It was obvious to stockholders that the $6,000,000 investment would readily repay itself.[79]

The ACM also made significant changes to its smelting operations, all aimed at improving recoveries of metals and reducing costs. Over time the company added significantly to the number of MacDougall roasting furnaces it used, which in turn added to the sulfur dioxide, arsenic, and flue dust it sent up the stack. The flue dust was especially rich in metal values, so the company experimented with ways it could collect much of that dust before it went up the flue to the stack. One of the early methods created very dusty conditions in the roaster building and would "act with corrosive effect upon the skin of the workmen."[80] The ACM continuously experimented with increasing the lengths of its reverberatory furnaces, fourteen of which were 50 feet in length when the smelter opened in 1902. In 1916, the smelter had but eight furnaces, each 143 feet in length. ACM experiments showed that each successive increase in length enhanced the fuel economy of the furnaces and reduced the copper content in the slag that was sent to waste.[81] When the Washoe smelter opened in 1902, it had seven blast furnaces, each with a hearth fifteen feet long. The ACM then experimented with installing a twenty-one-foot hearth between a pair of the furnaces, in effect creating a single furnace fifty-one feet long. The company was able to effect savings in fuel, labor, construction, and maintenance costs and to achieve more flexible operations with the larger blast furnace, so it retrofitted four of the remaining five furnaces, leaving the smelter now with three much larger furnaces with improved performance.[82]

FIG. 8.5. Man Standing inside a Reverberatory Furnace. When operating, calcined (roasted) concentrates were charged into the furnace from the openings in the vaulted ceiling to lie on the bed of the furnace, while combustion gases, flowing from the firebox at one end to the flue at the other, passed between the bed of material being smelted and the ceiling. The flue gases heated the charge to a melting temperature, allowing copper matte to settle and slag to float atop the matte. Each molten substance could then be drained from the furnace through openings at the appropriate level. Photo courtesy of the Montana Historical Society, cat. # PAc 82–62.1599.

Another way the ACM achieved overall fuel economies at the Washoe Smelter was to install waste-heat boilers atop the reverberatories. When the smelter opened, none of the reverbs was so equipped. The ACM had been experimenting with waste-heat boilers attached to some of the reverbs at the Lower Works during its last couple years of operation, but experimental boilers interfered too much with performance of the smelting to be considered practical. Experimentation continued at the Washoe Smelter, though, and by the end of its first year of operation, most of the reverbs had waste-heat boilers attached to them. The steam generated by the waste-heat boilers reduced the amount that had to be produced at the smelter's coal-fired powerhouse for operating air compressors.[83]

Blast furnaces and reverberatory furnaces produced copper matte containing about 50 percent copper (plus or minus 10 percent). The two byproducts were slag and smoke. The matte was sent to the converter department for bessemerization, wherein molten matte was charged into converters. Air was blown into the bottoms of the converters through openings called tuyeres. As the air passed through the molten matte, the remaining sulfides were oxidized, and the matte separated into two parts: molten blister copper, also called converter copper, which settled to the bottom; and slag, consisting of iron oxides and silica, which rose to the top. To facilitate the formation of ample slag to carry the iron, smelterworkers added low-grade siliceous ore to the converters while the charge was being blown. Because of the rather violent reactions occurring within the converters, there was not a perfect separation, and therefore converter slag was relatively high in copper content. For that reason, it was resmelted rather than being discharged to the slag dump. Smoke from the converters contained sulfur and flue dust. Because of the nature of the operation, a considerable amount of the converter smoke did not go into the overhead hoods to be conveyed to the dust chamber, flue system, and stack. The converter department was therefore one the greatest sources of fugitive emissions at the smelter. The blister copper produced by the converters was more than 99 percent copper and was cast into anodes, which were shipped to electrolytic refineries elsewhere for final refining to yield a product of pure copper.[84]

Changes in converting technology at the Anaconda Reduction Works during the 1910s were rather subtle. The two main changes were in the shape of the converter and in the composition of the lining. When the smelter went into operation in 1902, it had the barrel-type of converter (cylinder in a horizontal position). In 1915, the ACM switched to the Great Falls–type of converter (cylinder in an upright position), similar to, but much larger than, the converters first used at the Parrot smelter in 1884. In the barrel-type converters, silica for slag was derived from the lining, consisting of cupreous slimes mixed with siliceous ores. In the Great Falls–type converter, magnesite bricks were used to line the converter shells and, as already stated, siliceous ores were added to the charge for slag. The conversion to magnesite linings had actually occurred with the barrel-type converters at the Washoe smelter a few years prior to conversion to the Great Falls converter.[85]

FIG. 8.6. Converter Aisle, Pouring Matte, December 1906. The converters initially used at the Washoe smelter were of the horizontal type. Note the different orientation of the cylinder of these converters compared to those at the Parrot and MOP smelters (figs. 4.2 and 5.3). In this view, workers are pouring copper matte from two converters into ladles, which will haul the molten matte to the casting floor, where the converter copper will be cast into anodes for electrolytic refining. Photo courtesy of the Montana Historical Society, cat. # PAc 82–62.1595.

When the ACM ceased using its blast furnaces after World War I, it required a furnace for smelting converter slag, which still contained a considerable percentage of copper. The company built a reverberatory furnace, known as no. 9 furnace, at the west end of the converter department specifically for smelting converter slag, to which had to be added fluxing material (low-grade concentrates).[86]

The ACM's staff and management certainly did not act as if constrained by the fact that the reconstructed reduction works would magnify the smoke problem, even though the company was operating under an agreement with the U.S. government wherein the company agreed to, in the words of the 1911 agreement, "at all times use its best efforts to prevent, minimize and ultimately to completely eliminate the emission and distribution from its smelting works at Anaconda, Mont., of all deleterious fumes."

With ever-increasing volumes of material being treated at the Anaconda Reduction Works, and with the introduction of the new processes at the smelter during the 1910s, more gases, more dust, and more fumes were charged through the 1903 flue-and-stack system, choking it beyond its capacity so that it neither collected much of the solids it had been designed to recover nor supplied the draft to the furnaces it was intended to provide. Overloading the facility's "exhaust" system not only temporarily precluded the Anaconda company from complying with the 1911 agreement; it greatly diminished working conditions as well. A report describing conditions in 1913, when the company first became aware that it was exceeding the capacity of its flues and stack, but before flotation was introduced and the company decided to abandon the blast furnaces, said conditions in the blast furnace department were the worst. This was because gases leaving the furnaces, driven by the blast of air charged to the furnaces, were often at a pressure greater than the draft being supplied by the flue, causing smoke to back up. "This building became excessively smoky and the efficiency of the men was thereby enormously reduced."[87] There is little information to describe working conditions around the furnaces during World War I, when production was at its peak, but smoke, dust, and fume from the Washoe's overloaded flue system must certainly have caused even greater discomfort to the workers and damage to the surrounding environment.

Opportunity

While the ACM was improving the efficiency of its milling and smelting operations at Anaconda and ostensibly was working to comply with its 1911 agreement with the U.S. government to reduce and eventually eliminate air pollutants from the discharges of the smelter, the company also addressed the pollution controversy on another front. During the Bliss case, the ACM had suggested in court that smelter smoke was not the principal threat to agriculture in the Deer Lodge Valley. Rather, waterlogged soils, leading to alkalinity, was the true cause of diminishing crops.[88] By as early as 1910, the ACM had installed drainage on an experimental parcel of land, planted crops, and demonstrated that crop yields could be improved through drainage of Deer Lodge Valley soils.[89]

Under the direction of veterinarian H. C. Gardiner, the Anaconda company established subsidiaries, Deer Lodge Valley Farms and the Willow

Glen Stock Farm, to demonstrate on a large scale that improved agricultural techniques could address the problems facing area farmers and greatly increase yields. To enhance drainage of waterlogged soils, the ACM laid thousands of feet of clay and concrete tile on farms it had acquired from farmers devastated by smoke and tailings. Gardiner and his crews conducted scientific tests of the company's theory that waterlogged soils were the cause of crop problems. Crops did indeed improve, and the company was quick to demonstrate the efficacy of its agricultural methods at fairs and other exhibitions. Interestingly, the company also experimented with using slimes from the concentrator as a cementing agent for the manufacture of drainage tiles. This, along with the manufacture of bricks, was an alternative use the ACM explored for tailings.[90]

The Anaconda company also hoped to demonstrate that families of smelterworkers could find healthy lifestyles by living on small plots of land out in the valley, even under the plume of the stack. As early as 1911, Gardiner was soliciting advice from agricultural and rural-development specialists about subdividing drained swampland for establishing a model rural community. An impressive array of individuals inspected Gardiner's plans and the swamp he hoped to drain and subdivide. Evidently, he initially considered selling eighty-acre parcels, but his advisors recommended that smelterworkers would be more likely to buy homesites in the development if the plots were as small as one or two acres and no larger than ten acres. His advisors were confident that the land would be very productive (ironically, the one nutrient that soil samples showed to be somewhat deficient was phosphorous). They suggested further that the model community would both attract and foster a better class of smelter employees. Architect/contractor Fred Whiteside, a state senator from Flathead County, thought the development would go far in improving the Anaconda company's image in Montana. Gardiner and the ACM went ahead with the scheme, creating the model community of Opportunity, complete with service by the ACM-owned Anaconda Street Railway.[91] Parcels of land were sold in five- and ten-acre plots to individuals so that "the people of Anaconda can enjoy the benefits of country life while sharing in the activities of the industries located here. [Opportunity] affords a real solution for the high cost of livng [*sic*]. Little farms, with cows, pigs and chickens; gardens with vegetables, apple trees and small fruits, will supplant the city street and the crowded city lot."[92] In addition to draining the swampy land on which Opportunity was developed to make the soil suitable for agriculture, the ACM organized the Mill Creek Irrigation Company to supply water under contract accompanying the deed for a plot. Gardiner sought and received the services of Samuel Fortier, chief of the USDA's Irrigation Division and E. P. Mathewson's old classmate at McGill University, to design the irrigation system. Prior to joining the USDA, Fortier had been irrigation engineer for the Montana Experiment Station in Bozeman, where Gardiner had previously worked. Fortier was director of the Experiment Station from 1900 to 1903. To encourage good order at Opportunity, the ACM also built the Beaver Creek School, limited each individual to the purchase of one plot to discourage speculation, and prohibited establishing saloons or comparable "nuisance" businesses.[93] Opportunity remains to this day a vital but contaminated little rural community at the south end of the Deer Lodge Valley.[94]

Implementing the Smoke Commission's Recommendations

Recognizing that operating conditions had greatly changed at the Washoe smelter, the Smoke Commission still understood that it had two main tasks: to find means to reduce arsenic emissions, and to find means to reduce sulfur dioxide emissions. The solutions involved very different technologies, so each will be described in turn. Regarding arsenic, the Smoke Commission's efforts were mainly directed at the Cottrell electrostatic precipitators. Once the Cottrell treaters on the roaster no. 2 flue had been reconfigured, the research staff obtained satisfactory test results, and it became apparent that the Cottrell apparatus would be recommended among the processes and equipment to be installed in compliance with the 1911 agreement. At first, however, project engineers anticipated a design with treaters attached to the department's branch flues leading to the main flue, the reverberatory department excepted. Most of the arsenic came from the roasting furnaces (and the blast furnaces, but by now it was understood that they were to be eliminated). Although the converters discharged relatively little arsenic with their smoke, they did give off considerable metal-rich flue dust. It was in the ACM's economic interest to collect that flue dust in order to recover the metals it carried. The reverberatory furnaces gave off little arsenic, compared with the MacDougalls and the blast furnaces, and the amount of copper in the flue dust emanating from the reverbs was considered relatively insignificant, hence the initial belief that treaters would not be necessary for the flue serving the reverbs. The plan specified that a new tall stack be built near the juncture of the flues, obviating the need for the long main flue and the stack atop the hill.[95]

In 1916, the Smoke Commission made a preliminary recommendation to the ACM to eliminate the 1903 main flue and three-hundred-foot stack and to replace them with a stack more than six hundred feet tall to be located approximately at the confluence of the flues from the individual departments. The proposed stack would be sixty feet in diameter and twice as tall as the three-hundred-foot stack for two reasons, to give more draft than the existing stack did, and to make up for the fact that the proposed new location was at a lower elevation on the hill than that of the existing stack, standing as it did atop smelter hill. The flues connecting the roasting and blast furnace departments—these being the greatest contributors of flue dust and arsenic—would be fitted with electrostatic precipitators to enhance the recovery of flue dust and sublimated arsenic. In anticipation of the tremendous volume of brick that would be required to build the stack, the ACM immediately began expanding the capacity of its brick plant.[96]

After ACM management and engineers, in cooperation with the Smoke Commission, spent a year thoroughly studying the proposed system, the company in early 1917 devised an alternative plan that its engineers considered superior to the Smoke Commission's preliminary recommendation. The new plan called for using the existing main flue to convey gases to the top of the hill, where new Cottrell treaters would be built, along with a new stack, taller and of greater capacity than the existing one. This plan would allow the recovery of copper-laden dust given off by the reverberatory furnaces to be recovered by the new treaters along with dust from the roasting and blast furnaces, and the new treaters would be in one central location, rather than being located on individual connecting flues. The new plan consisted of two alternate schemes, one with twenty units of electrostatic precipitators and one with forty units. On January 23, 1917, Frederick Laist, by then manager of the smelter, requested that ACM vice president Con Kelley authorize construction of the twenty-treater alternative and that ACM management appropriate the necessary funds. Laist's main reason for recommending the twenty-treater alternative was that it projected a 17 percent return on investment, while the forty-treater alternative only showed a 10 percent return. Laist's other two reasons for recommending that Kelley authorize construction were: 1) the Smoke Commission was urging the company to do so; and 2) "the uneasiness constantly expressed by Dr. Gardiner, who is unqualified in his assertions that the damage is not only very much larger than formerly but considerably more intense." The Smoke Commission approved the ACM's selection of the twenty-treater plan, even though researchers estimated that it would recover only about 50 percent of the arsenic being given off by the smelter at wartime levels of copper production.[97]

Because of Gardiner's ongoing monitoring of the environment in the region around Anaconda, he was aware that environmental damage was intensifying in the Deer Lodge Valley during the 1910s because of the ACM's increased production and the changes made in operations. Gardiner was especially concerned that evidence of damage would become more apparent because Montana's climate, and in particular its records of annual rainfall, followed a recognizable pattern. He had become convinced that the Deer Lodge Valley was soon due for some dry years, which would exacerbate the smoke damage. As the ACM official in charge of monitoring the effects of smelter smoke on the environment surrounding the Washoe smelter, Gardiner had come to realize: 1) that emanations from the stack would cause more harm during dry years; 2) that the years immediately preceding the Bliss suit had been particularly dry, making the damage from smelter smoke acute; 3) that the years the scientific experts in the Bliss case conducted their fieldwork were rather moist, making it appear that damage from smelter smoke had ceased; 4) that discharges from the Washoe stack were greater in the mid-1910s than they had been when the smelter opened; and 5) that when dry years returned, damage from smelter smoke would again be very apparent, and the ACM risked being subject to another series of smoke-damage suits.[98]

Gardiner also knew from his field monitoring (described in more detail below) that the smelter-smoke damage had spread across the Continental Divide to the south of the smelter and into the Big Hole River drainage. Even after the new stack and treaters went into operation in 1919, Gardiner measured high levels of metals being deposited on vegetation downwind of the stack, a situation he attributed to the unusually dry climate of the period.[99]

Gardiner monitored environmental conditions in the vicinity of Anaconda both by employing his own staff and by contracting with outside experts. He had R. W. Thatcher keep him abreast of the technical literature concerning the effects on plants and animals of various pollutants, especially arsenic and sulfur dioxide. In 1911, when Thatcher was advising Gardiner on development of the community of Opportunity, he was director of the Agricultural Experiment Station at Washington State College, Pullman. In 1917, Thatcher was dean and director of the University Farm at the University of Minnesota. He sent Gardiner abstracts of articles published by government agencies, like the New Hampshire Agricultural Experiment Station *Bulletin*, and articles that appeared in periodicals like the *Journal of Industrial and Engineering Chemistry* and the *Journal of the American Society of Agronomists.*[100]

Another expert Gardiner retained was William Moore, head of the research section in economic zoology at the University Farm, University of Minnesota. Thatcher dispatched Moore to Anaconda specifically to determine the extent to which insects had injured forests around the smelter. He found a variety of insects that were injuring and killing trees, in some cases causing an appearance similar to that of trees killed or dying from smoke injury. Other causes of injury Moore identified included fire and fungus. He also described how the causes sometimes interacted and compounded each other in certain sections of the forest.[101]

Another of Gardiner's experts was E. P. Taylor, the director of Agricultural Extension at the University of Arizona in Tucson. In the summer of 1917, Taylor toured the Anaconda vicinity, taking soil samples and inspecting crops in fields and gardens. He found no damage to plants from smelter smoke other than perhaps some slight discoloration of leaves on small grains. He attributed any poor crop conditions to drought, poor drainage of waterlogged soils, and ground squirrels. Taylor did, however, report damage from smelter smoke to coniferous trees. Analyzing needles, he found there had been little injury in 1917 (due, he said, to limited operation of the smelter), but that there had been increasing injury over the previous four years, with needle growth from 1916 showing the worst damage. Taylor reported that smoke damage to trees on the national forest extended as far west as Georgetown Lake (about twenty miles) and as far south as the Big Hole River, on the other side of the Continental Divide (forty-one miles). He mentioned in passing that, when standing on a small hill near the Big Hole, he could see the smoke plume of the Anaconda smelter thirty-five miles to the north, even though the smoke from the smelter that day was blowing north and northeast. In addition to analyzing damage to trees from smelter smoke, Taylor observed and described other causes of injury to the forests around Anaconda, including insects, disease, and fire.[102]

Gardiner kept close watch, and reported directly to Kelley, on the effects of both arsenic and sulfur dioxide on the smelter's surroundings, and he responded accordingly to keep the ACM's legal exposure to a minimum.[103] As already mentioned, for example, the smelter saw reduced operation in 1917, but the year overall was quite dry in southwest Montana. Gardiner therefore reported to Kelley that "there was not the slightest evidence of SO_2 injury in any direction this year to any of the deciduous trees or growing crops." On the other hand, the dry conditions that year made grazing conditions in the Deer Lodge Valley very poor, despite the somewhat curtailed operation. Gardiner reported to Kelley that he had measured concentrations of arsenic on grasses in the valley of 1,200–1,500 parts per million and concentrations of copper of 100–450 parts per million. To head off any damage claims from stockgrowers, he bought about three hundred head of cattle from individuals who had sustained losses, a strategy he considered quite successful:

> I feel that we should consider ourselves very fortunate in having been able to go through this excessively dry year with the minimum amount of loss that has occurred, and I am certain that it could have been accomplished in no other way than to get right in and buy this livestock before it died, because, even at that, quite a number of the animals I bought were in such a weakened condition that we were unable to take them to the Big Hole, and if they had been allowed to go for a month to six weeks longer, conditions would have been very different.[104]

Gardiner also reported a technique he had developed to reduce the public's perception of the smoke's effects on grazing conditions. Because of his weekly monitoring of levels of metals on grasses, he had noticed that rain seemed to reduce the arsenic level appreciably (rain did not have the same effect on copper levels, for reasons he did not yet understand). Working from this knowledge, Gardiner would have the company-owned cattle graze on either side of a zone of pasture directly beneath the prevailing plume of smoke. After a substantial rain, he would move the cattle into the areas under the prevailing plume, thus allowing them to graze on grasses with reduced levels of arsenic. "In this way [we are] utilizing the feed," Gardiner told Kelley, "and not attracting any special attention to the fact that any part of the county was not being made full use of."[105]

In 1918, as the new stack and Cottrell treaters were under construction and the existing stack was still in use, Gardiner continued putting experts in the field. J. W. Blankenship, who had testified for the ACM in

FIG. 8.7. New Stack under Construction, Having Reached a Height of 478 Feet. As the ACM built the new stack, the smelter continued to operate, discharging smoke through the existing 300-foot stack. The steel-frame structures under construction at the base of the stack will house the Cottrell electrostatic precipitators. Photo courtesy of the Montana Historical Society, cat. # PAc 82–62.2807.

the Bliss case and then prepared a report on injury caused by smelter smoke for the Smoke Commission in 1911, had two stints in the fields and forests around Anaconda in 1918. His second stint of fieldwork was conducted in late August and early September, after rains and atmospheric conditions had combined with smelter smoke to be especially damaging. As Blankenship put it, "recent rains had furnished peculiarly favorable conditions for smoke damage, so that the various lines of 'smoke drift' showed up with unusual distinctness." Smoke damage was so severe in 1918, in fact, "that it is now possible to identify positively certain hitherto doubtful 'smoke paths,' though it must be said injury outside these usual directions is not infrequent and sometimes quite bad, depending on wind-direction, when the smoke 'drops down' in bad weather." The most severe "smoke path" extended southwest of the smelter, up Mill Creek and over the Continental Divide into the Big Hole River drainage. Blankenship said damage along this path was discernable as far as Wisdom, more that forty miles by air from the smelter. Another smoke path extended about twenty miles north and a little east of the smelter. A third ran slightly north of west from the smelter to Georgetown Lake. Blankenship named the others the "Butte Drift" (southeast toward Butte), "Lost Creek Drift" (northwest nearly to Philipsburg), and "German Gulch Drift" (south toward Feeley). He identified sulfur dioxide as the principal cause of injury.[106]

Blankenship noted that the data he collected in 1918 was particularly important, because the new Cottrell treaters and taller stack would be put into service at the Washoe smelter the next year. Future observations could be compared to the results of his fieldwork to see what effects the new equipment was having. Blankenship expressed his expectation that the treaters would recover more solids from the smoke, thereby improving livestock conditions, but the impact of the taller stack on sulfur dioxide emissions was less certain. He cited experiences in Germany, where taller stacks had reduced the intensity of damage near smelters but had expanded the area over which injury occurred.[107]

Construction of the new stack and the Cottrell electrostatic precipitators began with site preparation in July 1917. Masons began laying brick for the stack in May 1918 and finished that December. Construction of the treaters was delayed, however, by World War I; the government deemed that the steel necessary for the project was better allocated for the war effort. Finally, in May 1919, the first five Cottrell units were put in service, and by mid-June all twenty treaters were operating. They did not function as well as expected at first, so at the recommendation of the Smoke Commission, the ACM hired a member of the Research Corporation staff to fine-tune the treaters and their operation. By August 1920, after a year at Anaconda, the Cottrell specialist had the treaters recovering about 85 percent of the available copper and 70 percent of the available lead and arsenic. With postwar demand for copper down, however, the smelter was operating at less than full capacity, thereby improving the treaters' performance over what it would have been if operating at peak capacity. The commission anticipated that the ACM would have to install additional Cottrell units after the smelter returned to peak production in order to achieve the projected recovery of arsenic. This would not be difficult from a structural perspective. The flue connections to the new treaters and stack were built in 1918 to accommodate the anticipated additional treaters.[108]

In conjunction with the Cottrell treaters, the ACM had built a special reverberatory furnace for smelting the flue dust collected. The furnace produced copper matte from the flue dust, as well as slag and smoke. The furnace was equipped with its own specially designed Cottrell treaters for collecting condensed arsenic, smoke from the furnace being especially rich in arsenic fume. The dust collected by the special Cottrell units was about 80 percent arsenic. Some of it was sold in its crude form, and some was further refined at Anaconda to be sold as pure arsenic trioxide. Demand for the pure arsenic was excellent, and the ACM was able to sell all that it produced. Slag from the special reverb was also unusually high in arsenic.[109]

In 1923 and 1924, as the smelter returned to near normal operating conditions, tests conducted by ACM staff (with testing procedures inspected by Smoke Commission staff) showed that performance of the Cottrell treaters had improved. They were recovering more than 85 percent of the total dust and fume in the smoke, including about 94 percent of the copper, 78 percent of the lead, and 80 percent of the arsenic. The commission compared these figures with tests showing what had been leaving the three-hundred-foot stack in 1916. Total dust and fume losses had been cut to 6 percent of what they had been in 1916, copper losses had been cut to 2.5 percent of what they had been, lead losses to 15 percent of what they had been, and arsenic losses to 10.7 percent of former losses. During 1923, the ACM recovered $1,130,000 worth of materials from flue dust collected by the Cottrell electrostatic precipitators and related equipment. About $700,000 of that amount represented the gross value of the arsenic recovered. To recover those materials, it cost the company about $700,000 for operating and maintaining the Cottrell treaters, smelting the flue dust in the special reverberatory furnace, refining the arsenic, and so on. The operation thus yielded a net profit of more than $400,000. This figure represented 17 percent of the cost of installing the stack, treaters, and related equipment in 1917–1920. The commission estimated that during 1923, the ACM had recovered as marketable product about 24 percent of the arsenic present in the concentrates and ores smelted. The commission estimated that 12–16 percent was lost to the atmosphere, an equal amount was sent to the slag pile from the special reverberatory furnace, and the remainder was sent to the slag pile from the main smelting furnaces.[110]

Entering the 1920s, the Anaconda Smelter Smoke Commission seemed relatively sanguine with the reduction in arsenic and other solid particulate emissions from the Washoe smelter. Even though the Cottrells were not performing as well as had been expected and there was an anticipation that additional units would have to be built when the slump in the copper market ended and the smelter returned to full production, the Smoke Commission also stated in its 1920 report that it would probably not be necessary to eliminate arsenic emissions completely in order to protect agriculture in the area around the smelter. The commission recognized that completely eliminating arsenic from the smoke "would entail a very heavy burden of expense on the Anaconda Copper Mining Company." The commission was willing to await completion of studies by ACM field staff to determine how much arsenic could safely be deposited on the Anaconda environment. The commission also seemed content to await ACM determinations of what to do with other solids collected by the Cottrell treaters, such as lead and bismuth.[111]

In 1924, after the smelter was back to near normal operations, the commission seemed satisfied with the reduction in arsenic emissions the ACM had achieved. The commission's 1924 report stated:

> There remains for consideration then the matter of reducing the present loss of arsenic into the atmosphere. As shown above this is from 12 to 16 percent of the total arsenic entering the plant. In actual tonnage it is less than one third the quantity which it was believed in 1916–1917 probably could be discharged safely into the atmosphere without endangering any outside community interests. No conditions have arisen since then which would indicate that this arsenic output could constitute a nuisance to the outside surrounding community. Furthermore, a still greater proportion than in 1916–1917 of the agricultural area which could possibly be affected by the arsenic emanations has been bought or leased by the Anaconda Company.
>
> Hence, so far as the safeguarding of the community interests is concerned, the Anaconda Company has accomplished much more than what was thought to be necessary, and the Commission is of the opinion that all the danger of injury to outside interests by solid emanations has been removed.[112]

The report went on to explain why the commission recommended no further ACM expenditures to try increasing the percentage of arsenic recovered of the present Cottrell system.[113]

Although it appeared that the arsenic problem had largely been abated, the commission still had no plan to reduce and eventually eliminate the discharge of sulfur dioxide from Washoe smelter smoke. The commission had concluded that there was little likelihood of developing a market for either elemental sulfur or liquid sulfur dioxide. It was far more likely to find markets for sulfuric acid made from smelter smoke. Smoke Commission analysis showed that much of the smoke produced at the smelter, because of the concentration of sulfur dioxide, was amenable to sulfuric acid production. As the smelter had been operating in the mid-1910s, more than three-fourths of the sulfur being discharged came from the roasting and the blast furnaces. With the anticipated elimination of the blast furnace department, that 75 percent of sulfur would all be given off by the roasting department. Smoke from the roasters was sufficiently high in SO_2 that sulfuric acid could quite readily be made from it using available technologies. The Smoke Commission estimated that the Washoe smelter could theoretically produce 3,000 tons of 60° Be sulfuric acid daily from the roaster smoke alone (degrees Be stands for Baumé, a unit of measure for the relative concentration of an acid by measuring how much greater the density of the liquid is than the density of distilled water).[114]

As already mentioned, the ACM itself had already created a portion of the market necessary to use that acid. The tailings-leaching plant used sulfuric acid to extract residual copper from old gravity-concentrator tailings. The company had also signed an agreement to supply parties in Butte with 15 tons of acid daily and had identified promising markets in Idaho mining districts as well. In fact, by the close of 1915, the ACM had found that sales to outside customers were paying the entire cost of operating the 100-ton acid plant.[115]

By the following year, the ACM's new flotation mill was complete and the company had built its zinc concentrator, both of which increased internal demand for sulfuric acid. Correspondingly, the ACM expanded its acid plant in 1917 to a capacity of 180 tons per day. The acid plant was located on the flat west of the old slime ponds. Rather than using smoke from either of the existing roasting departments, however, the acid plant derived sulfur from smoke given off by special MacDougall roasting furnaces built in conjunction with the plant. This was necessary because the chamber process used to produce the acid required such a large facility that there was insufficient room within the smelter complex to build an acid plant adjacent to either of the existing roaster buildings. Therefore, coarse concentrates from the Washoe concentrator were hauled to the acid plant for roasting, and then calcines from the acid plant were hauled to the reverberatory department for smelting.[116]

Because the chamber process required such a great capital expenditure per ton of acid produced, and because the size of the equipment for the chamber process precluded construction of an acid plant near the main roasting furnaces, where most of the sulfur was given off, E. L. Larison, who had charge of the acid plant, successfully developed a new acid-producing method. After experimenting with a small 2-ton unit late in the decade, the ACM built a 25-ton acid plant according to Larison's design, wherein brick-filled towers replaced the lead chambers of the conventional acid plant. Larison's tower plant required about half the capital per ton of acid produced, and it occupied only an eighth of the ground. The development improved the chances that the ACM would be able to expand its acid-making capacity if additional markets could be found for ACM sulfuric acid. Meanwhile, Larison's addition brought the acid plant to a total daily capacity of about 200 tons. Note that this represented less than 10 percent of sulfuric acid calculated to be available from the roasting furnaces alone, disregarding all the sulfur given off by the other furnaces. In other words, at the close of the decade, the ACM's acid plant was daily using only about 60 tons of sulfur, out of the more than 1,000 tons of sulfur being given off each day the smelter operated at capacity.[117]

It was clear, then, that if the Smoke Commission was going to be able to recommend that the ACM install additional acid production capacity, and if that recommendation were to be commercially practical within the understanding of the 1911 agreement, other markets for sulfuric acid would have to be found. During negotiations leading to the 1911 agreement, Ligon Johnson had urged the ACM to consider using acid to produce phosphate fertilizer, but the ACM had objected that there were no phosphate deposits near enough to Anaconda to be economically shipped there and processed, and that there was no market for the fertilizer if it were to be made.

About the time the 1911 agreement was being signed, the U.S. Geological Survey had identified a large phosphate deposit in Idaho that was fairly close to Anaconda, and shortly thereafter a significant deposit was discovered in Montana, even closer. Nevertheless, the commission could not recommend that the ACM go into the fertilizer business unless it could demonstrate that there would be a market for the ACM's product. Conventional phosphate fertilizer then being produced in the U.S. used 50° Be sulfuric acid to make a product called superphosphate, containing about 16 percent available P_2O_5. At the commission's request, ACM researchers in 1916 explored the possibility of using the 60° Be sulfuric acid being made at the

Anaconda acid plant to produce an enriched product that could compete with fertilizers made elsewhere. The researchers found they could make a fertilizer, called treble superphosphate, containing 45–50 percent available P_2O_5. Investigation showed, however, that this product could not compete in the agricultural market of the southeastern states, where the largest market for phosphate fertilizer existed. The ACM also explored the possibility of using sulfuric acid to produce phosphoric acid from phosphate rock, ship the phosphoric acid to byproduct coke plants operated by the steel industry in the East, combine the phosphoric acid with ammonia generated by the coke plants to form ammonium phosphate, and sell ammonium phosphate as a fertilizer. Tests showed, however, that ammonium phosphate would be physically much harder for the steel industry to handle than was the ammonium sulfate fertilizer they were currently making from ammonia and sulfuric acid.[118]

In light of those setbacks, the Smoke Commission made a modest recommendation in June 1917 that the ACM build a relatively small experimental plant to process about 30 tons of phosphate rock per day and produce 17.5 tons of treble superphosphate fertilizer daily. The plant was intended to work out technical problems in fertilizer manufacture and to estimate construction and operating costs for a full-scale plant. Because of labor troubles and labor shortages during the war, however, the ACM opted not to implement the recommendation until June 1919, when the company began construction of the experimental phosphate plant in the old Bradley building on the flat near the old slime ponds.[119] During the entire process, L. D. Ricketts, the ACM's representative on the Smoke Commission, cautioned his fellow members not to pressure the ACM to move too quickly into business ventures aimed at utilizing sulfur recovered from smelter smoke. If there was a market for the resource, Ricketts reasoned, the ACM would recover the resource, because the company had a financial interest in doing so: "In my opinion they [the ACM] are fully alive to the situation and since they have this vital financial interest they can look after the investigation for a market and methods of manufacture better than the Smoke Commission which has no financial interest in the matter whatever."[120]

As the 1910s ended, the potential of the market for Anaconda phosphate fertilizer was still largely unknown. Early in 1920, Arthur Wells, a Bureau of Mines staffer furloughed to work for the Smoke Commission, prepared a thorough study of the phosphate market, with special attention to how the ACM could compete in certain facets of that market. Based on Wells's report, the Anaconda company agreed to build a plant capable of treating 400 tons of phosphate rock and producing 250 tons of treble superphosphate per day. Pursuant to those plans, the ACM also purchased a large phosphate deposit in Idaho and began mining phosphate rock from the deposit in Montana nearer the smelter. The company also established a department to begin marketing treble superphosphate fertilizer in the Midwest and West.[121]

In 1920, Gardiner offered the University of Arizona's E. P. Taylor a job with the ACM working on "phosphate extension work" and continuing the company's research into the impact of smelter smoke on agriculture, not only in the Deer Lodge Valley but also in the vicinities of the ACM's smelters at Great Falls, Montana, and Tooele, Utah. Taylor accepted, and took the title "agriculturist and forest specialist," working within the extension division of the ACM's phosphate department. By the end summer 1920, he had completed another field survey of injury to trees in the area, noting, for example, that there had been an increase, since his investigation two years earlier, in trees killed by smelter smoke along the south side of the Big Hole River.[122] Damage to vegetation from Anaconda smelter smoke was unabated.

Little actual plant construction happened at Anaconda, however, because of the serious postwar slump in the copper market. The copper industry was so depressed that the ACM actually closed the Washoe smelter from April 1921 to January 1922. At the same time, the fertilizer market in the U.S. was very unstable, and the ACM was having difficulty selling its treble superphosphate in 1922 and 1923, after the smelter resumed operations. Nevertheless, in late 1923, the ACM decided to build a large phosphate fertilizer plant, housed in the former leaching plant and capable of treating about 200 tons of phosphate rock and producing about 120 tons of treble superphosphate per day. This was about half the size agreed to in 1920. It consumed about 130 tons of 60° Be sulfuric acid each day, addressing one of the intents for the fertilizer business. It did not, however, necessitate the construction of addition sulfuric acid production capacity, because the ACM had ceased leaching old tailings, thus eliminating that demand for acid. Moreover, the company was about to discontinue its internal demand for sulfuric acid at the concentrator, as the switch from bulk flotation to selective flotation of Butte ores entailed a switch from acid circuits to alkaline circuits.[123] The entry of the Anaconda company into the fertilizer business as a large producer of treble superphosphate did not herald a solution to the sulfur problem in the Washoe smelter's smoke, as the Anaconda Smelter

Smoke Commission had hoped. The fertilizer business was a profitable one for the Anaconda company, however, and within a few years demand for the ACM's treble superphosphate exceeded the company's ability to meet it.[124]

The Anaconda company's entry into the fertilizer business was not at all unusual for a large corporation in the early twentieth century. As American corporations followed paths of both vertical and horizontal integration, they also typically broadened the kinds of products they made and marketed. Large steel corporations, such as U.S. Steel, for example, assembled numerous smaller corporations and came to own and operate not just iron-smelting and steel-fabricating facilities but also iron, limestone, and coal mines and railroad and shipping lines. They produced and marketed all varieties of steel, including sheet steel sold to manufacturers, finished structural shapes sold to the construction industry, and pipe sold to a variety of markets. They also sold metallurgical coke on the open market, sold byproducts of the coking process to the chemical industry, and even, as described earlier, sold ammonium sulfate as a fertilizer to agricultural customers.[125] The ACM fitted that diversified mold.

The ACM's Political and Physical Positions in Montana at the Close of the 1910s

In 1920, the Anaconda Copper Mining Company was one of the largest industrial corporations in the world, held incredible political power in Montana, dominated the scene in the Butte and Anaconda, and had incorporated much of the territory needed for the disposal of industrial wastes into its own industrial system. In the political realm, the ACM and its lobbyists dominated legislative affairs, allowing the company, for example, to operate without paying taxes, much to the displeasure of other important interests in the state, such as agriculture and timber. The ACM was able to implement this political power in part through its ownership of almost all the daily newspapers in Montana's major cities.[126]

It was in the context of world war and labor conflict that the ACM had agreed to implement the Smoke Commission's recommendations to build the 585-foot stack and the Cottrell treaters. And it was in that context that the ACM made major changes as well in the ways it managed the tailings it discarded. The former change was made under the provisions of the company's 1911 agreement with the United States government. The latter was made to forestall damage claims by downstream property owners.

The ACM's conversion to flotation in 1915 initiated the change in tailings disposal practices. Tailings from the flotation mill no longer carried enough copper to encourage the ACM to store the tailings for possible future treatment. Nevertheless, the company wanted to keep the tailings from flowing into the Clark Fork, so during the 1914–1916 period the ACM developed a new system of impoundments, known as the Opportunity Ponds. The first one was built east of the existing tailings dump and was known as A pond. It filled quickly and was followed in 1918–1919 by the B and C ponds, located between pond A and the Deer Lodge River. Ponds B-1 and C-1 were built in 1918, at which time the level of the dam impounding pond A was raised to store more tailings. Ponds B-2 and C-2 were built in 1919, by which time A pond was filled with tailings and its active use was discontinued.[127]

The two B ponds were about 4 miles east of the smelter; the two C ponds were about 1.5 miles further to the east and about fifty feet lower in elevation. Each impoundment had a spillway at both its north and south ends. Tailings were conveyed to the ponds via ditches and alternately discharged into the north side of a pond with the north spillway closed and then into the south side of a pond with the south spillway closed. By this means, water had to travel the entire width of the pond before it could flow out the spillway, and most of the solids were supposed to settle out along the way. Tailings water discharged from the upper ponds, having given up 75–80 percent of the solids, had to pass through one of the lower ponds, with more solids settling out, before it was allowed to flow into the Deer Lodge River near the old French Crossing Ranch. Water flowed via a ditch about two miles between the lower dams and the river. According to ACM reports from the period, water leaving the C ponds contained about 0.1 percent solids.[128]

When the ACM had entered the irrigation business in the early 1910s, it also attended more to the condition of the streams from which many irrigators in the Deer Lodge Valley drew their water. After the change from gravity concentration to flotation and development of the Opportunity Ponds, the Washoe smelter continued to discharge vast tonnages of tailings toward Silver Bow Creek and the Clark Fork. Those tailings presented a significant ongoing management problem to the Anaconda company. While the ACM was building its B and C ponds, it also developed a facility for final clarification of waters flowing into the Clark Fork. Known as the Warm Springs Ponds, the facility was placed downstream of the outflow from the Opportunity Ponds and just upstream of the confluence of

Warm Springs Creek and Silver Bow Creek. The Warm Springs Ponds gave the ACM an impoundment to collect fine solids overflowing the Opportunity Ponds and for tailings washing down Silver Bow Creek from Butte.[129]

The ACM built its Warm Springs Ponds following a general plan devised by Samuel Fortier, the USDA irrigation engineer who had helped E. P. Mathewson identify a plant pathologist during the Bliss case and had helped H. C. Gardiner design the ACM irrigation system for Opportunity. Fortier based his scheme for the Warm Springs Ponds on similar tailings impoundments in other western mining districts. He predicted that his plan would be an effective precaution against future damage claims from downstream water-users, especially farmers. He alluded to a claim already filed by J. E. Hoffman, but there had been others. Fortier also alluded to river conditions elsewhere in the West, where mining debris and tailings had been problematic. At Jackson, California, the Kennedy Mining and Milling Company had recently built, at considerable expense, a sophisticated multiple-arch concrete dam to keep its tailings out of the river. Fortier's cost estimates for the Warm Springs Ponds showed that the ACM could prevent solids from moving further downstream at considerably less cost per ton of material impounded than the Kennedy Mining and Milling Company had spent. After consulting with Gardiner and Fortier about the plan, and five days after Frank Little's murder in Butte, Washoe smelter manager Laist recommended to ACM vice president Con Kelley that the company implement the plan.[130]

Con Kelley was concerned with conditions along the Clark Fork River as far downstream as Drummond. He knew that the ACM was now liable for tailings discharged not only from the Washoe plant at Anaconda but from old tailings dumps at Butte as well, because of the incorporation of all the Amalgamated companies into the ACM in 1910. Kelley believed the ACM shared responsibility for injury caused by Butte tailings with the other companies still operating facilities there (Pittsmont, Butte and Superior, Timber Butte), so he hoped to enlist their participation in construction of the Warm Springs Ponds. Regardless of whether the other companies would participate, however, he favored building the Warm Springs Ponds so that they could capture tailings flowing out of both Butte and the Opportunity Ponds. John D. Ryan, who was president of both the ACM and the Montana Power Company at the time, arranged for Frank Scotten, a Montana Power engineer, to design the Warm Springs Ponds. Scotten had worked for the Boston and Montana on the construction of the Great Falls smelter in the early 1890s and then was involved in the design and construction of the Rainbow Falls dam in 1910. Construction of the Warm Springs Ponds was completed in 1918.[131]

With the Warm Springs Ponds built, the ACM could contain waterborne contaminants in the Deer Lodge Valley, and the company owned much of the land in the valley on which arsenic dust from the new 585-foot stack fell. If the company did not own the land in the southern end of the valley, then it likely held an easement to property, which the ACM believed freed it from any liability for damage to land, crops, or livestock. As the decade of the 1910s drew to an end, the ACM had implemented all the technical fixes that the Anaconda Smelter Smoke Commission had recommended, and as a consequence, the company was removing most of the arsenic from its smoke, but it was still discharging tremendous volumes of sulfur dioxide and related sulfur compounds from the stack. Even if the company no longer had to be concerned with the threat of damage claims from farmers in the valley, there was still the matter of damage that sulfurous smoke was causing to vegetation on adjoining National Forest lands.

Evidently, the federal agencies involved in the issue no longer had the heart to try to compel the ACM to abate its air pollution. The Roosevelt and Taft administrations, which had developed the legal case against the ACM and had negotiated the 1911 agreement, were long gone, and the administration of Woodrow Wilson, which had implemented the agreement through the war years, was about to leave office. By that time, it had come to light, in a paper prepared by Ligon Johnson for the American Institute of Mining Engineers, that the federal government had never intended to actually close the Washoe smelter, even if the ACM did not abate its pollution and cease its destruction of forest resources. After Johnson left his job as special assistant to the attorney general at the Justice Department, he went to work as "smoke or field counsel" for the American Smelting and Refining Company, or ASARCO. Often known as the "smelter trust," ASARCO owned lead and copper smelters throughout the western part of the United States and a giant copper refinery that was neighbor to the ACM's Raritan works in Perth Amboy, New Jersey. In the late 1910s, when he presented his paper to the AIME, he was defending ASARCO's Selby lead smelter in the San Francisco Bay area.[132]

Concerning the litigation he had pursued, seeking to enjoin the Washoe, Mammoth, Bully Hill, and Mountain Copper smelters from continuing to discharge harmful fumes, Johnson wrote, "While it was probably unknown to the smelters generally at that time, it was never the intention of the Government to close any plant. What it sought was the immediate and serious consideration of the fume problem and active effort to correct harmful conditions where such existed." As it had developed, he had been able to negotiate stipulations with each of the other smelters to conduct meaningful research and to install effective pollution-abatement methods; the ACM was the only company against which the government had filed suit. By means of the suit, Johnson wrote, the government was able to cooperate with what he called a "scientific commission," the board of experts consisting of Hammond, Ricketts, and Holmes. While Johnson seemed satisfied with the results that legal pressure and the Anaconda Smelter Smoke Commission had achieved, he asserted that smelter companies should not try to make the problem go away by paying damages, which he called "a little peace money." Such an approach, he warned, would only encourage more damage claims, and the price demanded of a smelting company would be much greater.[133]

The 1910s were a momentous decade for the Anaconda Copper Mining Company. It took control of nearly all the mining operations on the Butte hill, developing them into an integrated technological system. On a broader scale, the ACM became a global corporation and one of the largest industrial companies in the world. Along the way, the ACM faced a legal challenge to its monumental pollution of region around the Washoe smelter from a more formidable litigant, the U.S. Department of Justice, than it had faced in the Deer Lodge Valley farmers the previous decade. Faced with the prospect of going to court against the U.S. government, the ACM entered an agreement with the Department of Justice it would "at all times use its best efforts to prevent, minimize and ultimately to completely eliminate the emission and distribution from its smelting works at Anaconda, Mont., of all deleterious fumes, particularly those containing sulphur [*sic*] dioxide." Undaunted by the authority of the federal government or the agreement it had entered with the government, the ACM meanwhile completely reconfigured the methods it used to treat Butte ores at the Washoe smelter, the results of which meant that the smelter actually discharged more pollutants into the environment in the middle years of the decade than previously. As the smelter stood at the end of the decade, it featured facilities that greatly reduced, but did not eliminate, the arsenic billowing from the stack, but the smelter was discharging more sulfur than it had been at the beginning of the decade.

The 1910s left the landscape of the southern end of the Deer Lodge Valley with three prominent features that survive to this day, even though the smelter itself was demolished in the early 1980s. The Opportunity Ponds served the Anaconda smelter for as long as its concentrator operated, into the early 1960s. They remain a large man-made, barren feature on the landscape of the Deer Lodge Valley. The Warm Springs Ponds are still an important feature in the purification and clarification of the Upper Clark Fork, having recently been completely restructured as part of the work of the Superfund Project. And the 585-foot stack still stands atop Anaconda's smelter hill, now a state monument.

CHAPTER 9

Land Exchanges—The Environmental Controversies Draw to a Close

> The Forest Service cannot but be concerned with the continued and increasing damage caused by the smelter fumes in the Anaconda region. The exchange idea suggested seems to point a way of solving the matter to the mutual satisfaction of both the company and the Government.
>
> —Fred Morrell (U.S. Forest Service) to L. O. Evans (Anaconda Copper Mining Company), 1922[1]

At the beginning of the 1920s, Butte was no longer the world's unrivaled leading supplier of copper. Michigan's Keweenaw Peninsula had regained that position in 1919, and the Butte district dropped to third place, behind Arizona's Globe-Miami district. Butte regained the top position in 1920, but then it dropped to fourth place in 1921, producing only about 48,000,000 pounds of copper, and second place in 1922. In both those years the Globe-Miami district was the United States' leading copper producer. Butte was again the nation's leading copper producer for the four years of 1923 to 1926, but then Utah's giant Bingham Canyon open-pit copper mine permanently surpassed Butte in copper production in 1927.[2] The Anaconda Copper Mining Company's vertical and geographical expansion in the 1910s, however, had set the stage for even more consequential developments for the company in the early 1920s, with the acquisition of the American Brass Company, the largest fabricator of brass products in the country, and of the giant copper mine at Chuquicamata, Chile, the largest-known copper deposit in the world at the time.

After the copper market appeared to have returned to normal in the early 1920s, organized labor remained a defeated cause among Butte's miners, although the labor movement was still strong elsewhere in the community. As prosperity returned to Butte, many residents, especially women, turned their attention to fashioning the cultural infrastructure of a more permanent community, now that all the smelter smoke in Butte was largely gone, and to addressing the needs and desires of families, which increasingly characterized the demographics of the once-rugged mining camp formerly dominated by single men.[3]

Officials of the United States Forest Service turned their attention as well to a situation that appeared to be permanent: the ongoing damage, caused by smelter smoke, to forestlands adjacent to the Anaconda Reduction Works. The Forest Service initiated a new approach with the ACM for addressing the injury smelter smoke was causing to government-owned resources. The Washoe smelter had resumed nearly full-capacity operation, and sulfurous emissions from the giant stack neared previous high levels. The Forest Service's new approach to the problem called for the ACM to deed undamaged forestlands it owned elsewhere in Montana to the Forest Service in exchange for damaged National Forest lands in the vicinity of the Anaconda smelter. The idea arose with District Forester Fred Morrell as Congress was considering passage of the General Exchange Act of 1922. While inspecting the smoke-damaged areas on the Deerlodge National Forest, Morrell realized that the Forest Service's problem with ongoing injury by smelter fumes could be resolved by a land exchange. Implementation of the swaps of land that ensued employed the mechanisms for exchange enabled by the 1922 act and was accomplished without resort to the courts.[4]

The new approach did not continue the work the ACM and the U.S. had begun through the efforts of the Anaconda Smelter Smoke Commission to find technical means of reducing the harmful emanations from the smelter stack, nor did the land exchanges directly address the damages the United States had considered claiming against the ACM. For Forest Service officials, the land exchanges appeared to obviate the need to reduce the injurious emissions from the stack further and to postpone the need to address damages.[5]

By 1920, the Smoke Commission had seemed to be satisfied that, during periods of normal operation, the Cottrell treaters and accompanying apparatus would remove about 80 percent of the arsenic from the smoke, and the commission seemed resigned to the fact that no profitable technologies were available to recover any significant amounts of sulfur from the smoke. Meanwhile, the Anaconda company seemed satisfied, through cooperation with the Smoke Commission, to have installed facilities allowing the company to earn greater profits by better capturing metals heretofore being lost to waste and by recovering and selling byproducts of the treatment of resources extracted from the Butte hill, such as arsenic and a small amount of sulfuric acid.[6] More important, ACM officials must have been very pleased to have held the United States' 1910 suit in abeyance for a decade, during which the company had nominally complied with the commission's recommendations. The ACM had not been forced to curtail, in any way, its profitable smelting operations, nor had it been compelled to pay damages for the government property it had injured.

As the entity responsible for protecting and maintaining the government's forest resources, the U.S. Forest Service was intimately aware of the damage being sustained by trees and other vegetation in the region around the Washoe smelter. And given the course that events had followed at the smelter during the preceding decade, the Forest Service must have seen little prospect of a resolution to the problem through the legal mechanisms, either instituted or pending. By 1922, the Forest Service was ready to seek its own solution. Top Forest Service officials, both in Montana and Washington, D.C., turned their attention to the new idea of exchanging damaged National Forest lands for undamaged timberlands owned by the ACM. The ACM had purchased those timberlands from the Northern Pacific Railroad, which had received them as part of its land grant from the federal government in 1864, intended to help the Northern Pacific capitalize construction of its transcontinental railroad line. The Northern Pacific had received alternating sections of land on either side of its right-of-way, creating a checkerboard pattern of land ownership across Montana, a checkerboard pattern that continued through forest reserves when they were established in the 1890s.[7] Now, the ACM owned some of those alternating sections in the vicinity of the Northern Pacific's route through forested areas in western Montana near Missoula and to the northwest. Returning some of that land to federal ownership satisfied the spirit of the 1922 act, which was intended to consolidate National Forest lands.

Based on a preliminary inventory in March 1922, the Forest Service identified more than 150,000 acres on the Deerlodge National Forest it wanted to exchange. The assessment estimated the value of the damaged lands at about $1.3 million, including land, timber, and grazing values. There were other areas outside the preliminarily assessed lands that had also been damaged, but the Forest Service was prepared to start exploring this alternative avenue without a full accounting. On March 17, 1922, the acting forester of the Forest Service in Washington authorized Forest Service officials in Montana to approach officials of the Anaconda company with the idea of land exchanges, the purpose of the exchanges being "to eliminate the problem." Three days later, Congress passed the Act to Consolidate National Forest Lands.[8]

District Forester Morrell (the position he filled is now called regional forester) oversaw all the national forests in the region encompassing Montana. He was based in Missoula and was the first to broach the topic with the ACM. It occurred to Morrell that, rather than presenting the ACM with a claim for damages—which promised to continue, necessitating ongoing damage claims—the two parties could exchange lands so that the company would assume ownership of the lands sustaining ongoing damage by smelter smoke. After an informal meeting between Morrell and a low-ranking ACM official, and some preliminary letters from Morrell to the company, Morrell's assistant, M. H. Wolff, met in July 1922 with L. O. Evans, general counsel of the Anaconda Copper Mining Company, describing for him only in broad terms what the exchange might entail.[9]

Having initiated contact with the ACM, the Forest Service scrutinized the many issues it would have to resolve before any land exchanges took place. Because the damaged areas included mineralized lands, and the 1922 act precluded the exchange of mineralized lands, Forest Service officials anticipated they would have to seek special congressional action to enable the particular exchanges proposed in the area of the smelter. Forest Service officials also wondered how detailed their surveys of damages and land and timber values should be. Land exchanges, under the 1922 act, would not require fieldwork of excessive detail. The officials were aware, however, that the ACM might prove intransigent, in which case the Forest Service's eventual recourse would be to ask the DOJ to file formal claims for damages in court.[10]

Meanwhile, Forest Service officials in Washington were holding their own discussions about how to proceed. Assistant Forester L. F. Kneipp had visited the damaged area in July 1922, and later that month, the Forest Service's top official, Col. W. B. Greeley, visited with local Forest Service officials in Missoula. The local officials had tried to arrange a meeting between Greeley and ACM officials, but to no avail. Frustrated at the seeming lack of interest in the land exchange idea on the part of the ACM's local officials, Greeley recommended taking steps toward legal action to recover damages. Only in that way, he believed, could the company be forced to consider the option of the land exchanges. Toward that end, Greeley wanted Forest Service personnel in Montana to conduct thorough enough damage assessments in the vicinity of the smelter that the data collected would be serviceable in support of the suit. Such data, he believed, would also be effective in convincing the ACM of the need for the land exchanges. He proposed that the ACM be informed of the extent of damage it had caused in terms of acres, board feet of timber, and dollars. In addition to filing claims for damages already incurred, Greeley believed the ACM should be asked to pay rent to the United States for loss of productivity that would continue to occur. Greeley hoped the ACM, faced with such demands, would opt for the land exchanges.[11]

As Greeley and Kneipp were formulating their ideas in Washington, Forest Service officials in Montana continued their efforts to communicate with ACM officials. Late in July, Morrell wrote Evans again in the hope of scheduling another meeting. That letter includes the epigraph quoted at the head of this chapter. In the meetings that followed Morrell's July letter, Evans brought other ACM officials into the discussion, including John Gillie, general manager of ACM operations at Butte, and ACM veterinarian H. C. Gardiner, who was especially keen on the land exchange proposal.[12]

Gardiner's interest in remedies to problems of smoke damage are not surprising. As already described, Dr. Gardiner was the ACM official most intimately familiar with the damage smelter smoke caused. He supervised the company's several agricultural activities under the plume of the stack, and he was responsible for the company's program of monitoring environmental damage, described in the previous chapter. Gardiner's familiarity with damaged areas helped local Forest Service and ACM officials begin to discuss details concerning which damaged lands should be exchanged. Morrell urged that the more heavily damaged areas be exchanged first. Otherwise, if damage continued, he assured the ACM, the Forest Service would use legal means to seek an ACM response.[13] Morrell assumed that the ACM was as adept at litigating a case as the government was, so the idea of a land exchange offered both parties a less contentious means for addressing the problem.[14]

Morrell was sufficiently encouraged by his meetings with the ACM that he wrote his superiors in D.C. recommending continued local discussions as an alternative to the confrontational course proposed by the central bureaucrats. It was clear from the meetings that the key ACM officer was Con Kelley, ACM president, who would determine the company policy. Morrell counseled patience, urging his superiors to place their emphasis on the land exchange negotiations, saving the threat of a damage suit as a fallback position. He offered further arguments for his position: the fieldwork necessary for the legal case would be much more technical and therefore costly than that required for land exchanges; and if the United States did recover damages, the payments would go into an accounting category called "miscellaneous receipts." Using those funds to purchase undamaged lands to replace those ruined by smelter smoke would probably not be legal for the Forest Service.[15] In one of his letters to Washington, Morrell assured Greeley: "In conferences with Mr. Evans and Doctor Gardiner, I have made it clear that the only position the Forest Service can take is that the Government has been injured by the operation of the smelter, unless the Company is willing to acquire the injured lands at a valuation equal to what the lands would have without the smelter fume damage; and that unless the exchange can be made, the Service considers it necessary to proceed toward securing, if possible, a cash settlement."[16]

After several conferences had been held between various ACM and Forest Service personnel, L. O. Evans wrote Morrell summarizing his personal response, as opposed to an official company response, and suggesting some of concerns the company would bring to subsequent discussions. He cautioned that the ACM would be unwilling to have the Forest Service foist uninjured but nevertheless worthless lands, such as ridgetops, onto the company. He also cautioned the Forest Service against trying to transfer to the company timberlands previously damaged by fire or insects. Evans concluded by recommending that progress toward formulating a concrete proposal could be quickened by sending Forest Service representatives to go over the damaged areas with Gardiner so the two sides could compare, on the ground, their respective conclusions about causes of damage, whether smoke, insects, fire, or other.[17]

During the remainder of 1922 and the first six months of 1923, Forest Service personnel and ACM officials in Montana continued meeting, with each side anxious to learn more of the other's thinking. By June 1923, with local ACM representatives now apparently enthused by the idea, local participants for both parties were ready to formally seek the approval of their superiors to execute a land exchange for a portion of the lands injured by the ACM's smelter emissions. They had identified about seventy thousand acres of land they were ready to exchange in an initial deal, and they agreed that the two parties should proceed toward that exchange without any commitments or stipulations concerning future exchanges or legal actions, while agreeing informally that they would continue to work toward future exchanges as well.[18]

In August 1923, ACM's officials announced that they approved of the general idea of a land exchange, but they added a new wrinkle: the company now wanted to complete the entire land exchange for seriously damaged lands in a single transaction. Morrell advised his superiors that there might be some conflict between the parties in negotiating how many thousand board feet of timber should be estimated for the damaged areas, but he also recognized that the lands the ACM was offering were more productive than the lands on the Deerlodge National Forest (the latter being in semiarid southwest Montana, while the lands the ACM was offering were to the north, a region of Montana that gets more rain), so that would give the Forest Service some margin within which to negotiate estimates of timber values. Therefore, Morrell recommended accepting the ACM's proposed simple method of valuation.[19]

As discussions between the parties continued, it became clear that the ACM now wished the exchange to embrace about 130,000 acres of damaged area. In return, the ACM was offering a similar acreage, but because the company did not have enough cutover ground to equal that total, the company was offering a fairly large area still possessing merchantable timber. As the two sides continued communicating to each other what they would and would not accept, as measured in terms of land, timber, and stipulations of the bargain, they assured one another that neither side would be bound by the contents of any of the negotiations, but rather would be bound only by the final conveyance, when and if it came to pass.[20]

The next milestone in the negotiations occurred on October 9, 1923, when James R. Hobbins, a Butte-based ACM vice president, submitted a letter to Fred Morrell, accompanied by a map, proposing that the company convey specified timber lands elsewhere in Montana to the Forest Service in exchange for specified, damaged National Forest lands in the vicinity of Anaconda. The boundaries shown on Hobbins's map were still to be considered tentative, pending resolution of all details of the exchange, and he reiterated that either party could still withdraw from the transaction without obligation to the other party.[21]

As the likelihood of a land exchange appeared to grow more certain, the Forest Service and the USDA's solicitor began in autumn 1923 to explore the broader legal ramifications of the exchange. The forester was of the opinion that the only pending legal action was the 1910 suit, which had not developed further than the appointment of the Smoke Commission under the 1911 agreement. The Forester discussed the idea of a land exchange with H. Foster Bain, who, as director of the bureau, also served as a member of the Smoke Commission. Wondering whether the land exchange would cause the termination of the panel, Bain recommended the Forest Service contact John Hays Hammond, who chaired the Smoke Commission. The Department of Justice still had to be brought into the conversation. On behalf of the DOJ, William Hughes, who had worked with Ligon Johnson in preparing the United States' case against the ACM prior to 1910, suggested that the Forest Service seek Johnson's advice. Johnson was practicing law in New York City in 1923. He also recommended that the Forest Service consider what impact the U.S. dropping its suit would have on any suits farmers in the vicinity of the smelter might bring against the company. Deerlodge Forest Supervisor John Taylor investigated land holdings in the Deer Lodge Valley near the smelter and reported that there were at present no suits by farmers pending against the ACM for smoke trespass, nor was there a likelihood of such being filed.[22]

Once the ACM had made its proposal to exchange lands, acre for acre, board foot for board foot, it became clear how much effort it would take for the government to execute the exchange. Although he had been quite anxious to move the negotiations with the ACM's local officials quickly, Morrell did not respond to Hughes's requests for information until January 1924. Having received an encouraging response from Johnson in early March 1924, Acting Secretary of Agriculture C. F. Marvin formally asked Attorney General Harlan Stone for an opinion on the land exchange proposal. Stone responded that it appeared that a land exchange would remove the basis upon which the U.S. had filed its suit in 1910. In other words, if the U.S. did not own land that had been damaged or was being damaged by the smoke from the ACM's smelter, there would be no basis for a suit by the government against the ACM for damaging federal land. Stone approved of the land exchange.[23]

Meanwhile, the ACM's Gardiner seemed frustrated by the pace of progress, so he initiated discussions with Arthur Wells of the Bureau of Mines, asking that the Smoke Commission's attention to matters at the Washoe smelter be reactivated. According to Wells: "Dr. Gardiner expressed his opinion to me freely in a personal manner, but naturally does not want to be quoted as officially inviting the Anaconda Commission to resume the work which was temporarily held up during the interval of the last three years."[24] Wells reported that he had learned from Gardiner that the Cottrell treaters were not removing as much arsenic from the smelter smoke as Gardiner and the Smoke Commission had hoped they would. Wells recommended that he, on behalf of the Smoke Commission, should be authorized to help the smelter managers improve the treaters' performance. He also reported that the ACM had found his report on the phosphate industry to have been very helpful to the company in developing a market for the new product. Wells asserted that he and Gardiner shared the belief "that so long as the Anaconda Commission exists it should function." He went on to say: "The Commission has never before waited for an invitation of the Anaconda company or of the Department of Justice to function—It has taken the lead in outlining what is belived [*sic*] to be the work of the Commission. The negotiations for the exchange of forest lands with the Department of Agriculture are at a standstill and Gardiner says he has no assurance that they will go through."[25]

Upon Wells's recommendation to reactivate the Smoke Commission, Foster Bain wrote John Hays Hammond, who chaired the Smoke Commission, asking for his opinion on the matter. Hammond approved sending Wells to Anaconda to report on the operation of the Cottrell treaters, the arsenic plant, and other items of interest to the commission. Hammond went on to say that he had thought the commission had terminated. He would continue to serve as chairman, "if it is to the interest of the mining industry," but he expressed his desire to resign as soon as possible. Before heading to Anaconda, Wells met Gardiner in New York to go over some phosphate matters. While they were together, Gardiner convinced Wells that attending to the arsenic situation at Anaconda should take precedence over the phosphate work there. Wells and Gardiner left for Anaconda together in mid-February 1924. Following that and an April trip to Anaconda, Wells prepared the 1924 "Report of Anaconda Smelter Commission," the last report known to have been issued by the commission.[26]

Wells's last Smoke Commission report described the closure of the smelter, due to market conditions, for about nine months in 1922 and early 1923, but he said nothing about the prospective land exchanges. Wells reported that he found the testing methods being used by the ACM to determine arsenic emissions and arsenic recovery to be acceptable. He was also satisfied with the efficiency with which the Cottrell treaters were working. Wells found that 12–16 percent of the arsenic in the ores smelted at the Anaconda Reduction Works was still going up the stack and into the atmosphere, about 24 percent was recovered as a marketable product, and the remainder was in the slag sent to the dump. Wells reported that upon his recommendation the ACM was investigating various methods of recovering more arsenic from the flue dust, thereby reducing the amount lost to slag. On the other hand, he reported no efforts to reduce the amount of arsenic escaping into the atmosphere, but he did claim that arsenic emissions had been brought down to a third of what they had been during the peak of production in the 1910s. ACM control of agricultural lands in the vicinity of the smelter that could be harmed by arsenic, however, had virtually eliminated the threat to the company of damage claims from the surrounding community. The three smoke commissioners put their signatures to Wells's assessment of the arsenic situation at Anaconda. There was no discussion of Gardiner's concerns. Wells also reported on the developments of the phosphate fertilizer plant at Anaconda but said nothing about the amount of sulfur still going up the stack, which was more than 90 percent of the sulfur given off by the furnaces at the smelter.[27]

The commissioners' waning interest is evident in a handwritten note Bain sent Bureau of Mines staffer Dorsey Lyon about getting Hammond's signature on the 1924 report. Bain commented, "He [Hammond] never pays much attention to this matter but he always signs cheerfully. If he asks you can tell him we expect to be relieved of any further responsibility sometime this year." On the back of the note, Bain added the comment, "If Hammond wants to talk about the matter Wells can come down of course but he never has seemed that much interested." Despite the growing inactivity on the part of the Smoke Commission, the attorney general did take the trouble to name a replacement for Foster Bain after Bain resigned as director of the Bureau of Mines late in 1925. In November 1925, the attorney general appointed Scott Turner, Bain's successor at the bureau, to join Hammond and Ricketts on the commission.[28] No further Smoke Commission activity is known to have occurred. Based on the historical record, the most that can be said is that the work of the Anaconda Smelter Smoke Commission ended.

Despite Arthur Wells's apparently rosy assessment in the Smoke Commission's 1924 report of the arsenic situation at Anaconda, local Forest Service officials had other opinions. Deerlodge National Forest Supervisor John Taylor believed that the Anaconda company was knowledgeable of the arsenic problem but was careful not to give out much information. Two years after the last Smoke Commission report, for example, Taylor wrote that he believed that the ACM operated its Cottrell treaters in accordance with the arsenic market, and that in early 1926 the company was taking advantage of a slump in the arsenic market to cease operating some of the Cottrell treaters so repairs could be made. Taylor had also learned from Gardiner that arsenical poisoning of livestock was as dependent on the weather as it was on the ACM arsenic recovery rate. Favorable winds and ample precipitation reduced the likelihood of poisoning.[29]

Meanwhile, work on the land exchanges continued. In July 1925, several Forest Service officials visited with members of the ACM's lumber department at Bonner, near Missoula, to discuss the status of the land exchange. They informed the ACM that, following legislation passed by Congress earlier in 1925, there was nothing legislative to block the land exchange, that the U.S. would have to make mineral reservations on the lands it transferred to the ACM, and that the ACM could make similar mineral reservations on the lands it deeded to the U.S. if necessary. Much of the land the Forest Service intended to convey to the ACM, however, remained unsurveyed. It would take about three years to complete the necessary surveys so that patents for the land could be issued in the name of the ACM. For this reason, E. A. Sherman, representing the Forest Service headquarters in D.C., recommended that the exchange be executed in smaller blocks as they became ready for patent. Sherman suggested that it would be easier to get smaller blocks approved by the secretary of agriculture and secretary of the interior, both of whom would have to sign off on the deal.[30]

By April 1926, it appeared that the land exchange was getting close enough to being executed that the Forest Service would have to advertise the deal to the public, as required by the 1922 act. When Taylor discussed with Hobbins the publicity the Forest Service planned for the land exchange, the ACM executive responded that the company had hoped to keep the deal quiet. He was concerned that the Forest Service planned to highlight the injuries smoke had caused to government property and brag of having extracted payment from the company for those injuries. Taylor responded that the Forest Service was compelled to advertise the plan because the public's interest was at stake, but he also assured Hobbins that the Forest Service had no desire to circulate any publicity the ACM would find untoward. Taylor believed that the land exchanges would serve the public interest, even absent the smoke issue, and that there was no need to mention the smoke damage in any press release. Taylor also predicted that there was not a newspaper in the region that would oppose the land exchange.[31]

Small details still stood between the intentions of the two parties and an official execution of the proposed land exchange. One source of delay was the several approvals that were required on the government's part. The supervisor of the Deerlodge National Forest placed his signature of approval on the exchange of the first block in September 1926, but the regional forester, secretary of agriculture, and secretary of the interior did not approve it until March 1927. The ACM deeded about 22,700 acres to the United States, and approximately two years later all the necessary surveys and other necessities had been completed so that the United States could issue a patent to the ACM for the roughly 21,900 acres immediately west of the Anaconda smelter. In October 1927, the ACM deeded the United States another 9,100 acres, receiving a patent for the 11,300 damaged acres southwest of the smelter. And in December 1927, the ACM deeded the United States 10,200 acres, again receiving its patent for the 7,100 acres south of the smelter.[32]

As the Anaconda company and the Forest Service were executing the first three land exchanges, ACM metallurgists were implementing important changes at the Anaconda concentrator. In 1926, the ACM began to conduct tests on selective flotation, which would allow the company to better produce concentrates of the quality and quantity required by the various roasting and smelting departments. An important facet of the quality difference was the iron content in the concentrates. Slimes from Butte ores typically contained about the same percentage of copper as the original ores but only about half the iron. Correspondingly, "deslimed" ores had about the same copper content as the original ores but had a much higher percentage of iron. Because iron is present in Butte ores largely in the form of pyrite (iron sulfide) as well as in the forms of some of the complex copper sulfide minerals (e.g., chalcopyrite and bornite), sulfur content tended to vary in correspondence with iron content.[33]

Concentrates of the deslimed ore, produced on Wilfley tables, were needed at the Washoe smelter for use in the roasting furnaces associated with acid manufacture and for use as flux for converter slags, which were fed from the converter department back into the smelting furnaces. The Washoe smelter, however, did not need all the high-iron concentrate the mill was providing. By regrinding the deslimed ore, ACM metallurgists

FIG. 9.1. Typical Damaged Forest, August 1910. This photo appears in Lovejoy's 1910 report on damage to trees on the Deerlodge National Forest. The site is on the west side of the Continental Divide about eight miles southwest of Anaconda. The caption in the report reads, "The fir practically all dead; many of the pine badly injured, but few yet killed outright." The area was part of the second land exchange between the U.S. Forest Service and the Anaconda company. Photo courtesy of the National Archives and Records Administration, College Park, MD.

found they could use reagents to accomplish a selective flotation yielding a concentrate suitably high in copper but with reduced levels of iron pyrite. The net result of this change in concentration method was that the ACM was able to produce concentrates averaging more than 20 percent copper, twice the copper content of the concentrates under the old system. The change resulted in a reduction by half of the amount of material that had to be roasted and smelted.[34] This, in turn, had a significant influence on the performance of the smelting departments. For example, when the ACM adopted selective flotation in 1927, only half of the roasting furnaces in roaster no. 2 were needed for copper concentrates. The other fourteen furnaces were then converted to zinc roasters.[35]

Although the new system increased the loss of copper to tailings (old tailings had contained only 0.17 percent copper, while the new tailings contained 0.24 percent copper, still considerably less than tailings from gravity concentration), the loss was partially offset by a reduction in the loss of copper to slag, because less material was being smelted. The change had two other important consequences for the byproducts emanating from the Washoe smelter: 1) particle size of tailings remained relatively unchanged, but the iron and sulfur content of the tailings increased, meaning the iron and sulfur content of the material sent to the roasters decreased, and 2) nearly all the material roasted and smelted had been finely crushed. This change in concentrating technology may have further taxed the flue-dust collection system at the Washoe smelter and therefore increased the volume of flue dust passing the Cottrell treaters and escaping the stack. On the other hand, it meant less sulfur dioxide was being exhausted into the air by the roasting process, and therefore the smelter's smoke caused less damage to vegetation downwind of the stack.[36]

Through the 1920s, Gardiner continued to bring experts to the Anaconda vicinity to investigate damage due to smelter smoke, looking not only at trees and other vegetation but at the effects of smoke on animals as well. In 1928, he brought V. A. Moore out from Cornell to inspect the condition of livestock in the vicinities of the smelters at Anaconda and Tooele. Moore was one of the veterinary experts the ACM had retained to testify in the Bliss case more twenty years earlier. He was accompanied during his 1928 inspection by R. W. Thatcher, who by then was president of the Massachusetts Agricultural College at Amherst. Concerning the agricultural areas around Anaconda, Moore found crop conditions to be excellent and livestock to be thriving. There were few beef cattle in the Deer Lodge Valley, but dairy herds looked good west of Anaconda and near Warm Springs and Galen. He reported no signs of smoke injury to alfalfa and other crops, but he recognized that the generally good appearances in the area might have been due to the unusually high rainfall during July that year.[37]

Meanwhile, the ACM and Forest Service continued working to implement more land exchanges. Shortly after the exchange of the first block of damaged land, the Forest Service formally requested that the Department of Justice maintain suspension of the 1910 suit against the Anaconda company. The forester assessed the outlook for addressing the damage to government lands as follows:

> Each exchange is being put through upon its own merits without reference to what may follow. As each exchange is consummated the damage problem is correspondingly diminished. Until the entire area

> damaged has been acquired by the Company, I do not feel that the suit should be dismissed. Upon the other hand, as long as we are going forward with exchanges I do not feel that the suit should be pressed. If we reach a deadlock at some point in our exchange work, then the question of pressing the suit may properly be raised, but so long as we are making good progress, as I now feel we are, in bringing about a practical and businesslike settlement, I think legal action should simply remain suspended awaiting the outcome of our administrative efforts.[38]

Accordingly, the assistant attorney general instructed the U.S. attorney in Montana to take no further action on the 1910 suit against the Anaconda Copper Mining Company until further notice.[39]

One important facet of the land exchanges was that the ACM agreed, for the blocks of damaged lands it received in exchange, to maintain the land uses already permitted by the Forest Service. As part of the agreement, the Forest Service would administer the permitted land uses, including grazing, cutting firewood, and recreation, and would provide fire protection for the new ACM lands. The Forest Service and the company formalized their Cooperative Agreement on May 1, 1930.[40]

In such a cooperative atmosphere, the ACM and the Forest Service completed five land exchanges by early 1932, consummating the fifth on March 5. The fifth exchange included land more than twenty miles southwest of the Anaconda smelter, extending over the Continental Divide almost to the Big Hole River. In those five exchanges, the ACM received patents for a total of about 84,000 acres of former National Forest land and deeded to the U.S. about 95,000 acres of timber lands elsewhere in western Montana.[41]

On March 24, 1933, about a year after the fifth land exchange was completed, a clerk at the United States District Court for Montana entered the following statement on the docket for *United States v. Anaconda Copper Mining Company*, the suit that the federal government had filed against the ACM: "abandoned, so never called up for final order." [42]

There is no record of either the U.S. Department of Justice or the U.S. Forest Service being inclined to revive the old 1910 suit during the 1930s. There is also no record to explain under whose direction the court clerk recorded that the suit was abandoned. What is clear is that the suit had ended. Moreover, the Department of Justice in the 1930s appears to have been content to let the suit end, although there is no record of the court informing the DOJ that the docket had been closed. There is also no record of the court informing the ACM that the docket had been closed nor record of any official response on the part of the ACM to an end of the 1910 suit. Con Kelley and the others at the head of the company, however, must certainly have been satisfied. After several decades of wrangling with the U.S. government over damage that smoke from the company's smelter had caused to federal property, the ACM had succeeded in not having to go to court to defend itself against the government's claims. As the case ended, the ACM had succeeded in one other thing: the company had never been found by a court to be liable for the widely acknowledged damage to the trees and lands on the Deerlodge National Forest, nor had it admitted culpability in any out-of-court settlement.

By the early 1930s, the ACM had incorporated most of the dumping grounds for harmful discharges from its smelter into its company holdings. The ACM now owned a large buffer of forestlands in the mountainous vicinity of the smelter, and it had already acquired from farmers much of the private agricultural land in Deer Lodge Valley that had been severely damaged by smoke or tailings. In cases where the ACM did not actually own the property, it owned smoke or tailings easements stating that the company had the right to dump smelter wastes on the property.[43]

After the fifth land exchange, progress on subsequent exchanges grew more difficult. The land exchanges had begun to encounter opposition among farmers in the Deer Lodge area. Approximately 130 petitioners, about half of whom the Forest Service determined were neither landowners nor owned livestock, objected to the Forest Service plan to exchange lands west of Deer Lodge. The petitioners feared that once the public lands became private and owned by the ACM, they would no longer be able to graze livestock, cut firewood, and enjoy recreational opportunities on the land, and that, if the ACM chose to log the land, it would harm the watershed for downstream irrigation. Some petitioners were particularly concerned that once the ACM owned the land, all grazing rights would be transferred to Gardiner, who owned a large sheep operation and ran his sheep on much of the land in the vicinity of Anaconda owned by the company. The Forest Service assured the petitioners that, under the Cooperative Agreement signed by the ACM and the Forest Service in 1930, land uses in the areas to be conveyed to the ACM would remain as they had been.[44]

The Forest Service and the ACM proceeded to consummate the exchange of a sixth block of damaged lands in 1936, but not before a hint of the growing acrimony between the company and the government crept into the proceedings. The General Land Office (GLO), as staffed in the 1930s, was reluctant to continue processing the exchanges by the irregular means that had been instituted to please ACM officials in the 1920s. The ACM in turn refused to record the deed for the lands it was transferring to the Forest Service in exchange for the sixth block until the GLO approved the abstract of title. Anxious to complete the deal, the Forest Service succeeded in convincing the GLO to continue the precedent, but not without the GLO letting the Forest Service know it was not pleased to do so and that it would not do so with any exchanges other than those completed between the ACM and the Forest Service for smoke-damaged lands. The Forest Service was especially anxious to get the sixth exchange approved, because the ACM was also refusing to begin negotiations on the seventh exchange until the sixth was finalized. Forest Service officials were perplexed by the company's demand because negotiations for previous exchanges had proceeded without dependence on the status of the previous ones. The exchange for the sixth block was completed in 1937, with the ACM receiving a block of land in the Lost Creek area north and northwest of Anaconda.[45] It was the last exchange that would be consummated between the ACM and the Forest Service, bringing the total to 108,000 acres of smoke-damaged lands that the government exchanged to the ACM.

Following World War II, some officials in the Forest Service advocated that the government seek to reacquire some of the lands exchanged earlier. The last known reference to such an exchange came in 1951, when H. C. Gardiner contacted the Forest Service to propose an exchange for parcels of damaged Forest Service land which the ACM and the Forest Service had informally agreed to swap in 1923, but which had never actually been exchanged. By then, Gardiner was retired from the ACM, but he still had his own livestock business, and he grazed stock on most of the lands the ACM had acquired from the Forest Service in the Anaconda vicinity. The Forest Service evidently saw little likelihood of negotiations proceeding further, because Gardiner no longer worked for the ACM and the current ACM management had little interest in pursuing the matter.[46]

As far as the government lands damaged by smelter smoke in the early part of the century were concerned, both sides seemed content to let matters lie where they lay following a period of watching the seventh exchange go nowhere. The 1910 suit was no longer pending on the books of the U.S. District Court. In exchange for approximately 108,000 acres of the most severely injured forestlands in the vicinity of Anaconda, the Forest Service had received timberlands of comparable value elsewhere in western Montana. And it appeared that, as the Anaconda company was currently operating its smelter, remaining National Forest lands near Anaconda were no longer being injured. Both the U.S. and the ACM had avoided the cost, acrimony, and uncertainty of litigating the issue of smoke from the Washoe smelter having damaged United States property, and the ACM never had to formally admit culpability for the damage its smelter smoke had caused.

That situation became the environmental status quo in the upper Clark Fork watershed for three decades until the new regulatory regimes introduced by state and federal legislation around 1970 and by creation of the Superfund program in 1980. That status quo was able to persist because, until the legislation of the 1960s and 1970s, which was intent on protecting the more abstract entity called "the environment," the only legal recourse available to people or the government who complained that the environment was being harmed was tort law, which allowed them to ask courts to award them damages for harm to their property or to enjoin polluters from damaging their property. Once the ACM held the property rights to discharge its smelter wastes upon forests and farmland near the smelter, there was little that people or the federal government could do under existing law to stop the ACM from operating its smelter as it pleased for the next three decades.

CHAPTER 10

The Decades of Anaconda Company Hegemony

As American industry slowly began to emerge from the depths of the Great Depression in the mid-1930s, *Fortune* magazine ran a series of articles spotlighting some of the nation's greatest enterprises. The two-part article on the Anaconda Copper Mining Company, appearing in December 1936 and January 1937, chronicled the corporation's rise from the rugged nineteenth-century mining camp of Butte, Montana, to the position as the world's largest producer of copper and the world's largest fabricator of copper and copper-alloy products. *Fortune* was impressed with the vast extent of the ACM's horizontally and vertically integrated operations. The article analyzed the ACM's conservative management structure, ensconced in opulent corporate offices in New York City and presided over by Cornelius F. Kelley. *Fortune* appraised the financial health of the ACM, having survived the 1929 Wall Street crash, in which the company's stock fell from a high value of $175 per share to a low of $7 per share.[1] Using language similar to Lewis Mumford's assessment of mining, cited at the beginning of the first chapter, as "an unflinching assault upon the physical environment," akin to the ruthlessness of modern warfare, the *Fortune* article also waxed poetic about the physical setting from which the company's mineral riches sprang.[2]

By the 1930s, Butte had a well-established reputation as a subject on which journalists could test their descriptive skills. The anonymous writer of the *Fortune* article, too, took a turn. A first-time visitor to Butte, the writer predicted, "would find man, who has already been gutting the hill for two generations, still blasting his way farther into hotter and wetter depths, helmeted against the rock itself, naked against its heat, drenched with the sweat of his body and of the earth's." [3] In Anaconda, the visitor would see that "the dehydrated coppery slime goes through fire to be roasted, and into the hellish gas-burning reverberatory furnace for more of its dross to be burned away, and thence into the converters to boil until a pygmy man, staring up at the hissing nozzle mouth of the converter furnace thirty feet above him, decides by the color of its flame that the brew is ready to pour."[4] The descriptions offer vivid pictures of the environments in which miners, mill operatives, and smelterworkers labored and of the effects those labors had on the Earth, but the *Fortune* articles gives no hint that the ACM and associated companies in Montana had sustained several decades of opposition over the environmental consequences of the tremendous discharges of waste products from the mining and metallurgical processes in Butte, Anaconda, and Great Falls. By the time the *Fortune* article was written, such environmental consequences were apparently accepted as a given, much like the ore deposit in the Butte hill from which the mineral values were extracted.

A map accompanying the *Fortune* article shows the geographical extent of the Anaconda enterprise. In addition to its Montana operations—metal mines in Butte, smelter and refinery at Anaconda and Great Falls, the Butte, Anaconda and Pacific Railway, a lumber mill at Bonner, and coal mines near Red Lodge, it had its phosphate mine in Idaho and more coal mines in Wyoming. Its International Smelting and Refining subsidiary had smelters at Tooele, Utah, and Miami, Arizona, as well as interests in metal mines in Utah, Arizona, Nevada, and California. International also had a lead refinery near Chicago and the Raritan copper refinery at Perth Amboy, New Jersey. The Anaconda Wire and Cable subsidiary had copper-wire mills at Great Falls; Oakland and Orange, California; Hastings, New York; Pawtucket, Rhode Island; and elsewhere. Anaconda's American Brass subsidiary had brass plants throughout Connecticut's Naugatuck Valley as well as in Detroit, Buffalo, Toronto, and elsewhere. Outside the U.S., Anaconda had copper mines and smelters at Cananea, Mexico, and Potrerillos and Chuquicamata, Chile. The ACM even owned the Giesche zinc mining and smelting operation at Katowice, Poland, which it had acquired in 1926 (only to lose the property in 1939, after Nazi Germany invaded Poland).[5]

As mentioned in earlier chapters, the history of Montana's copper industry is not a story devoid of conflict. Journalists and historians have written extensively, sometimes in mythical or heroic style, of conflicts in Montana's copper industry. Many writers have attempted to describe the conflict in words as colorful as in the *Fortune* articles. Clark, Daly, and Heinze fought over economic and political control of Butte and the State of Montana; they bought elections and a seat in the U.S. Senate, and they generally besmirched the image of new state's political integrity in the eyes of the nation.[6] Butte's strong labor tradition and a monolithic employer, the ACM, provided the grist for many written accounts of the often bloody episodes of conflict between workers and capital.[7] There is no doubt that the outcomes of these episodes were contested. Interpreters of the conflicts, however, typically staged their dramas against the ravaged backdrop of Butte's contaminated environment as if it had been a given, perhaps like the sulfurous caldrons of Yellowstone National Park.

This book has shown that such a landscape in southwestern Montana, devastated by the copper industry, was not a given. The devastation was contested for a period of at least three decades, from about 1890 to the early 1920s, during which Butte reigned as the world's greatest copper producer. That story of conflict over environmental destruction should be added to our rendering of the history of "The Richest Hill on Earth," "The Battle for Butte," "The War of the Copper Kings," and "The Black Heart of Montana." The story of environmental conflict should be told not in the fabled manner of so much of Butte lore, but as a sober reflection on the cost that our material affection for copper, and the electricity it delivers, has exacted on the Earth and its inhabitants in a part of Montana still distant from the lives of most Americans, who enjoy the comforts of electricity, the convenience of modern communications, and the ease of automobility, without having to be confronted by their associated monumental environmental damage.[8]

As told here, an important part of the political and economic environment within which Butte's various stories of conflict took place at the turn of the century was the nationwide transition from competitive, proprietor (or entrepreneurial) capitalism to corporate (or managerial) capitalism. A key factor in this transition was the corporate reorganization of American industry, to use Martin Sklar's words.[9] This shift in the nation's political economy had its consequences for Butte in the form of the corporate history of the Anaconda Copper Mining Company. The extent to which the ACM was able to incorporate various fields of conflict into its own domain was the measure of the extent to which conflict on those several fronts either withered away or remained alive in the hearts and minds of the participants, their descendants, and outside observers. The ACM had eliminated potential conflict with railroad companies over hauling ore from Butte to Anaconda by building its own railroad in 1893. It eliminated considerable unfavorable publicity by controlling the daily newspapers in most of Montana's large cites. It eliminated potential conflict with equipment suppliers by owning its own foundry and fabricating plant in Anaconda. It eliminated potential conflict over environmental damage from its smelter by owning much of the farmland and timberland near the facility.[10] Political and labor conflict continued to erupt after the early twentieth-century period and the completion of the corporate reorganization of the Montana copper industry, however, because the ACM was not successful in thoroughly incorporating some of those matters into its own domain or system.

Most noteworthy, the ACM could not own the miners who worked its mines, mills, and smelters. After a period of relative quiet during the 1920s, labor unrest reappeared in the 1930s. Miners went out in May 1934, and an often-violent strike ensued for several months. The ACM vigorously resisted recognizing the Butte Miners' Union (BMU), but this time the miners had the backing of New Deal legislation passed that year, which gave legal protection to workers' efforts to organize unions. In September 1934, twenty years after the ACM had announced that it would no longer recognize any labor organization representing the miners, the ACM capitulated, again recognizing the BMU as the miners' representative in labor negotiations.[11] From then until the Anaconda Company ceased mining at Butte in 1983, unions would continue to present the company with formidable resistance.

Other arenas of conflict disappeared, or at least seemed to disappear, because of the ACM's ability to fully incorporate them under its control. As has been described in the foregoing chapters, environmental conflict nearly disappeared—until the era of more extensive government regulation, beginning in the 1960s—because the Anaconda Company had come to embrace most of the property, or at least key rights to that property, on which it dumped wastes from its smelter. Owning such a complex technological system that managed and manipulated environments, however, the ACM still had to contend with challenges.

Problems with Tailings

How the Anaconda Company came to own the properties and property rights for dumping dust and fumes from smelter smoke was described in the previous chapter about the land exchanges. The story regarding the tailings problem is still to be told. Most of the tailings generated at the Anaconda Reduction Works can be accounted for on the various tailings dumps established by the ACM. Some amount, of course, made its way downstream despite the efforts of the ACM to impound all tailings behind the

Opportunity and Warm Springs Ponds. A significant volume of tailings, however, blew away through a process of wind erosion the ACM called "dusting."[12] The ACM was aware of dusting as an environmental problem at least by the 1910s. A farmer in 1917 reported that tailings had been damaging his hay and pasture for about two years, which corresponded with the time the company began using flotation at the concentrator. One of the methods the ACM used to reduce the potential of damage claims being filed against the company by private property owners was to acquire easements. In the wake of the Bliss case, the company had begun acquiring "smoke easements." By 1917, the ACM routinely acquired tailings easements along with the smoke easements to accommodate blowing tailings.[13]

As tailings were being discharged into the Opportunity Ponds, the random meandering of streams of water through the pond area would leave dry islands, which were prone to dusting, even though a pond was

FIG. 10.1. City of Anaconda Looking East, 1925, showing the city of Anaconda at center, the 585-foot smelter stack at right, and a large cloud of airborne dust blowing off the Opportunity tailings ponds in the background. Photo courtesy of the Montana Historical Society, cat. # PAc 82–62.0283 (2 of 2).

largely wet. As early as the 1930s, the ACM considered dusting its most serious problem in terms of efforts to retain tailings on the tailings fields. One reason cited was the proximity of the ponds to the Montana State Mental Hospital at Warm Springs and the Montana State Tuberculosis Sanitarium at Galen, which had registered complaints against the dusting. It is unclear when the ACM began a concerted effort to control dusting, but in the 1920s, the ACM tried irrigating the tailings fields to keep them from drying out, tried erecting storm fences to discourage the blowing of fines, and tried covering tailings with granulated slag. Keeping the tailings wet proved impractical; the storm fences succeeded in causing some blowing material to form drifts behind the fences but did not keep the material from blowing off the fields when winds gained high velocities; and the slag, too, blew away under high winds. In the 1937, a portion of the water in Silver Bow Creek, contaminated because it flowed out of Butte, was diverted to the head of the Opportunity Ponds to keep more of the tailings field wet, thereby reducing dusting. Water in Silver Bow Creek was acidic, so another function the diversion served was to neutralize the water of Silver Bow Creek with the tailings being discharged from the concentrator at the Washoe smelter, which were alkaline. This reduced the amount of lime that had to be added to the Warm Springs Ponds to neutralize water flowing into the ponds from Silver Bow Creek. The diversion came to be called the "Yellow Ditch," because of the staining that resulted from the high level of dissolved iron in the water.[14]

One can appreciate the extent of metals content in old tailings by examining the actions the ACM took during World War II to reprocess tailings associated with the Butte smelters. As described in earlier chapters, each of the smelters had a concentrator associated with it, and the concentrators had discharged thousands of tons of tailings onto the floodplain adjacent to Silver Bow Creek as it flowed along the base of the Butte hill. When W. A. Clark had sold his Butte Reduction Works to the ACM in 1910, terms of the sale gave him the right to reprocess tailings left on the site. He reprocessed more than a million tons of Butte Reduction Works tailings at his Timber Butte mill between 1916 and 1920. The ACM excavated about 45,000 tons of slimes from the Butte and Boston tailings dump and hauled them to the Washoe smelter in the 1920s. The rest of the B&B tailings and other giant tailings deposits survived in Butte until the 1940s when, as a means of increasing copper output in support of the war effort, the ACM excavated and reprocessed about 1,400,000 dry tons of Butte slimes and sand tailings by hauling them to the Washoe smelter. That included about 660,000 dry tons of B&B tailings, about 313,000 tons of Colorado tailings, 189,000 dry tons of Parrot tailings, and 186,000 dry tons of tailings from the Montana Ore Purchasing Company smelter and concentrator. The ACM continued to reprocess old Butte tailings at these dumps into the 1950s.[15]

Even though the concentration of ores and the disposal of tailings had ceased at Butte in 1930, with the closure of the Timber Butte mill, the Warm Springs Ponds have continued capturing tailings being washed down Silver Bow Creek. Tailings deposited in earlier years along the creek upstream of the ponds still washed downstream during periods of high water, as the course of the creek channel changed, and due to other erosive factors. Even though the Warm Springs Ponds allowed solids to settle out of the waters

FIG. 10.2. Tailings along the Banks of Silver Bow Creek near Rocker, 1947. Silver Bow Creek flows from right to left (east to west) in this U.S. Department of Agriculture aerial photo. The very light material along the floodplain of the creek is tailings from the Butte mills. The Ramsay townsite is the rectangular grid of lots and house at top center. The Miles Ranch, shown in fig. 6.6, is at the left edge of the photo. Photo courtesy of the National Archives and Records Administration, College Park, MD.

of Silver Bow Creek before flowing down the Clark Fork, metals continued to flow past the dams that impounded the ponds. Those metals, including copper, arsenic, lead, zinc, and especially iron, were present mainly in dissolved form. Because of the high dissolved iron content of the water, it had a reddish-brown color. The iron did not precipitate out of the water until other tributaries added alkalinity to the flow. In general, and aside from a break in the dam of the upper pond (no. 2) in 1922, the Warm Springs Ponds were successful in capturing solids that otherwise would have washed downstream. Pond no. 3 was built in the 1950s because ponds 1 and 2 had nearly filled with sediment. Within ten years, pond no. 3 was also filled with sediment. The ACM considered building a fourth pond, but that action was never taken. Instead, the height of the dam for pond no. 3 was increased about five feet in 1969.[16]

With the construction of no. 3 pond, the ACM began adding lime to the water, as necessary, to reduce the acidity of Silver Bow Creek and the Warm Springs Ponds. The addition of lime served to precipitate dissolved iron and copper from the water as well. In the 1970s, the ACM also began experimenting with the addition of coagulants to the ponds to discourage the resuspension of solids during windy periods. Prior to that time, high winds often agitated the water enough to resuspend solids, which would then be discharged from the ponds with the normal overflow. Because of the high content of iron oxides in the settled solids, such resuspension would cause the water in the ponds to turn red, and the red water would be discharged from the ponds. The ACM called such conditions "red water" conditions.[17]

In about 1970, the company responded to a request by the Montana Department of Fish and Game by building a channel west of the ponds called the Mill-Willow Bypass. The channel was meant to convey water from Mill Creek and Willow Creek, which rose in the hills south of the Deer Lodge Valley and did not emanate in Butte. Fish and Game thought those streams were purer than Silver Bow Creek and wanted them to be able to flow into the Clark Fork without having to pass through the Warm Springs Ponds. The bypass proved problematic, however, because it also served to carry excess spring runoff from Silver Bow Creek if the volume flowing into the Warm Springs Ponds exceeded 900 cfs. The spring runoff carried tailings and other solids from Butte and the banks and bed of Silver Bow Creek, and it deposited them in the bypass, choking its channel by the end of the 1980s.[18]

Because the Warm Springs Ponds were kept filled with water, solids that settled to the bottoms of the ponds were usually kept beneath the surface of the water, except on rare occasions, and thus were not prone to drying and "dusting," the problem that plagued the Opportunity Ponds.[19]

The ACM continued to send tailings to the Opportunity Ponds until 1947, when the company built two new tailings impoundments just to the northeast of smelter hill and south of the Opportunity Ponds. Each pond had an area of about 240 acres and was designed to rise to a level 125–135 feet above grade. After 1949, the ACM sent nearly all its Anaconda tailings to the new ponds, called nos. 1 and 2. They were used until the Anaconda concentrator closed in 1964 (described below).[20]

The Situation since World War II

After World War II, many changes at the Anaconda smelter occurred because of new developments in mining at Butte. In the late 1940s, the ACM embarked on the "Greater Butte Project." The purpose of the undertaking was to make underground mining more economical so the company could afford to hoist lower-grade ores. The ACM developed the new Kelley mine, at the heart of Anaconda hill, to employ a technique called "block cave" mining. Much greater volumes of ore could be broken underground and hoisted with much less labor. The Kelley mine went into full production in early 1952, hoisting 15,000 tons of ore per day. Three years later, the ACM changed its name to the Anaconda Company and embarked on an even more revolutionary change in mining practice at Butte. The company opened the Berkeley Pit, and the era of large-scale surface mining in Butte began. When the Berkeley Pit opened, the ACM used haul trucks with capacities of 25 tons or less. Over the next quarter century, the capacity of the trucks the company used increased to 200 tons, although 150-ton trucks were found to be the most efficient. Production from the Berkeley Pit grew from 17,000 tons per day in 1955 to more than 50,000 tons per day in 1982, when the pit closed. In 1973, the company ceased all underground ore production, and all the remaining shafts were closed except the Kelley, which was retained for underground exploration and to service the pumps at the 3,900-foot level.[21]

Opening the Berkeley Pit brought profound changes to Butte. It marked a transition from large numbers of skilled miners, working in relatively isolated and unobserved pairs underground, to a considerably smaller force of truck drivers and equipment operators who worked in a surface environment where their every action could be observed by supervisors. Open-pit mining was dangerous, but it was not nearly as hazardous as underground mining had been. As the pit grew in size, it began to consume ground on which long-standing Butte neighborhoods had developed into tight communities. Expansion of the pit led some in the community to try to stop its advance, but the company held the economic power to continue its development of the pit, despite resistance, which was also muted because most people in Butte recognized that the economic base in mining could not survive without making the transition from underground to surface mining. As a result of the Berkeley Pit, Butte was able to retain its identity as a mining town, but it did so with a greatly reduced workforce of altered character.[22]

The year the ACM commenced excavating the Berkeley Pit, the Montana legislature passed the Water Pollution Act of 1955, signaling a growing public awareness of the importance of protecting clean water, but also demonstrating that the balancing doctrine was still shaping legal thinking. The act instructed the Montana Water Pollution Council to classify all the streams in the state according to water use: water supply without treatment, water supply with treatment, recreation and swimming, growth and propagation of fish and other aquatic life, or agricultural and industrial use. The act explicitly exempted Silver Bow Creek from classification because the stream had continuously been used primarily as a receptacle for industrial waste for more than thirty years. The exemption extended from Butte to the mouth of Warm Springs, at which point the stream was known as the Clark Fork, not the Deer Lodge River. Anaconda worked to try to improve water quality in the Clark Fork by attending more diligently to its discharges of water from Butte and Anaconda, incorporating virtually the entirety of Silver Creek into its water treatment system. Anaconda centralized its precipitation plant in Butte to recover copper from the daily average of 5,000–5,500 gallons of mine water pumped from the mines with a pH of 2.5–4.0.[23]

At the precipitation plant, the mine water was run through tanks filled with scrap iron (mostly "tin cans" collected from cities throughout Montana). The iron and copper traded places, with the iron going into solution and the copper settling as a sludge called cement copper, to be gathered and treated at the smelter. The effluent from the precipitation tanks had a pH of 3.2–4.5. Some of it was pumped back into the mines to produce more mine water for copper recovery, and some was discharged into Silver Bow

Creek, now with a decreased copper content and an increased iron content. By treating mine water by precipitation, Anaconda was able to produce as much as a million pounds of copper per month. While the industrial wastewater flowing out of Butte was acidic, the water being discharged with tailings from the concentrator at Anaconda was alkaline. After the solids in the tailings settled in the Opportunity Ponds, the alkaline water discharged into the river system was allowed to mingle with the acidic water flowing down from Butte, and the two streams largely neutralized each other as they flowed into the Warm Springs Ponds. Water flowing from there into the Clark Fork averaged a pH of 7.2–7.4, slightly more alkaline than neutral.[24]

Commenting on the accommodation the Montana legislature made in 1955 for Anaconda's polluting operation, the head of the Chemical Engineering Department at Michigan State University wrote, in a 1961 article on Anaconda's pollution control methods, "As awareness of the [Clark Fork] pollution grew, the State of Montana was, in this writer's opinion, neither backward nor unduly aggressive. The well-being of the state has always been largely dependent on its mining activities, and state officials recognized that these should be preserved."[25] The writer went on to describe a benefit of the industrial pollution. All of Butte's sewage, which included not only human waste but also the waste from a meatpacking plant, dairies, a brewery, and a cannery, flowed into Silver Bow Creek without treatment, but the stream was so acidic that it sterilized the organic wastes. Montana's attorney general classified the stream as an open sewer but considered the situation acceptable because alternative swimming facilities were available for children nearby. Similarly, the municipal waste of the city of Anaconda mixed with the tailings, which were extremely alkaline. Thus, neither city had to incur the expense of sewage treatment.[26]

There were at least two episodes in the postwar period of increased downstream discharge of dissolved metals because of strikes by Anaconda workers. During strikes, the unions would allow maintenance personnel to operate the pumps in the mines, but the unions considered the precipitation plant to represent production by the company, because it yielded copper. During a 1959 strike, the unions demanded that operation of the precipitation plant cease, resulting in the pumping of acidic mine water, laced with dissolved metals, directly into the head of Silver Bow Creek, killing thirty miles of the Clark Fork downstream. After the strike settled and Anaconda put the precipitation plant back into service, it took the river six years to recover. The situation arose again during a strike in 1967. Knowing what had happened in 1959, the State Fish and Game Department sought a restraining order from the court, and Anaconda briefly put the precipitation plant back into service, but the unions called the measure strikebreaking and threatened to prohibit maintenance men from tending the pumps, which would have flooded the mines. Anaconda again closed the precipitation plant and tried to store mine water in ponds until the strike ended.[27]

By the early 1950s, the ACM had realized that the relatively new ponds no. 1 and 2 at Anaconda would be filled more quickly than had originally been anticipated, because of the advent of mining methods at Butte that could economically extract large volumes of low-grade ore. First with the Kelley mine, which employed block-caving methods to extract large tonnages from underground, followed by the Berkeley Pit, the ACM greatly increased the volume of ore it shipped to the Anaconda Reduction Works. Moreover, concentration of those ores produced a higher proportion of tailings relative to concentrates (the concentration ratio was now about 20:1, ore to concentrates), further accelerating the rate at which tailings were discharged at Anaconda. Finally, some of the ores being produced from the Berkeley Pit could not be effectively treated using the methods employed at the Anaconda concentrator, and the company would have to develop new methods to maintain economical operations. The Anaconda Company therefore built a new concentrator at Butte. It was named for Clyde E. Weed, who had entered the Anaconda enterprise when he worked at Inspiration, Arizona, and moved to Cananea, in Mexico, before being named general manager of mines for the Anaconda group of companies in 1938, then vice president in charge of mining, and finally president of the Anaconda Company in 1956. When the Weed concentrator went into operation in 1964, the company closed the Anaconda concentrator permanently. Now, instead of hauling 20 tons of ore from Butte to the Anaconda smelter by rail to yield a given amount of copper, the company would have to ship only 1 ton of concentrate from Butte to produce that copper.[28]

The next significant changes in the Anaconda Company's structure occurred in the 1970s, and on two fronts. The company's financial situation received a devastating blow when Salvador Allende's government in Chile nationalized the copper mines there in 1971. Anaconda lost its Chuquicamata and El Salvador mines. Chuqui, as it was called, was the world's largest copper mine, and in most years it accounted for about two-thirds of Anaconda's profits. With support from the U.S. government, Augusto Pinochet mounted a violent coup in 1973, deposing Allende and installing a military dictatorship. A year later the new Chilean government paid some compensation to Anaconda for the nationalized properties, but that did not make up for having lost its most profitable ongoing operation.[29]

Reeling from the blow of losing its Chilean properties, the Anaconda Company became the target for a buyout. Crane Company and Tennaco made attempts to acquire the ailing company before an oil company, Atlantic Richfield (ARCO), acquired 27 percent of Anaconda's stock in March 1976. That summer ARCO announced that it would attempt to acquire the rest of Anaconda's stock. In November, Anaconda's stockholders approved a merger with ARCO. The merger was completed in January 1977, and the Anaconda Company became a wholly owned subsidiary of ARCO. As such, Anaconda continued to operate the Berkeley Pit, the Anaconda smelter, and the refinery at Great Falls, as well as Anaconda's remaining subsidiary operations. In 1979, ARCO began excavating a new copper-molybdenum surface mine at Butte, called the East Continental Pit. It is located at the base of the East Ridge, east of the Berkeley Pit.[30]

In 1980, another strike by Anaconda Company employees temporarily closed Anaconda's Montana operations. Several months later, ARCO surprised local management and strikers alike by announcing that both the Anaconda smelter and the Great Falls refinery would be permanently closed and demolished. ARCO briefly reopened the Berkeley Pit and began shipping concentrates to Japan for smelting. On April 22, 1982 (which happened to be Earth Day), ARCO unexpectedly decided to close the Berkeley Pit. ARCO's shipments of concentrates were being penalized by smelters because of the high arsenic content of the ore from the Berkeley Pit, making it uneconomical to operate the old open pit. ARCO continued operating the East Continental Pit, because arsenic was largely absent from its ore, and the molybdenum content enhanced the ore's value. An important part of the corporate action to close the Berkeley Pit was the decision to shut off the pumps at the Kelley mine in 1982. The pumps were located at the 3900 level of the Kelley, nominally 3,900 feet below the surface. The pumps served not only to keep dry the thousands of miles of underground mine workings above that level, but also to keep the Berkeley Pit dry. Immediately, the underground workings began to flood, and soon acidic mine water, laced with dissolved metals, began filling the Berkeley Pit from beneath. In June 1983, ARCO decided to close the East Continental Pit as well, and it offered its Butte properties for sale, thus ending nearly a century of Anaconda Company operations in Montana.[31]

Epilogue

When Ed Dobb's "Pennies from Hell" appeared in *Harper's Magazine* in 1996, mine water was rapidly rising in the Berkeley Pit, fed by groundwater in the underground workings that continued to rise up toward its "natural" level in the absence of pumping from the Kelley mine. As the water grew deeper, the Berkeley Pit increasingly emerged as a powerful symbol of modernity's failure to link humanity's consumption with industrialization's impacts on the environment. Dobb notes that most popular stories about industry are either "heroic valorization or naïve disdain," while paying little heed to impacts. "Especially regarding the use of such limited resources as timber, energy, and metals," he writes, "public debate has become so fragmented that it obscures the connections that tie all Americans to places like the Berkeley Pit, thereby precluding well-considered, honest responses to the uncomfortable questions they raise about desire and complicity, capitalism, and modern culture."[1] To put this dichotomy in moral terms, Dobb quotes Steve Blodgett, a Butte reclamation specialist: "The Berkeley Pit is the receptacle of all our sins."[2]

Thirteen years after Dobb published "Pennies from Hell" he gave the plenary address at the Vernacular Architecture Forum (VAF) annual meeting in Butte. VAF is a professional and scholarly organization that promotes the recording, analysis, and preservation of "everyday" (vernacular) buildings and landscapes. Butte has a wealth of such buildings, but in his plenary, Dobb chose to address the entirety of Butte's built environment: buildings as well as the industrial landscape of mines, waste rock, tailings, and the giant Berkeley Pit filling with contaminated mine water. The pit is the most visible emblem of a mining past that has culminated in the largest Superfund complex in the nation. He advocated for preservation and interpretation of the fulness of Butte's landscapes, the beautiful alongside the ugly, the valorizing of the city's hardworking population along with memorials to those who died in the mines, and yes, the contaminated landscape along with the monumental efforts to remediate the hazardous materials.[3]

Dobb's call fit well with the course some historic preservationists had begun to chart for Butte. When ARCO had decided to close its operations in Montana, the communities of Butte and Anaconda had to struggle to plot new directions. That task was made somewhat easier for Butte when another party purchased ARCO's Butte properties and decided to reopen the East Pit and the concentrator, but both cities have nevertheless had to find new identities to survive in the twenty-first century. Among those new identities are: 1) that, together, Butte and Anaconda form the largest Superfund complex in the nation, and 2) that, together, Butte and Anaconda form one of the largest National Historic Landmarks in the nation. This epilogue argues for the service that Butte and Anaconda can provide for the nation, and indeed the world, by wedding those two identities to help visitors ("pilgrims," to adopt Ed Dobb's term) interpret the deep history of environmental damage that lies behind the technological triumphs that the United States often boasts.

Superfund Comes to Town

In 1985, ARCO sold its Butte mining properties to Dennis Washington, a Missoula-based entrepreneur. ARCO was so anxious to unload its failed Anaconda acquisition that it agreed to sell Washington the Butte assets but to retain most of its environmental liabilities, such as they might come to be. The only environmental liabilities Washington would share with ARCO were those that arose in the areas around Butte that were under active mining permits, which included the Berkeley Pit, East Continental Pit, Weed concentrator, and Yankee Doodle Tailings. Washington formed a new company, Montana Resources (MR), to operate the East Continental Pit and Weed concentrator, and MR put them back into service in 1986, producing copper at the same rate ARCO had produced when it closed the Berkeley Pit in 1982. Anaconda/ARCO had maintained its union contracts until it laid off its remaining 700 workers in Butte in 1983; when MR resumed the Butte operations in 1986, it was with a workforce of only 350 and no union contract, a radical departure for miners in "the Gibraltar of Unionism." Workers agreed to come back to work at a wage 40 percent below what it had been in 1983, but MR included a profit-sharing plan in its new employment agreement. The price of copper was about sixty cents per pound, the break-even point, when MR resumed operations; by the end of 1986 it was up to $1.30 per pound. The price of copper has generally continued to rise, keeping the Butte operation profitable for both MR and its employees. The price of copper reached a record high of $4.86 per pound in May 2021, while the price of molybdenum jumped from an average $8.67 in 2020 to $19.80 in mid-2021. These prices were good news for the company and its workers, who were operating on an annual plan for 2021 based on copper at $3.00 per pound and molybdenum at $8.50 per pound.[4]

The return of profitable mining, however, was not the most profound development for Butte and its sister city, Anaconda. At about the time ARCO was closing its remaining Butte operations in 1983, the U.S. Environmental Protection Agency (EPA) was adding the Silver Bow Creek Superfund site and the Anaconda Smelter Superfund site to the National Priorities List. The areas would become parts of several "Operable Units" of the Clark Fork Superfund complex, the largest such complex in the United States.[5] These new terms, "Operable Units" and "Superfund," would soon be introduced to the common parlance of Butte's and Anaconda's citizens because of passage of the Comprehensive Environmental Response, Compensation, and Liability Act (CERCLA), passed by Congress in the waning days of 1980.

Americans in the 1970s had witnessed (and indeed advocated for) the inauguration of a new regulatory regime, at the state and federal levels, to protect the environment. Congress had passed the Clean Air Act, the Clean Water Act, and reinforcing amendments in the 1960s and 1970s. President Richard Nixon signed the National Environmental Policy Act in January 1970, establishing as public policy that corporations and government agencies had to undertake environmental impact studies for projects that involved federal participation, including such actions as licensing and permitting. By executive order in 1970, Nixon gathered together numerous federal agencies, which had been given regulatory responsibilities under the new legislation, heretofore scattered among many cabinet departments, and he placed them in the new cabinet-level agency, the EPA. This new attention to governmental protection of the physical environment was accompanied by the public in communities throughout the U.S. by celebrating the first Earth Day in April 1970.[6] The new attention also had an immediate impact on the Anaconda Company, whose emissions and discharges would now be regulated. For example, in order that discharges of wastewater from the Butte operations (Weed concentrator and the precipitation plant treating water pumped from the mines) could meet the requirements of the new National Pollutant Discharge Elimination System (NPDES), Anaconda altered its water treatment system at Butte and no longer relied on the Warm Springs Ponds to treat its Butte effluent.[7]

As the public, government agencies, and business enterprises were growing accustomed to the new regulator regimes, however, many people came to recognize that regulation of current industrial operations would not address the legacy of contamination left at sites of previous industrial activities. The dire situation was brought into full public view at Love Canal, near Buffalo, New York, where decades earlier Hooker Chemical Company had buried steel drums filled with dangerous chemicals. In 1953, Hooker Chemical had sold the land to the city for one dollar, without disclosing the dangers lying beneath the surface. The city in turn sold the land to a developer for a new residential community featuring a hundred houses and a school. A quarter century later, as the chemicals leached out of the corroded drums, parents in the community found that a significant number of their children were being born with birth defects. The EPA recognized the probability that there were similar conditions lurking at old industrial sites throughout the nation.[8]

Amid the public outcry, Congress began debating a suitable policy response. Questions included how contaminated sites should be identified, who should conduct the remediation work at contaminated sites, and who should be liable for remediation costs. Congress deliberated into the fall of 1980, with both Republicans and Democrats agreeing that legislation was needed but differing over details. When Ronald Reagan was elected president in November 1980, and Republicans took control of the U.S. Senate, members of Congress in both parties recognized that the needed legislation must be passed in "lame duck" session, for fear that Reagan, who was generally opposed to federal environmental regulations, would not sign any new bill if it were to be passed after he was inaugurated. Republicans and Democrats, members of the U.S. House of Representatives and the U.S. Senate, quickly compromised to pass CERCLA, and President Jimmy Carter, himself now a "lame duck" as well, signed the act in December 1980.[9]

The new law had several salient features. Thinking that most of the sites in the U.S. requiring remediation would be from the petrochemical industries, Congress imposed a tax on businesses in those industries in order to establish a trust fund from which the EPA could pay for remediation at sites for which the agency could not find responsible parties able to pay. This fund was called the "Superfund," and that name quickly was applied to the entire remediation program established under CERCLA. The act ordered the EPA to undertake a nationwide study, using a system for rating hazards, to identify sites requiring remediation and to place them on a National Priorities List (NPL). The EPA was then to look for Potentially Responsible Parties (PRPs) who were responsible for contaminating each site and who had the financial capacity to pay for remediation. Congress determined that the EPA would use the concept of "strict liability" when identifying PRPs, meaning a party could be found liable to pay for remediation whether or not it had been negligent. EPA could also use the concept of "joint and several" liability, meaning that, if a party was responsible for part of a problem requiring remediation, the party could be found liable for all of it, if it was not possible to distinguish that party's harm from the harm caused by other parties. If that party thought others should share the liability, it could take legal action to compel the other parties to pay their share, but it would not be the responsibility of the EPA to determine how to apportion liability. Finally, Congress decided that PRPs would not only pay for the remediation, but also they would perform the work, rather than reimbursing the EPA to do the work. The EPA would issue an order and set the standards to be met to accomplish the remediation, but the PRPs would design and implement the remediation, subject to approval by the EPA.[10]

Congress also gave the EPA enforcement authority, meaning if a PRP refused to pay, the EPA could sue and ask a federal court to compel the company to pay, or the EPA could conduct the remediation and then sue the PRP for "treble damages" (forcing the PRP to reimburse the EPA three times the cost of remediation). This was to encourage PRPs to participate with the EPA in the design and implementation of remediation from the outset, before the PRPs' legal liability was determined in the courts. In its early years, the Reagan administration seemed reluctant to enforce CERCLA, because of Reagan's broad deregulatory preferences. Meanwhile, the public and many state and local officials were recognizing the magnitude of the hazardous-waste problem nationwide, as the number of sites on the NPL grew to more than one thousand (Congress had initially required the EPA to list at least four hundred sites). Many Republican members of Congress favored reauthorizing the Superfund because of the magnitude of the problem nationwide and in their home jurisdictions. Entering a reelection year, President Reagan changed his position on the Superfund in his 1984 State of the Union address, pledging that he would support reauthorization (although he did not offer his own reauthorization bill). Over the next two years, members of both parties and both houses of Congress worked out compromises to strengthen the program. The Superfund Amendments and Reauthorization Act (SARA) of 1986 actually strengthened the Superfund program, giving it, where possible, an emphasis on permanent remediation designs and ensuring its continuity.[11]

These developments at the national level had profound consequences for the Clark Fork watershed, the communities of Butte and Anaconda, and ARCO. In September 1983, the EPA placed four sites in the upper Clark Fork on the NPL: Silver Bow Creek, Anaconda Smelter, Milltown Reservoir, and Montana Pole and Treating Plant. Silver Bow Creek was contaminated by tailings and other wastes from mining, milling, and smelting at Butte. The Anaconda smelter site was contaminated by milling and smelting at the three smelters that had operated at Anaconda from 1884 to 1980. The Milltown Reservoir, more than a hundred miles downstream of Butte and Anaconda and several miles upstream of Missoula, was impounded by the Milltown Dam, built by W. A. Clark in the early twentieth century as a hydroelectric facility to generate power for his nearby sawmill and to sell electricity to customers in Missoula. The reservoir created slack water, allowing solids that had washed downstream from Butte and Anaconda to settle. Analysis of those sediments and the groundwater adjacent to the reservoir showed elevated levels of arsenic. The Montana Pole site was contaminated by wood preservatives at an operation not directly related to the copper industry. The EPA added the Butte Area to the Silver Bow Creek site in 1987, because analyses showed that Butte remained an ongoing source of contamination to Silver Bow Creek. In 1992, the EPA added the Clark Fork, from the Warm Springs Ponds down to the Milltown Dam, to the Milltown Reservoir site. Each of the sites contains several operable units, which are geographical areas that have distinct contaminants and contaminated media that require distinct remediation treatments.[12]

The EPA identified ARCO as the PRP for all but the Montana Pole site. Although ARCO had owned the Anaconda Company as an operating subsidiary at Butte and Anaconda for little more than six years, ARCO was Anaconda's legal successor, which left ARCO to shoulder all the CERCLA liabilities for contamination that the ACM had caused since its creation in the nineteenth century. And through acquisitions and mergers, the ACM had been the legal successor to most of the mining companies that had operated in Butte prior to the ACM's corporate consolidation of mining in the early twentieth century. ARCO resisted accepting liability for the Superfund remediation at the several sites in the Clark Fork watershed, leading the U.S. Department of Justice (DOJ) to file a complaint in the

FIG. E.1. Streamside Tailings at Miles Crossing near Silver Bow Creek's Entry into Durant Canyon. Deposits of tailings along Silver Bow Creek comprised one of the significant Operable Units of the Silver Bow Creek / Butte Area Superfund Site. Tailings had filled the floodplain of Silver Bow Creek since the late nineteenth century, as shown in figs. 6.6 and 10.2. Shortly after the author made this photo, the streamside tailings were excavated from along Silver Bow Creek and hauled to the Opportunity Tailings Ponds. Photo by author, July 2007.

United States District Court for the District of Montana. After long negotiations, the United States and ARCO reached settlement agreements for each of the Superfund sites in the Clark Fork complex, with ARCO agreeing to pay more than $1 billion for remediation. After extensive analysis on each operable unit by both the EPA and ARCO, the EPA has issued a series of remediation orders, and ARCO has undertaken the work. As work is undertaken and the magnitude and fine detail of the contamination is better understood, the EPA and ARCO have continued to negotiate settlements, which have been lodged as consent decrees by the federal court overseeing the work.[13]

The Silver Bow Creek / Butte Area Superfund site has six operable units (OU). Remediation for the Streamside Tailings OU, for example, has entailed excavating tailings and mixed alluvium from the bed, banks, and floodplain of Silver Bow Creek between Butte and Warm Springs Ponds and hauling those materials to the Opportunity Ponds, now designated a permanent waste management area. Because excavation along seventeen miles of stream could cause the release of solids and dissolved contaminants into Silver Bow Creek, or could trigger an even more catastrophic release during a severe weather event, the EPA also ordered significant work at the Warm Springs Ponds OU. The Mill-Willow Bypass was improved, and the earthen embankments that impound the ponds were hardened to protect them during a severe weather event. Work on the Butte Priority Soils OU has excavated or capped considerable contaminant material along Silver Bow Creek, on the Butte hill, and on Timber Butte, and work is ongoing to develop treatment facilities in Butte, including stormwater retention ponds. One operable unit that will not have a permanent solution, in the conventional sense, is the Butte Mine Flooding OU, which includes the Berkeley Pit.[14] The Berkeley Pit is probably the most visible, and indeed notorious, feature of the many features that comprise the Clark Fork Superfund complex.

When ARCO turned off the underground pumps in the Kelley mine in 1982, the underground workings began to flood immediately, as groundwater in the rock formations beneath the Butte hill, which had been artificially depressed by a century of mining and pumping, began to return to its natural equilibrium, which had been near the Earth's surface. As the water rose in the mine workings, it inundated sulfide minerals, especially iron sulfide (pyrite, or fool's gold), in the walls of the shafts, drifts, and crosscuts that had long been exposed to the atmosphere and had thereby oxidized. The interaction of water with the oxidized minerals created sulfuric acid and put metals into solution. Dissolved metals include iron, copper, zinc, manganese, cadmium, and arsenic. Rising acidic water, laced with dissolved metals, reached the bottom of the Berkeley Pit about a year after the pumps had been turned off. Left unchecked, the water in the mine workings and Berkeley Pit would have continued to rise until it overtopped a groundwater divide separating the mine system from an alluvial aquifer and Silver Bow Creek. Such an occurrence would further pollute Silver Bow Creek and the shallow aquifer beneath much of the area of Butte south of the Butte hill. Recognizing this threat, the EPA added the Butte Mine Flooding OU to the Silver Bow Creek Superfund site in 1984. When Dennis Washington's Montana Resources (MR) purchased the Butte operation in 1985, MR accepted a share of the liability for the Berkeley Pit, which sits within the area with active mining permits MR acquired. MR now shares liability for the Berkeley Pit with ARCO.[15]

After studying the situation, the EPA concluded that the only practicable solution to the mine flooding and the toxic water in the Berkeley Pit would be at some point to begin pumping water from the pit before it reached the elevation at the top of the groundwater divide, and to build a water-treatment facility to neutralize the pumped water and remove the metals. After exploring possible methods that might recover the metals from the water in a manner that could generate revenue, the EPA concluded that the best solution would be the most conventional, to neutralize the acidic water with lime, allow the metals to drop out of solution as a useless sludge, and to dispose of the sludge, either back into the pit or to a suitable site on land. The EPA issued its formal Record of Decision (ROD) on the issue in September 1994.[16]

In August 2002, Judge Haddon of the U.S. District Court for the District of Montana lodged a consent decree documenting the settlement the EPA had reached with ARCO and MR (collectively called the Settling Parties). In addition to reimbursing the U.S. for the costs it had incurred in responding to hazards represented by the Mine Flooding OU, the Settling Parties agreed to continue their remediation design pursuant to the ROD, implement the remediation designs in compliance with the standards the EPA had established, and implement and pay for an ongoing monitoring program for the OU. The consent decree stipulated that the Settling Parties would begin pumping and treating water from the Berkeley Pit by such time as to prevent the level of the pit water from reaching what is called the "Protective Water Level," an elevation of 5,410 feet. That level is about 50 feet below the groundwater divide separating the mine aquifer system from the Silver Bow Creek groundwater and surface water system. The Settling Parties would dispose of the sludge created by the water treatment either into the Berkeley Pit or to an approved surface disposal site. Treated water produced by the Settling Parties could be discharged either into Silver Bow Creek or used for industrial processing by MR. The Settling Parties and the EPA agreed that ARCO and MR would continue this treatment of Berkeley Pit water in perpetuity.[17]

The Settling Parties built the Horseshoe Bend Water Treatment Plant along the northeast edge of the Berkeley Pit in 2003. It began treating the Horseshoe Bend surface water (the name given the water that had been seeping from the ground and flowing toward the Berkeley Pit from the north, where the headwaters of Silver Bow had once been and where the giant Yankee Doodle tailings impoundment now sits. Although the Horseshoe Bend water does not come from the Berkeley Pit, it is still contaminated with metals, because it flows from the old Berkeley Pit waste-rock pile. The treatment plant raises the pH of the water to 11 (a pH of 7 being neutral, and below 7 being acidic) and removes metals before sending the water to MR's concentrator (the former Weed concentrator of the Anaconda Company). Over the ensuing years, the Settling Parties also tested the plant with water pumped from the Berkeley Pit. In 2019, the parties finally began pumping and treating water from the Berkeley Pit. Three million gallons per day of pumped water first goes to copper-recovery cells, which remove dissolved copper by precipitation (adding to the copper MR produces at its Butte operation). Then the water is piped to the Horseshoe Bend Treatment Plant for pH adjustment and removal of other metals before sending it to the concentrator to process ore. The water becomes part of a tailings slurry that is pumped up to the Yankee Doodle Pond, where the fine solids are allowed to settle, adding to the height of the tailings impoundment, and leaving relatively clear water in a lake at the north end of the tailings pond. The water is largely free of dissolved metals but has a highly alkaline pH. A portion of this high-pH water, which is pumped back to the concentrator, is piped to a polishing plant along the southwest edge of the Berkeley Pit. The polishing plant uses filters and reverse osmosis to further clean the water so that it is fit to be discharged into Silver Bow Creek.[18]

The Berkeley Pit covers 675 acres (a bit more than one square mile) and is 1,780 feet deep, from its south rim to the bottom, which is at an elevation of 4,263 feet above sea level. When the pit started flooding, the surface area of the pool at the bottom was relatively small. Because the pit is cone-shaped, the surface area of the pool increases as the water rises, meaning it takes more water to bring it up each additional foot. Moreover, as the surface area of the pool grows, it also increases the rate of evaporation. By 1987, the pool had risen to an elevation of 4,850 feet, and by the end of the twentieth century it had risen another 350 feet to about 5,200 feet. Since then, it has risen only about 150 feet more to an elevation of 5,356 feet (as of September 2021). Current estimates are that the Berkeley Pit now holds about forty billion gallons of contaminated water.[19] It will remain a permanent threat to the environment in the upper Clark Fork, which is why the Settling Parties have agreed to treat the water in perpetuity.

Despite the monumental threat posed by the Berkeley Pit, the

FIG. E.2. Berkeley Pit as Seen from the International Space Station. An astronaut took this photo from the International Space Station on August 2, 2006, when the water in the pit was about 1,000 feet deep. The view shows the Berkeley Pit, East Continental Pit, Weed Concentrator, Yankee Doodle Tailings, and a portion of the city of Butte. Since 2006, the water in the Berkeley Pit has increased in depth about 100 feet. Photo courtesy of National Aeronautics and Space Administration, photo # ISS013-E-63766. Labels by author.

remediation of that problem has not been ARCO's most costly in the Clark Fork complex of Superfund sites. ARCO has spent millions of dollars on other OUs, including the Streamside Tailings OU, in which 6.1 million cubic yards of tailings and contaminated alluvial material along Silver Bow Creek from Butte to the Warm Springs Ponds have been excavated and hauled to the Opportunity Ponds Waste Management Area for permanent disposal; on the Butte Priority Soils OU, where tons of material have been removed and other improvements made in efforts to prevent contamination from the Butte hill reaching Silver Bow Creek; and the Warm Springs Ponds OU, where that facility has been enhanced to prevent contamination that continues flowing down Silver Bow Creek from getting into the Clark Fork. Three decades after Butte and Silver Bow Creek were placed on the NPL, and after many of the major remediation projects have been completed, new problems associated with the vast contaminated area continue to be recognized. For example, in the 2010s, groundwater testing around the location of the former Parrot smelter and its tailings dump revealed considerable contaminated material just below the surface and a plume of some of the most contaminated groundwater known to exist in Butte. Water samples taken from monitoring wells were turquoise in color because the water contained so much dissolved copper. The problem led to contentious negotiations among ARCO, EPA, state government, local government, and community groups about what should be done about the problem. The parties negotiated resolutions to those new problems, including excavating the buried, residual Parrot tailings and disposing of them into the Berkeley Pit, and the parties signed the Butte Priority Soils Consent Decree in May 2020.[20]

The Anaconda Smelter Superfund site covers three hundred square miles at the south end of the Deer Lodge Valley, including the site of the Washoe smelter and the sites of the previous two Anaconda smelters, the Upper Works and the Lower Works. This Superfund site has five OUs. The largest is the Anaconda Regional Waste, Water, and Soils OU, which includes contaminated groundwater, the contamination in Willow, Mill, Warm Springs, and Lost Creeks, smelter hill, the Opportunity Ponds, Anaconda Ponds, and the giant piles of granulated slag along the north side of smelter hill. The Old Works / East Anaconda Development Area OU includes the surviving ruins of the Old Works' foundations, and various collections of smelter wastes on the bottomland on both sides of Warm Springs Creek adjacent to the old smelter sites. Rather than excavate all that material, ARCO and the EPA developed a plan to cap the materials in situ, establish vegetation, and manage the area as a golf course, in order to prevent precipitation from percolating down through the contaminated material and into the groundwater. To design the golf course, ARCO contracted with Jack Nicklaus, the famous professional golfer who, in retirement, had formed a successful golf-course-design firm. This novel remediation plan, the Old Works Golf Course, which opened to the public in 1997, became one of the more widely publicized of the remediations implemented in the Clark Fork Superfund complex. A novel feature of the golf course is that its bunkers are filled with granulated black slag rather than conventional sand, giving the golf course its distinctive appearance. ARCO also helped to develop a paved interpretive trail that allows walkers to travel along the hillside north of the Upper Works for better views of the smelter ruins.[21]

The Milltown Reservoir Superfund site has received perhaps the most monumental movement of material among the three sites in the Clark Fork complex. After the discovery of excessive levels of arsenic in wells serving the community of Milltown, adjacent to the reservoir, and a period of considerable advocacy by community groups in Missoula (about seven miles downstream of the dam), the EPA and ARCO agreed that ARCO would implement a plan to demolish the Milltown dam and powerhouse, dewater the sediments (tailings and alluvium) in the reservoir behind the dam by temporarily diverting the river around the site, excavate and transport those sediments ninety miles by rail to the Opportunity Ponds, and permanently dispose of them there. The dam destruction and sediment removal began in 2006 and was completed in 2008. About 2.3 million cubic yards of sediment were transported from the Milltown Reservoir to the Opportunity Ponds. Two years later, after the site was prepared to receive a free-flowing river, the Clark Fork was diverted back into the area where its channel had once flowed. The other OU in this Superfund site is the Clark Fork River OU, in which streamside tailings along about 120 river-miles of the Clark Fork are being excavated and transported to the Opportunity Ponds.[22]

As of this writing, word has it that ARCO has spent more than $1.5 billion on the three sites in the Clark Fork Superfund complex associated with the mining industry, and there is still remediation to complete and costs for ARCO to incur.[23] Some of the remediation has created areas suitable again for commercial development, agricultural uses, or recreational activities. Other sites, like the Berkeley Pit, can never be left alone and will require perpetual human attention to prevent environmental devastation expanding beyond the underground workings in Butte and the confines of the giant open pit. The legacy of contamination initiated by a small gold rush to Butte in the 1860s will live on forever.

Butte and Anaconda: National Historic Landmarks

The U.S. Secretary of the Interior designated Butte a National Historic Landmark (NHL) in 1961 for its significance under the theme of Westward Expansion (Mining Frontier of the Trans-Mississippi West). This last section of *Smoke and Tailings* argues for adding more than that theme to Butte's (and Anaconda's) significance as a National Historic Landmark. I echo Ed Dobb's assertion, in the opening epigraph of this book, that Butte should be added to Concord, Gettysburg, and Wounded Knee as places to which Americans make pilgrimage to understand our nation's past more deeply. As citizens, we need to know the legacy of environmental devastation that stands behind our technological achievements, and the largest Superfund complex in the U.S. is an excellent place to do that. As Jeremy Zallen's *American Lucifers* shows us, human suffering and oppression helped form the foundation of America's quest for enhanced artificial lighting. Likewise the landscapes of Butte and Anaconda show us the environmental devastation at the base of industrial society's appetite for metals like copper.

The federal government had been designating national landmarks for decades before designating Butte in 1961, and Butte's designation also occurred before 1966, when Congress passed the National Historic Preservation Act, under which the National Park Service developed standards for documenting the boundaries of historic districts. As a national landmark, Butte had as yet no definitive boundary. It also had no detailed analysis of the features within the boundary that contributed to the landmark status and those that did not.[24] Nevertheless, Butte stood in a remarkable state of preservation entering the 1970s, because the community had undergone a gradual but steady process of economic decline since its peak population during the First World War. Consequently, Butte's Uptown central business district and many neighborhoods on the Butte hill had been preserved through the process of benign neglect.

In the early 1970s, however, Butte's Uptown central business district (CBD) and some neighborhoods near it faced a new potential threat. As the Berkeley Pit grew in area and depth, many of the city's residents believed the expansion would inexorably swallow the Uptown, and the Anaconda Company encouraged the notion. Interest groups in the community, including city government, the Anaconda Company, and prominent Uptown businesses formed a commission to study the issue, with an eye toward an orderly abandonment of the historic Uptown and the construction of a new, planned CBD somewhere on the Flats, hoping to enlist significant federal aid in the undertaking. The plan was called Butte Forward. By the time the Anaconda Company announced in 1975 that it probably would not consider mining the ground under the Uptown for at least twenty years (and even then it would probably be by underground methods and not by an expansion of the open pit), Butte Forward had gained enough momentum that its proponents continued to press for relocating the CBD. In mid-1976, Butte Forward was ready to release its preliminary plan, including an announcement of its preferred site on the Flats for the new CBD. Owners of a few small businesses in the Uptown opposed the plan, however, as did much of Butte's populace. The city council entertained the proposal and held public hearings on the matter, but in July 1976, the council voted resoundingly to reject the proposed relocation.[25]

Another important transaction was completed in 1976 as voters approved a consolidation of the city and county governments into a new legal entity, Butte-Silver Bow, governed by an elected chief executive and a twelve-member council of commissioners. After the Butte Forward proposal was rejected, Mayor Mario "Mike" Micone, who then became the first chief executive of Butte-Silver Bow, determined that it was in the community's best interest to revitalize the Uptown, after years of neglect and catastrophic fires that had destroyed some prominent business blocks in the district. That effort was continued by Micone's elected successor, Don Peoples, who had been director of Butte's Public Works Department. While attending to revitalization efforts in the Uptown, Peoples also began some publicly sponsored efforts to diversify Butte's economy. Under Peoples's direction, Butte-Silver Bow sponsored a summer "Rehab Action Team" of the Historic American Engineering Record (HAER, a small unit within the National Park Service), which brought a team of about a dozen architects, planners, and historians to Butte in 1979 to document historic buildings in the Uptown and to propose adaptive reuses, which could be partially financed by federal tax credits now available for rehabilitating historic structures under the Tax Reform Act of 1976. Building on the HAER work, the local government established a tax-increment financing district for the Uptown, whereby any additional tax revenues generated for the local government by revitalization, and the resulting increase in taxable value, would be invested directly back into public projects to enhance the Uptown district. Butte-Silver Bow created the Urban Revitalization Agency (URA) to administer the tax-increment financing plan, and Peoples hired Janet Cornish, a local resident who had worked on the HAER team, to be the URA's first director. Within a few years, the council of commissioners also passed a historic district ordinance to protect the NHL district and specifically the fourteen iconic headframes that stand on the Butte hill.[26]

The Butte Historical Society contributed as well. Bill Walker, a local nurse and society member, proposed in 1979, with society support, that the local government establish a public archives and house it in the old Quartz Street Fire Station (built in 1900), which was no longer used and which the local government had slated for demolition in order to expand a parking lot. Butte-Silver Bow accepted the proposal, and Peoples hired Walker to work part time as director of the Butte-Silver Bow Public Archives. Walker and subsequent directors, working with Butte-Silver Bow and volunteers, continued to make improvements to the old fire station and to expand the collections of local government records, Anaconda Company records, newspapers, other business records, and manuscript collections. In 2007, Ellen Crain, director of the archives, proposed that Butte-Silver Bow put a $7.5 million bond issue before the voters, to pay for restoration of the fire station and to build an addition, nearly doubling the floor area of the facility and furnishing it with state-of-the-art archival equipment. The voters approved the bond issue by the widest margin ever in a Butte-Silver Bow bond election, and the resulting building is the finest archive facility in the state (pending expansion of the Montana Historical Society's building in Helena, which is currently underway). The Butte Archives attracts a wide audience of local researchers, students, and educators—as well as scholars from around the world.[27]

Because the Butte NHL was established before documentation standards had been created under the National Historical Preservation Act of 1966, no formal determination had been made of which buildings in Butte merited the protections from federal actions afforded by the act, or which were eligible for the tax credits under the Tax Reform Act of 1976. To document Butte's historic district to those standards, the Butte Historical Society applied for and received grants from the Montana State Historic Preservation Office (SHPO) in 1981–1983 to hire a single employee each summer to survey a particular neighborhood. In 1984, the SHPO awarded the Butte Historical Society a larger grant that allowed the society to hire nine summer workers, who completed the architectural and historical analysis of the NHL district. The following year, the SHPO awarded the Butte Historical Society a matching grant (with local matches provided by the Butte Historical Society, the Anaconda Historical Society, and ARCO) to retain a local historic preservation firm, Renewable Technologies Inc. (RTI) to develop a master plan for the preservation and interpretation of historic industrial sites in Butte and Anaconda and along the rail corridor between the two cities. Inspired by the National Park Service's development of the historical park system at Lowell, Massachusetts, which focuses on sites there associated with the textile industry, the plan for Butte and

Anaconda suggested a range of preservation and interpretation projects for sites ranging from headframes in Butte to the 585-foot stack, which still stands atop Anaconda's smelter hill but cannot be visited by tourists. RI's master plan identified seven themes for organizing interpretation, one of which was environmental degradation and remediation.[28]

Both projects of the Butte Historical Society bore further fruit. Preservation advocates in Butte continued to press ARCO and the EPA to heed the historical significance of many of the cultural resources at sites requiring remediation under the Superfund program. In 1992, ARCO joined with the EPA and state and local governmental entities to sign a programmatic agreement under which ARCO would fund a study intended to develop a Regional Historic Preservation Plan (RHPP), which would guide the parties as they prepared plans to implement the numerous Superfund remediation projects needed for the area. Numerous people from agencies and community groups worked together to prepare the plan, with Janet Cornish serving as project manager. The RHPP was issued and adopted by the parties to the programmatic agreement in 1993. It included many of the preservation and interpretation ideas first proposed in the RTI master plan, but because the RHPP enjoyed the direct participation of ARCO and the EPA, it became the road map for much of the treatment that cultural resources associated with the Superfund remediation would receive.[29]

To date, several of the proposals offered by RTI's master plan, and forwarded by the RHPP, have been implemented, including a viewing stand and memorial overlooking the headframe of the Granite Mountain mine, scene of the deadly 1917 disaster that killed 165 miners; the Old Works Interpretive Trail in Anaconda; and a state-park viewing stand at the base of Smelter Hill that interprets the Washoe smelter stack, now owned by the State of Montana as a state monument. Numerous other sites have been made accessible to the public and interpreted. Butte and Anaconda jointly created a program, called Montana's Copperway, to create a logo and a network of consistent signage at historic sites, such as the park and trails in the vicinity of the Mountain Con mineyard and the trails at the Warm Springs Ponds, managed as a fish and wildlife refuge. The grounds of the Original mineyard have been improved to serve as a kind of amphitheater, with the stage located under the headframe. The facility serves as the main stage for the Montana Folk Festival and other outdoor concerts, and during those events the hoist house at the Original is often opened for tours.[30]

The challenges of integrating preservation of a mining landscape with the remediation of that same landscape are immense, as analyzed by Alison LaFever. A historic preservationist in New York City, she analyzed the preservation issues Butte faces in her 2012 master's thesis at Columbia University. She cites Ed Dobb in emphasizing the importance of preserving the mining landscape, in the face of the driving force that environmental remediation exerts in influencing decision-making: "It's also true that, in the attempt to repair the ravaged land, the poisoned water—a noble and necessary task, certainly—we run the risk of burying or erasing and, therefore, systematically forgetting what we most need to remember. Reclamation [is] a kind of amnesia."[31] She identifies sluggish administrative mechanisms as one of the major hurdles blocking significant preservation of the mining landscape, and she suggests that Butte's mining landscape needs stronger institutional support to overcome those hurdles.[32]

One small step the preservation community has taken to gain institutional backing has been the designation of Butte and Anaconda as a National Historic Landmark. The Butte Historical Society's survey of historic structures in the Butte NHL laid an important foundation for an expansion of the landmark district to encompass the communities of Walkerville and Anaconda, as well as the rail corridor between Butte and Anaconda. The Montana Preservation Alliance (now called Preserve Montana) coordinated an effort funded and supported by the National Park Service, the Montana State Historic Preservation Office, the Butte-Silver Bow Public Archives, and the local governments of Butte-Silver Bow and Anaconda-Deer Lodge. After fourteen years of effort and the work of more than fifty contributors, the U.S. Secretary of the Interior officially approved the expanded NHL, now called the Butte-Anaconda National Historic Landmark District, in April 2006.[33] Groups in Butte and Anaconda continue working toward designation of a Butte-Anaconda national historical park system, patterned after the Keweenaw National Historic Park in Michigan.

The epigraph from Ed Dobb (1950–2019) at the beginning of this book is from a short essay in which, after summarizing the staggering environmental ruin caused by the copper industry over the past century and a half, he suggests that Americans should make pilgrimage to Butte, to see where our nation comes from, as much as it has come from Concord, Gettysburg, and Wounded Knee. Dobb's words comprise the introductory essay to the catalog and exhibition guide for *Extraction: Art on the Edge of the Abyss*, a series of exhibitions, performances, installations, readings, and other events that he, together with Peter Koch and Samuel Pelts, organized to occur at galleries, stages, and other venues around the U.S. in 2021 to spotlight, through art, the magnitude of the impacts that extractive industries have had on the planet. The Berkeley Pit was the inspiration for the undertaking. Sadly, Ed Dobb died before the exhibitions and other events took place.[34]

That same year, art historian Betsy Fahlman curated an exhibition at the Phoenix Art Museum titled *Landscapes of Extraction: The Art of Mining in the American West*. The exhibition is mostly a collection of paintings—but also contains photographs—depicting landscapes and work environments created by mining and petroleum extraction. The artists represented, from across the twentieth century and into the twenty-first, have been inspired by the awesome beauty of landscapes scarred by industry on such large scales, and by the stark realities of the human communities that live in landscapes of extraction.[35]

Both *Extraction* and *Landscapes of Extraction* are stunning accomplishments, in part because of what David Nye calls the "technological sublime." He uses the word "sublime" not in the way we have come to use it colloquially, as in piece of chocolate cake that we might call sublime, but rather as a way of describing something that fills us emotionally with a sense of awe, by both its magnitude and the sense of horror it can instill. We find nature sublime when we behold the Grand Canyon. Nye argues that Americans have come to find certain aspects of technology, both in its functioning and in its consequences, also to be sublime. And that applies to the negative aspects of technology, like the atomic bomb, as well as the positive. Developing an appreciation for the technological sublime, according to Nye, has helped to shape the American identity.[36] The technologies employed in the extractive industries are part of the array that comprise the American character. We must acknowledge and learn from the destruction they have wrought in addition to appreciating the material benefits they appear to have given us.

Philosopher Susan Neiman's book *Learning from the Germans* analyzes how, in recent decades, many Germans have learned to talk about—to work through—the meaning of the fact that the Nazi era is part of their past, part of their legacy and heritage. She commends the Germans for having done that and for continuing to work on it, and she uses the Germans as a model for assessing how we have fared in the U.S. in working through the facts of our own history of slavery, of Jim Crow, and of lynchings. For example, she commends the State of Mississippi for its development of the Mississippi Civil Rights Museum in Jackson, which opened in 2017. The museum does a remarkable job of interpreting the state's background in the institutions of slavery, of Jim Crow, of voter suppression, of educational inequities for children, and it interprets in that context and in great detail the several facets of the Civil Rights Movement in Mississippi.[37]

Many Germans, especially those who lived through the Second World War, would rather not have talked about their Nazi past, but now many Germans are engaging that history to help them negotiate the new realities that face their nation, such as the many refugees that seek entry into their country. Despite the rise of a new nationalism in Germany, that nation has been among the most welcoming of refugees, in part of because of an awareness of its past. Many Americans, especially those who worked to maintain the institutions of Jim Crow, would rather not discuss the horrible realities of our nation's racist past, but now many Americans are working to address the ways that history is embedded in our institutions in the form of systemic racism. Both of these contemporary examples exhibit how people in Germany and the U.S. are drawing moral connections between their history and the present as they seek to shape their future. Expanding on that theme, books like Zallen's *American Lucifers* draw moral connections between the technologies that enhance day-to-day lives, like artificial lighting, and the people who toiled at the base of the supply chains for that lighting. How can we draw comparable moral connections between the technologies that we take for granted and the environments that we sacrifice to sustain those technologies?

I do not mean to draw a direct moral equivalence between our abuse of the environment and the Holocaust, or the institution of chattel slavery and Jim Crow, but we must also examine the moral reverberations that extend from what our extractive industries have done to environments like Butte, Anaconda, and the Clark Fork. The extraction of copper, and of iron, coal, petroleum, natural gas, and the array of other minerals we have learned to exploit, has provided one of the bases from which together we are pushing Earth's climate out of the "Goldilocks Zone" that helped sustain the development of civilization as we know it.[38] As the planet warms and weather becomes more extreme, nations and communities that did little to contribute to global warming are the first to suffer, while we who have enjoyed the fruits of industrialization can easily remain oblivious, much as one hundred years ago people in large cities, enjoying the artificial light brought to them by copper, could remain oblivious of environmental harm caused in southwest Montana by the mining and processing of that copper.

It is in this vein that the Butte-Anaconda NHL can contribute to the maturing of our collective memory and national character. The Butte-Anaconda NHL is more than a marvelous collection of neighborhoods, business blocks, churches, and industrial structures like headframes, hoist houses, and smokestacks, although it certainly is all those things. It is also a cultural landscape that has been ravaged by Americans' insatiable appetite for copper and other metals. It is a cultural landscape that exhibits the power that giant corporations have exerted to produce profits and generate wealth for distant shareholders while destroying the lives of plants and animals, mountains and watersheds, and the agricultural communities that sought to husband those organisms. It is also a cultural landscape now shaped by a national policy to try to remediate some of the legacy of environmental devastation left in the wake of industrial extraction, only to find that some of the environmental wounds cannot be healed and will require treatment in perpetuity.[39]

Fig. E.3. Anselmo Mineyard. There are thirteen steel headframes surviving in Butte. Several stand alone, and several still have an associated hoist house standing. The Anselmo is the most intact mineyard in Butte. This view to the northwest shows the headframe (steel-truss structure at the left), a warehouse (gabled building at center), and the main hoist house (gabled building at the right). The grassy slopes represent the way Superfund remediation has altered much of Butte's landscape by replacing bare surfaces with grass. Photo by author, July 2013.

FIG. E.4. Berkeley Pit, as Seen from the Berkeley Pit Viewing Stand. The Berkeley Pit has carved away a significant portion of the east flank of the Butte, which forms the background of this view. Note the headframes visible atop the north rim of the pit. Water that has flooded the pit is visible in the lower right corner. Photo by author, July 2013.

FIG. E.5. Silver Bow Creek Flowing through a Slag Canyon Built by the Butte Reduction Works in the Early Twentieth Century. As shown in fig. 6.8, workers at the Butte Reduction Works built these slag walls by pouring molten slag into steel forms. The slag walls at the former site of the Butte Reduction Works are the only landscape feature surviving from the several smelters that once operated in Butte. Photo by author, July 1995.

Fig. E.6. (*Left*) Warm Springs Ponds. Pond 2 is to the right and Pond 3 is to the left, with the earthen embankment that separates them running across the center of the photo. Note the 585-foot stack of the Anaconda smelter in the center of the background. Photo by author, July 2013.

Fig. E.7. (*Below*) Stack of the Anaconda Smelter. Built in 1918, the stack is 585 feet tall. The two slim openings in the concrete base of the stack admitted flue gases after they had passed through the Cottrell electrostatic precipitators, which were demolished along with the rest of the Anaconda Smelter in the early 1980s. The stack stands today as a Montana State Monument. Photo by author, November 1984.

FIG. E.8. Old Works Golf Course, Anaconda, Montana. Designed by Jack Nicklaus, the golf course was built over the old tailings of Anaconda's Upper Works to control surface water and prevent it from passing through the old tailings and into the groundwater. Note the black hazards, made of granulated slag rather than the traditional beach sand. This photo to the southeast was taken from the Old Works Trail, a paved interpretive trail that allows walkers to view the ruins of the Upper Works furnaces and flues. Photo by author, June 1997.

The Butte-Anaconda NHL is a monument to all those unseemly aspects of our history. We need to look them in the face and to talk about them. The Berkeley Pit and Yankee Doodle Tailings, the slag walls at the site of the Butte Reduction Works, the restored banks, bed, and floodplain of Silver Bow Creek, Warm Springs Ponds, Opportunity Ponds, the Anaconda stack, and the Old Works Golf Course all bear witness to conflicting truths of the American past. Interpretation of those and other sites may be able to shape hearts and minds of Americans as together we try to chart more sustainable paths into the future.

Notes

Introduction

1. Ed Dobb was born in Butte in 1950, worked in journalism elsewhere, and returned to Butte in the mid-1990s as the public was becoming more aware that the Berkeley Pit was filling with highly contaminated water, and aware of the environmental ramifications of that fact. He wrote for *Harper's Magazine* while living in Butte. He was teaching journalism at the University of California, Berkeley, when he died in 2019. Prior to his death, he began organizing, with others, a series art exhibitions called *Extraction: Art on the Edge of the Abyss*. Mounted at galleries around the United States in 2021, the exhibitions, featuring works by various artists, depict the monumental scale of impacts the extractive industries have had on landscapes.

2. Fiege, *Republic of Nature*. Mark Fiege narrates nine iconic episodes in American history, ranging chronologically from the Salem witch trials to the oil crisis of the early 1970s, showing how intrinsically and dynamically the episodes were intertwined with their natural environments. The argument in *Smoke and Tailings* is that Butte, as an iconic place in American history, can also be best understood through its environmental history.

3. Zallen, *American Lucifers*.

4. "Epilogue," in Zallen, *American Lucifers*, 256–271.

5. Dobb, "Pennies from Hell," 39–54. In this essay also, Dobb calls for Butte to be added to Concord, Gettysburg, and Wounded Knee as places to which Americans should make pilgrimage.

6. The Granite Mountain disaster is described in chapter 7 of this book. Butte has long been recognized as a rough place to live, and numerous articles have been written about the various facets of rugged social life; see, for example, Murphy, "Bootlegging Mothers and Drinking Daughters," 174–194; Scheidler, "Failure to Protect," 29–40.

7. Howard, *Montana: High, Wide, and Handsome*, 85.

8. Malone, *Battle for Butte*.

9. Marcosson, *Anaconda*.

10. Calvert, *Gibraltar*; Emmons, *Butte Irish*; Murphy, *Mining Cultures*; Punke, *Fire and Brimstone*; Swibold, *Copper Chorus*; Finn, *Mining Childhood*; Basso, *Meet Joe Copper*; Leech, *City That Ate Itself*.

11. Mercier, *Anaconda*.

12. Smith, *Mining America*; LeCain, *Mass Destruction*; Curtis, *Gambling on Ore*.

13. MacMillan, *Smoke Wars*; Wirth, *Smelter Smoke in North America*; Bloom, *Murder*

of a Landscape; Maysilles, *Ducktown Smoke*; Morin, *Legacy of American Copper Smelting*; Mix, *Leaded*; Sullivan, *Tainted Earth*; Snow, *Living with Lead*.

14. McNeill and Vrtis, *Mining North America*. The Reserve Mining Company dumped tailings from its taconite (iron) plant into Lake Superior, leading to a monumental lawsuit because of the apparent presence of asbestos in those tailings. Nonferrous metals, on the other hand, are typically derived from sulfide minerals, which oxidize when exposed to the atmosphere. As a consequence, sulfide mining operations can generate acid mine drainage flowing from abandoned mines and acid rock drainage flowing from old waste-rock piles and tailings deposits associated with mines.

15. Smith, *Mining America*, 67–73, 117–119, 129–132; Kelley, *Gold vs. Grain*; Rohe, "Man and the Land," 299–338.

16. Bakken and Bakken, "Goldfish Died," 38–51; Sandlos and Keeling, "Claiming the New North," 5–34.

17. Hardesty, "Evaluating Site Significance in Mining," 42–51; White, "Heads, Tails, and Decisions In-Between," 47–66; Gohman, "It's Not Time to Be Wasted," 5–22; Quivik, "Historical Significance of Tailings," 35–54; Quivik, "Nuisance, Source of Wealth," 41–64; Quivik, "Renaturalization and Industrial Heritage," 202–221.

18. Rosenberg, *Inside the Black Box*.

19. Lankton, *Cradle to Grave*.

20. This paragraph and the next are a brief synopsis of Hugh Gorman's excellent book on the subject, *Redefining Efficiency*.

21. Aiken, "Environmental Consequences," 132–164.

22. Finding sinks for waste is a concept analyzed by Tarr in *Search for the Ultimate Sink*. The concepts of sinks for hazardous wastes and of internalizing the costs of waste disposal are elaborated in chapter 1.

Chapter 1. Historical Context

1. Kirsch, *Mining Capitalism*, 1.

2. Brands, *T.R.: The Last Romantic*, 189, 415, 622–623; Miller, *Gifford Pinchott*, 155–157.

3. Brands, *T.R.: The Last Romantic*, 434–439; Goodwin, *Bully Pulpit*, 344–346, 441–443, 669.

4. Gordon, *Rise and Fall*, 27–61; Hughes, *American Genesis*, 1–12.

5. McNeill, "First Hundred Thousand Years," 16–19; Radkau, *Nature and Power*, 195–205.

6. Sieferle, *Subterranean Forest*, 19–27.

7. Sieferle, *Subterranean Forest*, 9–31.

8. Friedel, *Culture of Improvement*, 1–11, 225–230; Hounshell, *From the American System*, 4–10; Sieferle, *Subterranean Forest*, 34–37, 133–134; Stearns, *Industrial Revolution in World History*, 6–8.

9. Cowan, *Social History of American Technology*, 36–39; Hills, *Power in the Industrial Revolution*, 1–31.

10. Sieferle, *Subterranean Forest*, 19–27, 133–134.

11. Sieferle, *Subterranean Forest*, 78–137; Malm, *Fossil Capital*, 121–222.

12. Sieferle, *Subterranean Forest*, 19–27, 89; Malm, *Fossil Capital*, 121–222.

13. Sieferle, *Subterranean Forest*, 133, notes the ways coal, iron, and steam interacted symbiotically in the development of the railroad to greatly enhance transportation, in particular, and industrialization generally. On the transition from wood to coal in the United States, see Johnson, *Carbon Nation*, 9–23; Jones, *Routes of Power*, 15–21, 59–87.

14. McNeill and Engelke, *Great Acceleration*. McNeill and Engelke analyze the tremendous increase in industrial production and environmental impacts since the Second World War, characterized by an acceleration in global burning of fossil fuels in the postwar period. Because they claim the name "The Great Acceleration" for the period after World War II, we can refer the period after 1880 as "the previous great acceleration."

15. "World's Metal Production for 40 Years," *E&MJ* 95 (12 April 1913): 742.

16. Sources for global coal production: Parker, "Coal," in *Mineral Resources of the United States, 1910*, 62; Parker, "Coal," in *Mineral Resources of the United States, 1913*, 784.

17. *Mineral Resources of the United States, 1910*, 165.

18. The graph is based on data derived from *Mineral Resources of the United States, 1882*, 306–307.

19. *Mineral Resources of the United States, 1887*, 69. I am indebted for this insight to Kent Curtis, *Gambling on Ore*, 121–122, 141–143.

20. Robert Peele, *Mining Engineers' Handbook* (New York: John Wiley & Sons Inc., 1918), 94.

21. Day and Tylecote, *Industrial Revolution in Metals*.

22. Andrews, *Killing for Coal*, 125–153. Andrews calls the working spaces in mines "workscapes." They were largely inorganic (mineral) confines, but they had to support organisms (workers) so they could perform their duties. Other organisms (microorganisms, mice and rats, mules, etc.) occupied those spaces as well.

23. Coal may be an "organic" mineral resource, in a chemical sense, but it is classified a mineral fuel because geological processes, across geological time, converted the residue of the organic realm into useable fuel buried in Earth's crust.

24. Tarr, *Search for the Ultimate Sink*.

25. The concept of competing groups of people working to convert a particular environment into the place they imagined it to be is developed in West, *Contested Plains*. West shows how different groups of people on the central plains of Kansas, Nebraska, and Colorado used technologies and other methods to realize their visions of those plains, of what they imagined the plains truly were. The groups, including various Indian peoples and successive groups of white settlers, imagined the plains to be a particular place, and they succeeded and failed in realizing their visions of that place because of both environmental factors and the actions of competing groups. In this book we also see competing visions for what Butte and the Deer Lodge Valley were and therefore would become.

26. Hughes, *Networks of Power*, 6. Hughes writes: "Those parts of the world that are not subject to a system's control, but that influence the system, are called the environment. A sector of the environment can be incorporated into a system by bringing it under system control."

27. Malone, *Battle for Butte*, 41–42.

28. *Mineral Resources of the United States, 1905*, 345.

29. The legal actions of farmers in the Deer Lodge Valley are described later in this book. The belief in the legal system out of which they acted is nicely described in Novak, *People's Welfare*; and Freyfogle, *Land We Share*.

30. Baldwin, *Antitrust and the Changing Corporation*, 25–39; Sklar, *Corporate Reconstruction of American Capitalism*, 146–154, 170–172, 238–242, 419–430; Freyer, "Legal Restraints on Economic Coordination," 183–201.

31. The giant trust was the Amalgamated Copper Company. Its rise and influence over Montana have been analyzed by several historians, notably Toole, *Twentieth-Century Montana*; and Malone, *Battle for Butte*.

32. Freese, *Industrial Strength Denial*, 1.

33. Freese, *Industrial Strength Denial*, 10.

34. Diamond, *Collapse*, 38–40.

Chapter 2. Background to Mining at Butte

1. Mumford, *Technics and Civilization*, 69.

2. Mumford, *Technics and Civilization*, 74–75.

3. Mumford, *Technics and Civilization*, 67.

4. Mumford, *Technics and Civilization*, 69.

5. McClernan, "Sixty Million Years of History," 16–19; Czehura, "Butte: A World Class," 16–17.

6. Each of these minerals is in turn made up of individual elements. Quartz is a glassy compound of silicon and oxygen, for example, while feldspar is white in appearance and consists of sodium, calcium, potassium, aluminum, silicon, and oxygen.

7. This deposit is found in small, closely spaced veins beneath what is now known as the Berkeley Pit, Butte's first large open-pit mine. The copper and molybdenum are found in the form of chalcopyrite ($CuFeS_2$) and molybdenite (MoS_2).

8. The descriptions in the ensuing paragraphs about the geologic formation of the Butte area are from McClernan, "Sixty Million Years of History," 18–19; and Czehura, "Butte: A World Class," 14–17.

9. Snyder, *Scenic Routes and Byways Montana*, 48.

10. Goetzmann, *Exploration and Empire*, 3–35.

11. White, *"It's Your Misfortune,"* 191–193.

12. Many of Virginia's first settlers at Jamestown were mainly interested in finding gold; see Woodward, *New American History*, 27, 31; Schlebecker, *Whereby We Thrive*, 3.

13. Bining, *Pennsylvania Iron Manufacturing*, 14–17, 24, 49–64; Mayer and Riley, *Public Domain, Private Domain*, 20–24; Larry Lankton, *Hollowed Ground*, 7–15. America's "copper frontier" is generally described in Young Jr., "American Copper Frontier, 1640–1893," 4–15; Hyde, *Copper for America*, 3–48.

14. For example, Iowa grew from 43,112 in 1840 to 192,214 in 1850, and in the same period Wisconsin grew from 30,945 to 274,446; Pomeroy, *Pacific Slope*, 32.

15. Pomeroy, *Pacific Slope*, 26–36; Smith, *Pacific Visions*, 11–18.

16. Pomeroy, *Pacific Slope*, 42–48; Rohrbough, *Days of Gold*, 202–203.

17. Paul, *Far West Plains*, 24–45; Bryant, "Entering the Global Economy," 197–208; Curtis, *Gambling on Ore*, 20–29. For a general description of Butte's place within that broader context, see Malone, *Battle for Butte*, 5–10.

18. Malone and Roeder, *Montana: A History*, 50.

19. Malone and Roeder, *Montana: A History*, 50–55; Malone, *Battle for Butte*, 3–8; Curtis, *Gambling on Ore*, 22–25, 54.

Chapter 3. Beginnings of Butte's Mining Industry

1. *Montana Post* (7 January 1865), letter to the editor from Butte City, copied in W. J. Wilcox, "His Record of Anaconda," typescript, ca. 1936, Records of the Anaconda Copper Mining Company, #AW W021, Butte–Silver Bow Public Archives, Butte. In his lengthy "Record," Wilcox copied excerpts from many old newspapers, including the *Montana Post*, Montana's first newspaper, published in Virginia City. In cases where the citation is to his copies of early newspapers rather than to the original, it is so noted.

2. *Montana Post* (9 December 1865), from Wilcox, "His Record of Anaconda."

3. West, *Contested Plains*.

4. Fogarty, "A History of Butte, Montana," (n.d., n.p.), 1, 6, Butte Public Library Collection, #AW F014, Butte–Silver Bow Public Archives, Butte.

5. William Cronon, in a 1992 essay, suggests the distinction between the organic and the inorganic in environmental history, but he does not explicitly delineate the distinction; see Cronon, "Kennecott Journey," 28–51.

6. Gordon, *Rise and Fall*, 27–61.

7. Johnson, *Carbon Nation*, xviii–xix.

8. For an excellent history of failure in promoting mines, see Safford, *Mechanics of Optimism*. Curtis, in *Gambling on Ore*, 69–114, describes the gamble involved in trying to invest profitably in ores. A vivid account of the ways miners and geologists used mine maps to better understand what they could not yet see underground is Nystrom, *Seeing Underground*. Malone, *Battle for Butte*, is still the best account of the miners, bankers, and investors who were most successful in Butte's early years.

9. Several scholars have written about the gendered nature of Butte, especially in its early years; see Murphy, *Mining Cultures*; Finn, *Mining Childhood*; Basso, *Meet Joe Copper*.

10. Early descriptions of the discovery of gold at Butte include Leeson, *History of Montana*, 916–917; Fogarty, "History of Butte, Montana"; Freeman, *Brief History of Butte*, 10; Charles S. Warren, "The Romance of Butte" (speech delivered 4 July 1876 and reprinted in *The Butte Miner*, Semi-Centennial Number, 5 July 1926): 3.

11. Weed, *Geology and Ore Deposits*, 17–18; Fogarty, "History of Butte, Montana," 2–3.

12. Richards, *Text Book of Ore Dressing*, 1.

13. Fogarty, "History of Butte, Montana," 4; *Montana Post* (18 February 1865, 25 February 1865, 4 March 1865, 11 March 1865, 20 May 1865, from Wilcox, "His Record of Anaconda").

14. Malone and Roeder, *Montana: A History*, 55–61; *Montana Post* (11 February 1865, 25 February 1865, from Wilcox, "His Record of Anaconda").

15. Piatt, ed., *Story of Butte*, 28.

16. *Montana Post* (19 January 1867, 16 November 1867, from Wilcox, "His Record of Anaconda"); *Mining & Scientific Press (M&SP)* 18 (17 April 1869): 247; Piatt, ed., *Story of Butte*, 7; Leeson, *History of Montana*, 917; Browne, *Report on the Mineral Resources*, 504; Raymond, *Statistics of Mines and Mining*, 278; Smith, *History of the Early Reduction*, 1–2.

17. Piggot, *Chemistry and Metallurgy of Copper*, 286–287; Douglas, "Historical Sketch of Copper Smelting," 269–280.

18. Douglas, "Historical Sketch of Copper Smelting," 280; Fell, *Ores to Metals*, 10–49; Egleston, "Boston and Colorado Smelting Works," 276–298.

19. Browne, *Report on the Mineral Resources*, 504.

20. Malone, *Battle for Butte*, 8–10; Leeson, *History of Montana*, 917–918; Raymond, *Statistics of Mines and Mining*, 274; *New Northwest* (Deer Lodge, Montana), 11 November 1870, 27 May 1871, 24 June 1871, from Wilcox, "His Record of Anaconda"; Lou P. Smith, testimony in *Hugh Magone v. Colorado Smelting and Mining Company, et al.*, No. 222 in Equity, Circuit Court of the United States, Ninth Circuit, District of Montana, 1903, typed transcripts in Records of the Circuit Court of the United States, Ninth Circuit, District of Montana, Record Group 21, National Archives, Seattle Branch (hereafter cited as MvC), vol. VII, 2735–2736. The prevalence of Chinese miners in worked-over areas is nicely analyzed in Rohe, "Chinese River Mining," 14–29.

21. Malone, *Battle for Butte*, 15–16; *Butte Miner*, 3 June 1876; Warren, "Romance of Butte," 4.

22. Malone, *Battle for Butte*, 15–17; Smith, *History of the Early Reduction*, 3; *Butte Miner* (13 June 1876): 3, (17 June 1876): 3, (1 July 1876): 3, (20 July 1876): 3, (25 August 1876): 3; *New Northwest* (4 February 1876): 3, (10 March 1876): 3, (28 April 1876): 3, (16 June 1876): 3; "Indenture between William L. Farlin and William A. Clark," document dated 31 January 1876, and "Indenture between William L. Farlin and William A. Clark," document dated 23 February 1876, Transcribed Deed Book 3, pp. 337–346, Clerk and Recorder's Office, Butte–Silver Bow Courthouse, Butte.

23. *New Northwest* (28 April 1876): 3.

24. Warren, "Romance of Butte," 5; *Butte Miner* (8 June 1876): 3, (17 June 1876): 3, (27 June 1876): 3, (8 July 1876): 3; *Engineering & Mining Journal (E&MJ)* 22 (15 July 1876): 45; Malone, *Battle for Butte*, 16.

25. Malone, *Battle for Butte*, 17.

26. Warren, "Romance of Butte," 6.

27. Safford, *Mechanics of Optimism*, 141–146.

28. *E&MJ* 22 (16 December 1876): 397.

29. *E&MJ* 23 (2 June 1877): 379; 24 (15 September 1877): 207, (3 November 1877): 327; 25 (23 March 1878): 205; "Deed Made between William A. Clark and Joseph K. Clark," document dated 6 October 1877, Transcribed Deed Book 6, pp. 169–173, Clerk and Recorder's Office, Silver Bow County Courthouse, Butte; "Deed Made between William A. Clark and Rolla Butcher," document dated 6 October 1877, Transcribed Deed Book 6, pp. 186–190, Clerk and Recorder's Office, Silver Bow County Courthouse, Butte.

30. Leeson, *History of Montana*, 923; listing for John How, R. G. Dun & Co. Collection, Montana vol. I, Historical Collections, Baker Library, Harvard University, Cambridge, MA; Malone, *Battle for Butte*, 16–17; Smith, *History of the Early Reduction*, 3; *E&MJ* 21 (10 June 1876): 565; *New Northwest* (10 March 1876): 3, (28 April 1876): 3, (9 August 1878): 3; *Butte Miner* (13 June 1876): 3, (1 July 1876): 3, (11 July 1876): 3; Raymond, *Statistics of Mines and Mining*, 730–733.

The map of the Mary Ann lode accompanying its patent to Lee Mantle shows a mill building located in the southwest corner of the claim; see "Patent for Mary Ann Lode," document dated 15 July 1882; field notes for the survey of the Mary Ann lode, "Survey No. 799," unpublished field notes dated 10 June 1880, Mining Claim Records, Recorder's Office, Butte–Silver Bow Courthouse, Butte. Location of How's Centennial mill is also corroborated in W. W. de Lacy, "Field Notes of the Subdivisional Lines of Township 3 North, Range 8 West of the Principal Meridian in the Territory of Montana" (unpublished survey notes dated 13–17 April 1877), 6–7, U.S. Dept. of the Interior, Bureau of Land Management, Butte.

31. Leeson, *History of Montana*, 918, 923, 954; *Montana Post* (4 January 1868, from Wilcox, "His Record of Anaconda"); Raymond, *Statistics of Mines and Mining*, 277; Malone, *Battle for Butte*, 12, 16; listing for Charles Hendrie, R. G. Dun & Co. Collection, Montana vol. I, p. 218, Historical Collections, Baker Library, Harvard University, Cambridge, MA; Smith, *History of the Early Reduction*, 2–3; *M&SP* 18 (17 April 1869): 247; *E&MJ* 23 (3 March 1877): 140, (21 April 1877): 261, (2 June 1877): 379; 26 (17 August 1878): 118; 30 (10 July 1880): 29; (31 July 1880): 75; 32 (3 September 1881): 152; *New Northwest* (14 April 1876): 3, (19 October 1877): 3; *Butte Miner* (13 June 1876): 3; Sanborn Fire Insurance Map for "Butte City" (New York: Sanborn Map & Publishing Co., Ltd., 1884, hereafter cited as "Sanborn Map for Butte" and date), Digital Collections, Library of Congress, Washington, D.C.

32. There are no detailed records of the production of these early mills. Estimates here of the ore they treated and the tailings they produced are based on certain assumptions and analyses of historical descriptions detailed in Fredric L. Quivik, "Expert Report" (unpublished report prepared for the U.S. Department of Justice, dated 15 August 1997, and submitted in the case *U.S. v. ARCO*; the Montana Historical Society's library holds a copy of the expert report).

33. Quivik, "Expert Report" (15 August 1997), iv.

34. Malone, *Battle for Butte*, 16–20.

35. *New Northwest* (19 October 1877): 3, (1 March 1878): 3; (23 August 1878): 3; *E&MJ* 23 (3 March 1877): 140; 24 (7 July 1877): 8, (11 August 1877): 110, (10 October 1877): 296, (8 December 1877): 421; 26 (27 July 1878): 63, (17 August 1878):

118; 29 (28 February 1880): 155, (6 March 1880): 172–173; Blake, "Rainbow Lode," 71. The wet and dry processes of stamping quartz ores are described in Emmons and Becker, "Amalgamating Mills," 242–285.

36. Emmons and Becker, "Amalgamating Mills," 269–273.

37. Blake, "Rainbow Lode," 74–75; Emmons and Becker, "Amalgamating Mills," 269–273; Raymond, *Statistics of Mines and Mining*, 732–733, 738, 748–755.

38. Lingenfelter, *Hardrock Miners*, 132, 183–188; Frisch, "'Gibraltar of Unionism,'" 73–95, 113–114.

39. Smith, *History of the Early Reduction*, 4; James A. Talbott, testimony in MvC, vol. VII, 1634–1637; John Gillie, testimony in MvC, vol. IX, 3502; *New Northwest* (10 January 1879): 3; *E&MJ* 24 (8 December 1877): 421; *E&MJ* 26 (17 August 1878): 118; *E&MJ* 27 (22 February 1879): 128; *E&MJ* 28 (12 July 1879): 26; *E&MJ* 29 (15 May 1880): 342, (12 June 1880): 411, (26 June 1880): 442; *E&MJ* 30 (11 September 1880): 176, (9 October1880): 240, (30 October 1880): 289; Leeson, *History of Montana*, 957–958; "Sanborn Map for Butte" (1884), sheet 8; "Sanborn Map for Butte" (1888), sheet 14.

40. Quivik, "Expert Report," iv.

41. *E&MJ* 51 (20 June 1891): 721; listing for Clark & Donnell, R. G. Dun & Co. Collection, Montana vol. I, p. 240, Historical Collections, Baker Library, Harvard University, Cambridge, MA; Malone, *Battle for Butte*, 13–15.

42. *E&MJ* 51 (20 June 1891): 721; Fell, *Ores to Metals*, 51; Malone, *Battle for Butte*, 15. In the *E&MJ* biographical sketch of Clark cited here, he recalled that his shipping costs to Baltimore were $75/ton. An article in *E&MJ* 23 (28 April 1877): 272–273, reviewing Butte's 1876 mining season, reports that costs averaged about $65/ton. The latter figure is used here, being closer to the time when the costs were incurred.

43. *New Northwest* (28 April 1876): 3; *Butte Miner* (27 June 1876): 3; *E&MJ* 22 (15 July 1876): 45; *E&MJ* 23 (3 March 1977): 140; Smith, *History of the Early Reduction*, 5.

44. Cornish rolls are a pair of steel rolls that operate like the wringer of an old washing machine to crush ore more finely as it passes between the rolls. Jigs operate on the principle that more-dense, metal-bearing mineral particles will sink in agitated water, while less-dense particles (largely gangue, and devoid of metals) will remain suspended in that agitated water. Jigs work best when they are treating particles that are about the same size, so crushed and ground ore is usually classified by particle size before the particles are fed to jigs. Jigs are effective on coarsely ground ore but are not effective on particle sizes of sand or smaller (something that is hard to avoid when crushing and grinding ore). Concentrating tables treat the finer particle sizes, which are unsuitable for treatment in jigs. Concentrating tables employ such features as textured surfaces and shaking motion to encourage the more-dense particles to move one direction in a stream of water and encourage the less-dense particles to move in another. For both jigs and tables, the more-dense particles are collected as concentrates (with a higher metals content than the raw ore), and the less-dense particles are considered tailings, either to be discarded as waste or to be processed further in an effort to recover more concentrate. Both kinds of machines are used in the process called "gravity concentration," because they use gravity to separate

the metal-bearing particles from the gangue. Neither jigs nor tables are 100 percent effective, meaning there is still gangue in the concentrates, and there are still metals in the tailings. Varieties of these kinds of equipment are described in detail in texts like Richards, *Text Book of Ore Dressing*.

45. *New Northwest* (5 May 1876); *Butte Miner* (13 June 1876): 3, (27 June 1876): 3, (22 July 1876): 3, (2 October 1877): 3; *E&MJ* 23 (28 April 1877): 272–273; *E&MJ* 24 (3 November 1877): 327, (8 December 1877): 421; Surveyor General's Office, "Township No. 3 North Range No. 7 West of the Principal Meridian Montana," map dated 27 June 1877, Bureau of Land Management, Butte; Smith, *History of the Early Reduction*, 5; Malone, *Battle for Butte*, 16–17.

46. *E&MJ* 27 (15 March 1879): 185, (21 June 1879): 452; *E&MJ* 28 (5 July 1879): 7.

47. Fell, *Ores to Metals*, 27–54, 136–141; *E&MJ* 51 (20 June 1891): 721.

48. *E&MJ* 74 (19 July 1902): 82; Newell, "Atmospheric Pollution," 658.

49. Fell, *Ores to Metals*, 137–141; "Certificate of Incorporation of the Colorado and Montana Smelting Company," document dated 1 June 1879 (Y000077–078); *E&MJ* 27 (21 June 1879): 452; 28 (5 July 1879): 7; Douglas, "Metallurgy of Copper," in *Mineral Resources of the United States* (Washington, D.C.: Government Printing Office, 1883), 265.

50. *E&MJ* 27 (21 June 1879): 452; *E&MJ* 28 (5 July 1879): 7, (25 October 1879): 302.

51. *E&MJ* 28 (6 September 1879): 163, (12 October 1879): 302; *E&MJ* 29 (10 April 1880): 256; *E&MJ* 30 (31 July 1880): 75; Leeson, *History of Montana*, 1357.

52. *E&MJ* 27 (26 April 1879): 299, (31 May 1879): 390; *E&MJ* 28 (6 December 1879): 418; *E&MJ* 29 (1 May 1880): 307, (19 June 1880): 426; *E&MJ* 30 (3 July 1880): 10, (9 October 1880): 240.

53. Emmons and Becker, "Amalgamating Mills," 249.

54. "Montana Mining Notes," *E&MJ* 29 (14 February 1880): 119.

55. Kirchhoff, "Copper: The Copper Industry of the United States" (1883), 215–216, 221.

Chapter 4. Amazing Growth of the 1880s

1. Peters, "Mines and Reduction Works," 384.

2. Deciphering how people of another age understood a word or concept can be a useful exercise for learning the values of that age. Adam Rome has performed a similar exercise on the word "pollution"; see Rome, "Coming to Terms with Pollution," 6–28.

3. Coxe, Rothwell, and Coryell, "Call to a Meeting," dated April 1871 and reprinted in *Transactions of the American Institute of Mining Engineers* 1 (1871–1873): 3.

4. Hewitt, "Century of Mining and Metallurgy," 164–191.

5. Hays, in *Conservation and the Gospel of Efficiency*, has written about the turn-of-the-century conservation movement and its efforts, through the application of science and technology to the extraction of the nation's resources, to maximize efficiency and eliminate waste as much as possible. He makes it clear that when conservationists and federal policymakers talked of waste, they were referring to resources

that went unused. This is a very similar understanding of waste to that of mining and metallurgical engineers when they spoke of waste in the 1870s and 1880s.

6. Coxe, "Presidential Address," 126–127.

7. The classic reference is Agricola, *De Re Metallica*, translated by Hoover and Hoover, 279. Note especially Hoover's extensive footnote on pp. 279–283. See also, for example, Peters, *Practice of Copper Smelting*, 32–33; Richards, *Text Book of Ore Dressing*, 1.

8. Goodale, "Concentration of Ores," 601; Rudolph Siever, testimony in Circuit Court of the United States, Ninth Circuit, District of Montana, *Hugh Magone v. Colorado Smelting and Mining Company*, et al., No. 222 in Equity, 1903, MvC, vol. IV, 409; W. A. Kidney, testimony in MvC, vol. VIII, 2924–2925, 2929, 2944; David T. Gilbert, testimony in MvC, vol. VIII, 2963.

9. Goodale, "Concentration of Ores," 601.

10. Curtis, *Gambling on Ore*, 117.

11. Kirchhoff, "Copper Industry in the United States" (1885), 329; Butler, "Copper," 142–143, 165; Lankton, *Hollowed Ground*, 85–125.

12. Kirchhoff, "Copper" (1885), 336–340.

13. Kirchhoff, "Copper: The Copper Industry in the United States" (1886), 208, 225; Kirchhoff, "Copper: The Copper Industry in the United States" (1887), 124; Kirchhoff, "Copper" (1890), 66.

14. Knowles, *Economic Development in the Nineteenth*, 201; Chandler, *Visible Hand*, 159–165; Lamoreaux, *Great Merger Movement*, 1; Bryant, "Entering the Global Economy," 213–224; Malone and Peterson, "Politics and Protests," 28–53.

15. Rickard, *History of American Mining*, 147–170; Cronon, *Nature's Metropolis*, 65–68; Lankton, *Hollowed Ground*, 33–34; Malone and Roeder, *Montana: A History*, 55–57.

16. Cronon, *Nature's Metropolis*, 55–259; Robbins, *Colony and Empire*, 162–183.

17. Malone and Roeder, *Montana: A History*, 131–132; Myers, "Butte Rail Connection," 31. Malone, in *Battle for Butte*, 11–39, provides a nice overview of Butte's early investors and of the national and international markets for copper that drove Butte's rise.

18. Singer et al., *History of Technology*, 76–77; Kirchhoff, "Copper: The Copper Industry of the United States" (1883), 215, 252; Kirchhoff, "Copper: The Copper Industry of the United States" (1885), 327, 356.

19. Hildebrand and Mangum, *Capital and Labor in American*, 94; Singer et al., *History of Technology*, 218–230; Blake-Coleman, *Copper Wire and Electrical Conductors*, 143–149, 185–208.

20. Nye, *Electrifying America*, 28–37; Hughes, *Networks of Power*, 40–43.

21. Kirchhoff, "Copper" (1883), 213–215, 232–236; *Mineral Resources of the United States, 1910*, 165; Herfindahl, *Copper Costs and Prices*, 70–71, 243–255; Hildebrand and Mangum, *Capital and Labor in American*, 100.

22. Kirchhoff, "Copper" (1883), 213–214, 221–225.

23. "Leonard Lewisohn," *E&MJ* 73 (15 March 1902): 375; "Leonard Lewisohn," in *Jewish Encyclopedia*, vol. VIII, 69–70; *E&MJ* 28 (12 July 1879): 26; *The Anaconda*

Standard (25 November 1900): 22; "Certificate of Incorporation of Montana Copper Company," document dated 16 October 1879 (Quivik Collection). *E&MJ* refers to a New York Copper Company in early references to development around the Colusa. Note that there is no legal record of a New York Copper Company owning or operating the Colusa property in Butte. It is likely that *E&MJ* mistakenly referred to the Montana Copper Company as the New York Copper Company because the investors were based in New York.

24. *E&MJ* 28 (8 November 1879): 339; *E&MJ* 29 (3 January 1880): 10, (14 February 1880): 134, (10 April 1880): 256, (19 June 1880): 426; Howe, *Copper Smelting*, 99–103; Peters, "Mount Lincoln Smelting Works," 310–314; Morin, *Legacy of American Copper Smelting*, 2–4, 13–14.

25. *E&MJ* 29 (19 June 1880): 426; *E&MJ* 30 (2 October 1880): 224; *E&MJ* 31 (26 February 1881): 149, (26 March 1881): 210, (18 June 1881): 418; *E&MJ* 32 (29 October 1881): 289; *E&MJ* 33 (28 January 1882): 56; Goodale, "Concentration of Ores," 601; Michael Curnow, testimony in MvC, vol. VIII, 3169; John Gillie, testimony in MvC, vol. IX, 3442.

26. *E&MJ* 33 (28 January 1882): 56; Goodale, "Concentration of Ores," 601; Richards, *Ore Dressing*, vol. II, 644–651.

27. John Gillie, testimony in MvC, vol. IX, 3523.

28. David T. Gilbert, testimony in MvC, vol. VIII, 2968. Gilbert did not specify that he worked at the B&M's upper works, as opposed to the lower works. This can be inferred, however, by the fact that he located the B&M's concentrator where he worked about three-quarters of a mile above the Liquidator, and he elsewhere referred to the lower works as Clark's.

29. Smith, *History of the Early Reduction*, 7; *New Northwest* (23 June 1876): 3; *E&MJ* 22 (15 July 1876): 45; 32 (10 September 1881): 172, (5 November 1881): 308; "Articles of Incorporation of the Parrot Silver and Copper Company," document dated 21 May 1880 (Quivik Collection).

30. *E&MJ* 35 (31 March 1883): 182; *E&MJ* 37 (8 March 1884): 186, (5 April 1884): 261; Goodale, "Concentration of Ores," 601; Hofman, "Notes on the Metallurgy," 260–261; Ed McCool, testimony in MvC, vol. VIII, 3057; George F. Bartlett, testimony in MvC, vol. VIII, 3098; John Gillie, testimony in MvC, vol. IX, 3444.

31. Douglas, "Recent American Methods and Appliances," 834, 839; Hofman, "Notes on the Metallurgy," 261; *E&MJ* 38 (9 August 1884): 94; Douglas, "Historical Sketch of Copper Smelting," 284; Peters, *Modern American Methods of Copper*, 330–335; Peters, *Practice of Copper Smelting*, 475–478, 484–486.

32. "Edward Dyer Peters," *Transactions of the American Institute of Mining Engineers* 60 (1919): 735–740; Peters, "Mount Lincoln Smelting Works," *AIME* 2 (1872): 310–314; Peters, *Modern American Methods of Copper*.

33. George F. Bartlett testimony in MvC, vol. VIII, 3099; *E&MJ* 37 (3 May 1884): 318; *E&MJ* 38 (12 July 1884): 27, (9 August 1884): 94, (30 September 1884): 147, (8 November 1884): 320; *E&MJ* 39 (7 February 1885): 94, (14 March 1885): 180, (4 April 1885): 232, (18 April 1885): 268, (2 May 1885): 304, (9 May 1885): 322, (27 June 1885): 446; *E&MJ* 40 (18 July 1885): 43, (15 August 1885): 117, (17 October

1885): 273; *E&MJ* 43 (8 January 1887): 19; *E&MJ* 47 (29 June 1889): 596; *E&MJ* 49 (7 June 1890): 649; *E&MJ* 51 (10 January 1891): 63–64; Ed McCool, testimony in MvC, vol. VIII, 3057; John Gillie, testimony in MvC, vol. IX, 3444; Goodale, "Concentration of Ores," 601; "Sanborn Map for Butte" (1884), sheet 9; "Sanborn Map for Butte" (1888), sheet 13.

34. Smith, *History of the Early Reduction*, 7; *E&MJ* 29 (6 March 1880): 173, (17 April 1880): 270; *E&MJ* 30 (10 July 1880): 29, (28 August 1880): 144, (2 October 1880): 224, (16 October 1880): 256, (6 November 1880): 300; "Butte Gives the Glad Hand to an Old-Timer," *Anaconda Standard* (25 November 1900): 22.

35. Peters, "Mines and Reduction Works," 395–396; Douglas, "Historical Sketch of Copper Smelting," 284; *E&MJ* 33 (4 March 1882): 124, (25 March 1882): 159; *E&MJ* 34 (19 August 1882): 95–96. On the debate among metallurgists over the best furnace for smelting copper-sulfide ores, see Peters, *Practice of Copper Smelting*, 406–422.

36. Peters, *Practice of Copper Smelting*, 115–203, 406–422; Hofman, *Metallurgy of Copper*, 138–189; Austin, *Metallurgy of the Common*, 360–377.

37. Tylecote, *History of Metallurgy*, 149, 155–157; Peters, *Practice of Copper Smelting*, 311–422; Hofman, *Metallurgy of Copper*, 236–288; Austin, *Metallurgy of the Common Metals*, 387–399.

38. *E&MJ* 32 (2 July 1881): 2; *E&MJ* 34 (19 August 1882): 95–96; *E&MJ* 35 (3 March 1883): 123, (28 April 1883): 244, (12 May 1883): 273, (26 May 1883): 306; *E&MJ* 36 (14 July 1883): 24, (21 July 1883): 40; *E&MJ* 37 (3 May 1884): 336; *E&MJ* 38 (2 August 1884): 79; 40 (25 July 1885): 65, (15 August 1885): 117; *E&MJ* 43 (11 June 1887): 426; *E&MJ* 44 (26 November 1887): 402; Smith, *History of the Early Reduction*, 7; Leeson, *History of Montana*, 950–953, 957.

39. *E&MJ* 38 (12 July 1884): 27; Quivik, "Expert Report" (15 August 1997), B-3, B-55, B-69, B-77, C-10, C-109, C-159.

40. *E&MJ* 30 (21 August 1880): 128. Hickey's reminiscences are recounted in numerous histories of the Anaconda company; see for example, Malone, *Battle for Butte*, 24; and Marcosson, *Anaconda*, 31, who state respectively that it would be McClellan's army and Grant's army that would surround Lee. Incidentally, Winfield Scott's plan to blockade the South with ships along the Coast and gunboats along the Mississippi was called the Anaconda Plan, likely giving rise to Greeley's use of the term; see McPherson, *Battle Cry of Freedom*, 333–334.

41. Malone, *Battle for Butte*, 25–29; Marcosson, *Anaconda*, 32–46; Brechin, *Imperial San Francisco*, 209–231; *Report of the Anaconda Copper Mining Company* (1895), 8.

42. Malone, *Battle for Butte*, 28–31; Marcosson, *Anaconda*, 46, 51–52.

43. *E&MJ* 37 (16 February 1884): 129; *E&MJ* 38 (20 September 1884): 200, (18 October 1884): 272; Malone, *Battle for Butte*, 29–31; Marcosson, *Anaconda*, 46–52.

44. Malone, *Battle for Butte*, 29–31; Marcosson, *Anaconda*, 46–52; *E&MJ* 34 (19 August 1882): 100, (23 September 1882): 165, (2 December 1882): 299, (16 December 1882): 325.

45. *E&MJ* 35 (9 June 1883): 338; *E&MJ* 37 (26 April 1884): 318; Marcosson, *Anaconda*, 47.

46. *E&MJ* 35 (12 May 1883): 273; *E&MJ* 37 (16 February 1884): 129; *E&MJ* 38 (23 August 1884): 132, (30 August 1884): 147.

47. Maxwell, "Mountain Railroads of the Mining," 64–65.

48. *E&MJ* 37 (16 February 1884): 129, (21 June 1884): 469; *E&MJ* 38 (4 October 1884): 236, (25 October 1884): 288; Kirchhoff, "Copper: The Copper Industry of the United States" (1884), 338–339; Goodale, "Concentration of Ores," 609; Charles C. Rueger, testimony in MvC, vol. VIII, 3216.

49. Kirchhoff, "Copper: The Copper Industry of the United States" (1885), 338–339.

50. Thomas Parke Hughes is often credited with introducing the concept of technological systems to the field of history of technology in *Networks of Power*. As Hughes showed, systems builders were intent not only on developing efficient and effective systems within a particular industrial plant but within the entire enterprise as well. Systems builders sought to incorporate such facets as transportation, marketing, finance, and energy supply into a single well-ordered and well-managed, integrated technological system. Marcus Daly and his Anaconda enterprise were especially adept at this kind of system building, as this book shows.

51. *E&MJ* 39 (10 January 1885): 28, (14 March 1885): 180; *E&MJ* 40 (15 August 1885): 117; Goodale, "Concentration of Ores," 609; Hofman, "Notes on the Metallurgy," 265–266; Charles C. Rueger, testimony in MvC, vol. VIII, 3207–3208.

52. Ulke, "Characteristic American Metal Mines," 520–522; Goodale, "Concentration of Ores," 609–610; Repath, "Mechanical Engineer," 143; Kirchhoff, "The Copper Industry" (1885), 338; Charles C. Rueger, testimony in MvC, vol. VIII, 3207–3210, 3213–3214, 3216; John Gillie, testimony in MvC, vol. IX, 3455–3456.

53. *Anaconda Weekly Review*, 29 December 1887, 16 August 1888, 30 August 1888; *Report of the Anaconda Copper Mining Company* (1895), 52.

54. Testifying at trial in 1907, concerning smoke from the new smelter, Nick Bielenberg stated that he had been forced to abandon grazing land north of the Old Works because smoke had destroyed his range and killed his sheep. Ephraim Staffanson testified at that trial that his land, which was close to the Lower Works, had been destroyed by the smelter, although he didn't specify whether it was smoke or tailings that had destroyed his land. Writing to President Woodrow Wilson in 1913 about the smoke problem, Harry J. Quinlan stated that, for twenty years prior to the 1905 suit in which Bielenberg and Staffanson testified, owners of livestock harmed by smoke from the Old Works had tried to get the Anaconda company to acknowledge that arsenic in the smoke was damaging their farm operations. See N. J. Bielenberg, testimony in *Fred J. Bliss v. The Washoe Copper Company and the Anaconda Copper Mining Company*, transcripts of records for the United States Circuit Court of Appeals for the Ninth Circuit (San Francisco: Filmer Bros. Co. Printers, 1909, hereafter cited as BvA), vol. XII, 4528–4537; Ephraim Staffanson, testimony in BvA, vol. VIII, 2912, 2941–2946; Harry J. Quinlan to Hon. Woodrow Wilson, letter dated 21 July 1913, box 886, DOJ Central File, Records of the U.S. Department of Justice, Record Group 60, National Archives, College Park, MD.

55. *E&MJ* 37 (26 April 1884): 318; *E&MJ* 38 (2 August 1884): 79, (30 August

1884): 147, (18 October 1884): 272, (25 October 1884): 288, (6 December 1884): 384; *E&MJ* 39 (20 June 1885): 430; *E&MJ* 40 (18 July 1885): 43; *E&MJ* 41 (13 February 1886): 118; "Sanborn Map for Butte" (1884), sheet 9.

56. Rudolph Siever, testimony in MvC, vol. IV, 396–400; *E&MJ* 39 (27 June 1885): 447; *E&MJ* 40 (18 July 1885): 43, (15 August 1885): 117; *E&MJ* 41 (9 January 1886): 27, (30 January 1886): 79, (April 24 1886): 299, 309; Hofman, "Notes on the Metallurgy," 262–263; Kirchhoff, "Copper Industry of the United States" (1886), 216; Douglas, "Historical Sketch of Copper Smelting," 275.

57. Hofman, "Notes on the Metallurgy," 263; "Separate Answer of Colusa-Parrot Mining and Smelting Company," in *Hugh Magone v. Colorado Smelting and Mining Company*, 12, 17, National Archives, Seattle Branch; *E&MJ* 45 (12 May 1888): 347; *E&MJ* 51 (20 June 1891): 721; MacMillan, "History of the Struggle," 83; A. H. Wethey, testimony in MvC, vol. IV, 2–3, 40–42; Frank Glaser, testimony in MvC, vol. IV, 474; James Doull, testimony in MvC, vol. IV, 531, 553; "Sanborn Map for Butte" (1888), sheet 9; "Sanborn Map for Butte" (1890), sheet 31; "Sanborn Map for Butte" (1891), sheet 50.

58. Myers, "Butte Rail Connection," 31–32; Malone, *Battle for Butte*, 40–41.

59. Rothwell, *Mineral Industry, 1892*, vol. I, 109; Ledux, "Future of Copper Production," in Rothwell, ed., *Mineral Industry, 1893*, vol. II, 241; *E&MJ* 40 (18 July 1885): 43; *E&MJ* 41 (5 June 1886): 407, (19 June 1886): 441, 445; *E&MJ* 42 (24 July 1886): 57; *E&MJ* 44 (13 August 1887): 110; *E&MJ* 45 (7 January 1888): 5; Malone, *Battle for Butte*, 35–37; Lankton, *Hollowed Ground*, 153–154.

60. *E&MJ* 39 (31 January 1885): 78, (18 April 1885): 268; *E&MJ* 40 (15 August 1885): 117; *E&MJ* 41 (30 January 1886): 65, (6 March 1886): 166, (24 April 1886): 299, (19 June 1886): 445; *E&MJ* 43 (1 January 1887): 3; *E&MJ* 44 (23 July 1887): 65–66, 68, (30 July 1887): 83–84, 86; Toole, "Anaconda Copper Mining Company," 318–320; Lankton, *Hollowed Ground*, 139.

61. *E&MJ* 43 (1 January 1887): 3; *E&MJ* 44 (23 July 1887): 65–66, 68, (30 July 1887): 83–84, 86.

62. *E&MJ* 44 (30 July 1887): 86, (13 August 1887): 122, (27 August 1887): 155, (3 September 1887): 174, (26 November 1887): 402, (24 December 1887): 476, (31 December 1887): 492; *E&MJ* 45 (7 January 1888): 25, (25 February 1888): 150; *E&MJ* 46 (1 September 1888): 180, (15 September 1888): 224, (10 November 1888): 400; *E&MJ* 48 (30 November 1889): 481.

63. *E&MJ* 46 (15 September 1888): 224, (22 September 1888): 245, (27 October 1888): 356; *E&MJ* 48 (6 July 1889): 13, (27 July 1889): 79, (17 August 1889): 145, (21 September 1889): 253, (19 October 1889): 346, (21 December 1889): 551; *E&MJ* 49 (7 June 1890): 649; "Certificate of Incorporation for Butte and Boston Mining Company," document dated 22 August 1888 (Quivik Collection); James A. Talbott, testimony in MvC, vol. VII, 1640; Arthur M. Day, testimony in MvC, vol. VIII, 2774–2776, 2784–2801; Henry Webking, testimony in MvC, vol. VIII, 2861–2862; W. A. Kidney, testimony in MvC, vol. VIII, 2926; John Gillie, testimony in MvC, vol. IX, 3445.

64. *E&MJ* 49 (25 January 1890): 117, (7 June 1890): 649, (14 June 1890): 688; *E&MJ* 50 (13 December 1890): 698–699; "Sanborn Map for Butte" (1890), sheet

31; James A. Talbott, testimony in MvC, vol. VII, 1640; Arthur M. Day, testimony in MvC, vol. VIII, 2778; Henry Webking, testimony in MvC, vol. VIII, 2873; John Gillie, testimony in MvC, vol. IX, 3446–3449. Gillie testified in the Magone case that the initial capacity of the B&B concentrator was 300 tons/day, but other sources suggest it was 400 tons/day.

65. Richards, *Ore Dressing*, vol. I; Richards, *Ore Dressing*, vol. II. These two volumes are the most complete early twentieth-century text on methods and equipment for concentrating ores. Part IV of vol. II includes an invaluable chapter, "Summary of Principles and Outlines of Mills," which includes summary descriptions and flow sheets for ninety-four mills and concentrators treating a wide variety of ores for precious and base metals in the U.S. and around the world, including six copper concentrators in Montana (Butte, Anaconda, and Great Falls).

66. Peters, "Mines and Reduction Works," 384; Richards, *Ore Dressing*, vol. II, 644–729.

67. Richards, *Ore Dressing*, vol. I, 318; Richards, *Ore Dressing*, vol. II, 1121–1122.

68. U.S.G.S., "Butte Special Map," in U.S. Geological Survey, *Butte Special Folio, Montana*, Geologic Folio 38 (Washington, D.C.: Government Printing Office, 1897); Doull and Pratt, "Deposition of Tailings," 4 (Quivik Collection); James Doull, testimony in MvC, vol. IV, 565–567; A. H. Wethey, testimony in MvC, vol. IV 12, 25, 69–70; Hugh McLeod, testimony in MvC, vol. VI, 1461–1462, 1469–1470.

69. W. A. Kidney, testimony in MvC, vol. VIII, 2929–2930, 2944.

70. "Sanborn Map for Butte" (1884), sheet 9; "Sanborn Map for Butte" (1888), sheet 14; "Sanborn Map for Butte" (1891), sheet 48; Rudolph Siever, testimony in MvC, vol. IV, 409; John Gillie, testimony in MvC, vol. IX, 3442, 3523; W. A. Kidney, testimony in MvC, vol. VIII, 2924–2925; David T. Gilbert, testimony in MvC, vol. VIII, 2963; A. H. Wethey, testimony in MvC, vol. IV, 16.

71. Goodale, "Concentration of Ores," 601–602, 605–606, 615–617; Richards, *Ore Dressing*, vol. II, 982–984; Peters, "Mines and Reduction Works," 386; Ed McCool, testimony in MvC, vol. VIII, 3058.

72. *E&MJ* 51 (11 April 1891): 452; Robert A. McArthur, testimony in MvC, vol. VIII, 3045, 3053; Ed McCool, testimony in MvC, vol. VIII, 3059–3061, 3074–3075; John M. Evans, testimony in MvC, vol. VIII, 3087–3088, 3094–3095; George F. Bartlett, testimony in MvC, vol. VIII, 3100; Arthur M. Day, testimony in MvC, vol. VIII, 2816; Henry Webking, testimony in MvC, vol. VIII, 2892.

73. Ed McCool, testimony in MvC, vol. VIII, 3061–3062; Evans, testimony in MvC, vol. VIII, 3088–3089; George F. Bartlett, testimony in MvC, vol. VIII, 3100–3101; Richards, *Ore Dressing*, vol. II, 1103.

74. Ed McCool, testimony in MvC, vol. VIII, 3061–3064, 3072–3073, 3077–3080; Evans, testimony in MvC, vol. VIII, 3088–3089; George F. Bartlett, testimony in MvC, vol. VIII, 3099–3101, 3104–3106; Robert A. McArthur, testimony in MvC, vol. VIII, 3047–3048, 3051; Lou P. Smith, testimony in MvC, vol. VII, 2745–2746; "Sanborn Map for Butte," (1884), sheet 9; "Sanborn Map for Butte," (1888), sheet 13.

75. A. H. Wethey, testimony in MvC, vol. IV, 5, 50, 384–387; Frank Glaser, testimony in MvC, vol. IV, 476–477; F. A. Pratt, testimony in MvC, vol. IV, 486, 507;

James Doull, testimony in MvC, vol. IV, 530–536, 541–542, 546–548; J. R. Wharton, testimony in MvC, vol. IV, 370–371; Charles W. Goodale, testimony in MvC, vol. VIII, 2832.

76. A. H. Wethey, testimony in MvC, vol. IV, 24, 50–51, 58, 102–103; James Doull, testimony in MvC, vol. IV, 535–536, 540, 565–567, 616; Rudolph Siever, testimony in MvC, vol. IV, 410.

77. *New Northwest* (14 November 1884): 3; *Weekly Missoulian* (28 November 1884): 2; "View of Anaconda: A Description of the Great Smelter at That Place," *Butte Daily Miner*, 30 November 1885.

78. Goodale, "Concentration of Ores," 609–610; Repath, "Mechanical Engineer," 143; Charles C. Rueger, testimony in MvC, vol. VIII, 3207–3210, 3213–3214, 3216; W. A. Kidney, testimony in MvC, vol. VIII, 2933–2934. Testifying at the Magone trial about Anaconda's efforts to treat the fine materials more intensively in order to recover more copper, Rueger concluded, "In other words the additional saving did not cover the additional expense."

79. *E&MJ* 42 (26 October 1886): 282; *E&MJ* 43 (28 May 1887): 381; *E&MJ* 49 (26 April 1890) 470; Charles C. Rueger, testimony in MvC, vol. VIII, 3212–3213; Goodale, "Concentration of Ores," 610–613; Repath, "Reminiscences of Early Years," 111; Richards, *Ore Dressing*, vol. I, 113–130; Lankton, *Hollowed Ground*, 87–90.

80. *E&MJ* 67 (15 April 1899): 450; John Gillie, testimony in MvC, vol. IX, 3508–3509, 3515; Frank A. Jones, testimony in MvC, vol. IX, 3429; S. J. Gormly, testimony in MvC, vol. VIII, 3310–3312. Maps include: "Map Showing Railroad Connections at Works near Anaconda" (n.d.); Butte Anaconda & Pacific Ry., "Map Showing Connections with Upper and Lower Works, Anaconda" (n.d.); Butte Anaconda & Pacific Ry., "Map Showing Connections with Works at Anaconda" (2 July 1895, revised 7 January 1902); all held in the map case at the Butte–Silver Bow Public Archives, Butte.

Chapter 5. Beginnings of Conflict over Smelter Smoke in the 1890s

1. William Andrews Clark, address to the Montana Constitutional Convention in *Proceedings and Debates of the Constitutional Convention 1889* (Helena: State Publishing Company, 1921), 754, quoted in Malone and Roeder, *Montana: A History*, 150.

2. *Butte Daily Miner* (14 July 1885), and cited in MacMillan, *Smoke Wars*, 26.

3. Thomas Couch's perspectives are documented in his letter press books, containing copies of frequent and very detailed letters he sent to Boston and Montana Consolidated Copper and Silver Mining Company officials in Boston and New York, especially A. S. Bigelow. Couch's letter press books are in the possession of Jim Combs, a collector of memorabilia who has been generous in granting me unrestricted access to the volumes. Copies of the Couch letter press books are in the Quivik Collection. See Couch to A. S. Bigelow, letter dated 1 January 1889.

4. Aiken, "Not Long Ago a Smoking Chimney," 81–82; Tarr, "Searching for a 'Sink,'" 15; Chicago Association of Commerce, *Smoke Abatement and Electrification*,

27–83. See also Stradling, *Smokestacks and Progressives*; Thorseim, *Inventing Pollution*; Uekoetter, *Age of Smoke*.

5. Kirchhoff, "Copper" (1901), 143–144. The government statistics for Montana copper production differ slightly from another reliable source; see Rothwell, ed., *Mineral Industry*, vol. I (1893), 111; Struthers, ed., *Mineral Industry*, vol. X, 182.

6. Stradling, *Smokestacks and Progressives*, 200, 237.

7. These estimates for the six Butte smelters in1890 are derived from Quivik, "Expert Report" (15 August 1997), C-11, C-35, C-46, C-86, C-109, C-143. On the percentage of sulfur in Butte ores and the percentage of that sulfur driven off by roasting and smelting during the nineteenth century, see Peters, *Practice of Copper Smelting*, 62–63.

8. Peters, *Modern American Methods of Copper*, 48–77; Peters, "Roasting of Copper Ores," 283–287; Peters, *Practice of Copper Smelting*, 578–597.

9. Peters, *Modern American Methods of Copper*, 65.

10. Peters, *Modern American Methods of Copper*, 52.

11. MacMillan, *Smoke Wars*, 36–42.

12. *E&MJ* 44 (2 July 1887): 14; Fell, *Ores to Metals*, 49–51; Peters, *Modern American Methods of Copper*, 50–52; Schramm, "Experts in the Smelter Smoke," 196–198.

13. Malone, *Battle for Butte*, 58.

14. MacMillan, *Smoke Wars*, 38–40; Swibold, *Copper Chorus*, 28–33, 43–46.

15. *E&MJ* 50 (6 December 1890): 658, (13 December 1890): 711; MacMillan, *Smoke Wars*, 39–43.

16. *Anaconda Standard*, 27 November 1890, 6 December 1890, cited in MacMillan, *Smoke Wars*, 40–41.

17. MacMillan, "Butte 'Smoke Messiahs,'" 48–55.

18. MacMillan, *Smoke Wars*, 53–57, 265n30.

19. *E&MJ* 44 (23 July 1887): 65–66, 68, (30 July 1887): 82; Sanders, *History of Montana*, vol. II, 1260–1261; Gates, *Michigan Copper and Boston Dollars*, 71; Lankton, *Cradle to Grave*, 59–60.

20. Thomas Couch to A. S. Bigelow, letters dated 14 February 1891, 17 February 1891, 21 February 1891, 7 March 1891, Thomas Couch Papers, Quivik Collection. Couch's incoming and outgoing correspondence are in the hands of several private collectors and the Butte–Silver Bow Public Archives. The letters cited here are from Jim Comb's private collection. The author has made archival copies of all correspondence in private hands and donated them to the Butte–Silver Bow Public Archives.

21. *E&MJ* 46 (1 September 1888): 180, (15 September 1888): 224; *E&MJ* 48 (13 July 1889): 36, (30 November 1889): 481; *E&MJ* 50 (18 October 1890): 461, (29 November 1890): 626.

22. *E&MJ* 45 (11 February 1888): 115; MacMillan, *Smoke Wars*, 67–78; Harvey, *Rio Tinto Company*, 132–133.

23. Thomas Couch to A. S. Bigelow, letters dated 9 November 1891, 24 November 1891, 30 November 1891, 4 January 1892, Quivik Collection. On the Irish-nativist conflict, see David M. Emmons's chapter, "Church, Party, and Fraternity: The Irish and Their Associations," in *Butte Irish*, 94–132. Couch was also anti-Irish. See his

letter to Bigelow dated 23 July 1888, Quivik Collection, in which he calls W. A. Clark an "Irish Mick" with senatorial aspirations.

24. MacMillan, "Butte 'Smoke Messiahs,'" 49; Smith, *Mining America*, 74–75, 79–80.

25. Jensen, *Heritage of Conflict*, 54–56; Lingenfelter, *Hardrock Miners*, 132, 172–173; Wyman, *Hard Rock Epic*, 172–173; Emmons, *Butte Irish*, 184; Frisch, "'Gibraltar of Unionism,'" 132, 264–267.

26. *E&MJ* 51 (7 March 1891): 295, (11 April 1891): 449.

27. *E&MJ* 51 (6 June 1891): 664, (13 June 1891): 703; *E&MJ* 53 (13 February 1892): 213, (19 March 1892): 335, (23 April 1892): 459, (14 May 1892): 529, (4 June 1892): 601; *E&MJ* 54 (2 July 1892): 15, (24 September 1892): 291, (8 October 1892): 351; Leonard, *Who's Who in Engineering*, 1171–1172.

28. *E&MJ* 54 (8 October 1892): 343, (26 November 1892): 507; "Great Falls Water Power and Townsite Company," report on file at the Records Center, box 060906, The Montana Power Company, Butte, n.d., pp. 1–2; *Great Falls and Cascade County*, 13–15, 19–20, 26.

29. *E&MJ* 54 (8 October 1892): 351; *E&MJ* 56 (23 September 1893): 327, (7 October 1893): 379; *Mineral Resources of the United States, 1910*, 165.

30. *E&MJ* 50 (29 November 1890): 626; *E&MJ* 53 (5 March 1892): 286; *E&MJ* 55 (8 April 1893): 327, (17 June 1893): 567.

31. Malone, *Battle for Butte*, 50–53, 141–148; *E&MJ* 54 (29 October 1892): 422; "Certificate of Incorporation of the Montana Ore Purchasing Company," document dated 23 February 1893 (Quivik Collection); McNelis, *Copper King at War*, 16–19. The October 1892 report in *E&MJ* is an example of the Heinze brothers doing business as the MOP prior to the incorporation.

32. Malone, *Battle for Butte*, 50–53, 141–148; *E&MJ* 55 (28 January 1893): 86; *E&MJ* 61 (4 April 1996): 334; *E&MJ* 63 (17 April 1897): 386, (24 April 1897): 409; *E&MJ* 64 (7 August 1897): 166–167, (4 September 1897): 288; *E&MJ* 65 (15 January 1898): 82; *E&MJ* 68 (9 September 1899): 317.

33. *E&MJ* 38 (23 August 1884): 132; Malone and Roeder, *Montana: A History*, 129–140; and Myers, "Butte Rail Connection," 30–37.

34. Marcus Daly to J. B Haggin, 21 July 1890, Dispute with Montana Union Railway and Union Pacific Railroad, Correspondence, 1889–1890, Folder 12, box 63, Montana Historical Society Archives, Anaconda Copper Mining Company Records, Helena.

35. Anonymous, "The Anaconda Electrolytic Copper Refinery," *E&MJ* 62 (19 September 1896): 271–274; Ulke, "Characteristic American Metal Mines," 527; Ulke, *Modern Electrolytic Copper Refining*, 2, 15–19.

36. Marcosson, *Anaconda*, 50–58, 88–91; Malone, *Battle for Butte*, 45–47; *Report of the Anaconda Copper Mining Company* (1895).

37. Sklar, *Corporate Reconstruction of American Capitalism*, 14–20; Lamoreaux, "Sklar's 'The Corporate Reconstruction,'" 36–39.

38. Chandler, *Visible Hand*, 381–382, 415–416; Malone, *Battle for Butte*, 45–47.

39. Rohrbough, *Aspen*, 211–221; Malone, *Battle for Butte*, 54–55; *E&MJ* 56 (1

July 1893): 2–3, (8 July 1893): 38, (22 July 1893): 87, (25 November 1893): 553, (2 December 1893): 601; *E&MJ* 57 (21 April 1894): 374; *E&MJ* 59 (23 February 1895): 182, (2 March 1895): 206; *E&MJ* 60 (13 July 1895): 28; *E&MJ* 62 (12 December 1896): 566, (26 December 1896): 620; *E&MJ* 63 (16 January 1897): 74; *E&MJ* 64 (23 October 1897): 497; *E&MJ* 65 (29 January 1898): 143, (26 March 1898): 369, (14 May 1898): 592; Alice Gold and Silver Mining Co., *Annual Report from January 1, 1894, to January 1, 1895*; Alice Gold and Silver Mining Co., *Annual Report from January 1, 1898, to January 1, 1899*. The two annual reports of the Alice Company and those for the intervening years are in the private collection of Al Hooper, Butte, and copies are in the Quivik Collection.

40. Rothwell, ed., *Mineral Industry*, vol. III (1895), 158–161.

41. Rothwell, ed., *Mineral Industry*, vol. VIII (1900), 160, 164; *Report of the Boston and Montana Consolidated Copper and Silver Mining Company for the Year Ending December 31, 1897*, 15 (Quivik Collection); *Report of the Boston and Montana Consolidated Copper and Silver Mining Company for the Year Ending December 31, 1898*, 13 (Quivik Collection).

42. Repath, "Mechanical Engineer," 143–145; Weed, *Geology and Ore Deposits*, 67; Hovis and Mouat, "Miners, Engineers, and the Transformation," 429–456.

43. Sanborn Fire Insurance maps for Butte in 1900 show that all the smelters had tall stacks. Photographs of the period show that fact as well. See, for example, the photos of each of Butte's smelters in Freeman, *Brief History of Butte*, 83–85.

44. Rohrbough, *Aspen*, 225–230.

45. A. C. Snow to Mr. and Mrs. Major & Ella C. and Capt. Geo S., letter dated 11 October 1897, private collection of Lon Johnson, Columbia Falls, Montana (copies in Quivik Collection).

46. A. C. Snow to Ella Collins, letter dated 19 January 1898, private collection of Lon Johnson, Columbia Falls, Montana (Quivik Collection).

47. *Daily Inter Mountain* (3 February 1898): 5.

48. *Daily Inter Mountain* (12 December 1898): 3; *Butte Miner* (13 December 1898): 6.

49. *Daily Inter Mountain* (12 December 1898): 3.

50. *Daily Inter Mountain* (14 December 1898): 6.

51. *Daily Inter Mountain* (15 December 1898): 5.

52. *Daily Inter Mountain* (13 December 1898): 2, (14 December 1898): 2.

53. *Butte Miner* (28 December 1898): 8, (29 December 1898): 5; *Daily Inter Mountain* (24 December 1898): 5.

54. *Butte Miner* (31 December 1898): 8; *Daily Inter Mountain* (31 December 1898): 5.

55. *Butte Miner* (8 January 1899): 5.

56. *Butte Miner* (8 January 1899): 5, (15 April 1899): 4, (2 May 1899): 5; *Daily Inter Mountain* (8 January 1899): 7.

57. *Daily Inter Mountain* (10 May 1899): 3; *Butte Miner* (3 May 1899): 6, (11 May 1899): 5, (14 May 1899): 5.

58. Lawson, *Frenzied Finance*, 237–252; Malone, *Battle for Butte*, 131–135; *E&MJ* 66 (27 August 1898): 258, (17 September 1898): 355; 67 (29 April 1899): 494.

59. *Boston Globe* (26 April 1899): 9.

60. *Boston Globe* (28 April 1899): 1 (morning) and 9 (evening); *E&MJ* 58 (24 November 1894): 494, (22 December 1894): 590; *E&MJ* 67 (29 April 1899): 516; Malone, *Battle for Butte*, 132–140; Walker, *Copper Mines of Butte*, 8–9; *Montana Supreme Court, MacGinnis, Respondent, v. Boston & Montana Consolidated Copper & Silver Mining Company, et al., Appellants*, case no. 2,001 (1 February 1904) 430–433, 448–453 (Quivik Collection).

61. Lawson, *Frenzied Finance*, 301–306; *E&MJ* 67 (8 April 1899): 420; *E&MJ* 68 (22 July 1899): 108, (2 September 1899): 289.

62. *Butte Miner* (10 November 1899): 5–6.

63. *E&MJ* 77 (14 April 1904): 620; *E&MJ* 78 (24 November 1904): 842, (15 December 1904): 962; *E&MJ* 79 (20 April 1905): 784, (27 April 1905): 833, (11 May 1905): 930, (15 June 1905): 1159; *E&MJ* 81 (17 February 1906): 350, (24 February 1906): 390–391, (3 March 1906): 445; *E&MJ* 82 (7 July 1906): 37, (21 July 1906): 130, (28 July 1906): 176. Joel Tarr and other urban historians have described a typical U.S. trend of smoke industries tending to move into suburban or exurban settings, in part to improve urban conditions; see Tarr, "Urban Environmental History," 81–84. All the evidence here points to the smelters in Butte closing because Anaconda/Amalgamated could run a more efficient copper mining and smelting enterprise in Montana by consolidating all smelting operations at Great Falls and Anaconda, and eventually at Anaconda only.

64. *E&MJ* 89 (18 June 1910): 1290; *Mineral Resources of the United States, 1924*, Part I—Metals (Washington, D.C.: Government Printing Office, 1927), 364.

Chapter 6. Corporate Consolidation and the Farmers' Suits

1. "The 'Smoke' Meeting," *Silver State*, 15 April 1903.

2. Novak, *People's Welfare*; and Freyfogle, *Land We Share*.

3. Steinberg, *Nature Incorporated*, 24–49, 99–239.

4. Struthers, *Mineral Industry*, vol. X, 176.

5. Leeson, *History of Montana*, 1077; Malone and Roeder, *Montana: A History*, 11–113, 179.

6. Lamoreaux, *Great Merger Movement*.

7. Dobb, "Pennies from Hell," 39–54.

8. Malone, *Battle for Butte*, 154; Swibold, *Copper Chorus*, 101.

9. Schmitz, "World's Largest Industrial Companies," 87, presents a table showing the world's one hundred largest industrial companies, with the Anaconda listed fifth, and as the largest copper-mining company.

10. Richter, "Amalgamated Copper Company," 387–407, and Malone, *Battle for Butte*, 131–206, describe Amalgamated's acquisition of the several companies, the consolidation of all Amalgamated's companies into the Anaconda Copper Mining Company, and the dissolution of Amalgamated. On the legal and political battles focusing on Amalgamated's looming control of the state, see also Toole, *Montana: An Uncommon Land*, 186–210; McNelis, *Copper King at War*. On the legal battles over Amalgamated's damage to the Deer Lodge Valley, see MacMillan, *Smoke Wars*,

83–143; Bakken, "Was There Arsenic," 30–41; Quivik, "Tragic Montana Career," 32–47, 92–94.

11. *E&MJ* 61 (13 June 1896): 561; *E&MJ* 62 (29 August 1896): 193, (14 November 1896): 470; *E&MJ* 63 (23 January 1897): 85; *E&MJ* 66 (12 November 1898): 587.

12. *E&MJ* 67 (29 April 1899): 494, 516, (27 May 1899): 630; 68 (2 September 1899): 288, (2 December 1899): 678, (23 December 1899): 769, (30 December 1899): 798; 69 (20 January 1900): 89, (27 January 1900): 120, (10 March 1900): 298–299, (16 June 1900): 720; 70 (11 August 1900): 169, (25 August 1900): 229; Hutchinson, "Sampling Spitzlutten Products" (Quivik Collection). The latter report is one of several by Hutchinson, all of which show results of Hutchinson's tests to see how various kinds of rolls, mills, and tables performed in sequence with each other, how they worked on materials of various particle sizes, and how much slimes they produced. The specific report cited here directly addresses how particular equipment being tested at the Upper and Lower Works might be used at the new works. The use of the old works as test facilities for machinery and processes for the new works is not explicitly mentioned in documents describing the history of the construction of the new works. Rather, one may surmise that the old works were used for that function by the facts that: 1) the equipment initially installed at the new works was installed at the old works during the planning stages of construction for the new works, as indicated by such documents as the Hutchinson reports, despite the fact that such new equipment at the old works would only be used for about two years before the old works closed, and 2) Frank Klepetko, who supervised the design and construction of the new works, was also in charge of the installation of new equipment at the old works.

13. *E&MJ* 66 (29 October 1898): 527, (5 November 1898): 558, (12 November 1898): 587; *E&MJ* 67 (7 January 1899): 7, (29 April 1899): 516, (27 May 1899): 630; *E&MJ* 68 (1 July 1899): 8–9, (2 September 1899): 288, (2 December 1899): 678, (23 December 1899): 769; *E&MJ* 69 (10 March 1900): 298–299, (16 June 1900): 720; *E&MJ* 70 (11 August 1900): 169, (25 August 1900): 229; Repath, "Mechanical Engineer," 143; Hutchinson, "Sampling Spitzlutten Products."

14. *E&MJ* 68 (23 December 1899): 769; *E&MJ* 69 (20 January 1900): 89, (10 March 1900): 298–299, (16 June 1900): 720; Laist, "History of Reverberatory Smelting," 51–52.

15. *E&MJ* 70 (11 August 1900): 169; *E&MJ* 72 (19 October 1900): 505, (9 November 1901): 614, (23 November 1900): 678; *E&MJ* 73 (1 February 1902): 191; *E&MJ* 74 (12 July 1902): 62; Laist, "History of Reverberatory Smelting," 51–52.

16. Ligon Johnson to U.S. Attorney General, letter dated 13 February 1909 (p. 6), box 277, DOJ Central File, Records of the U.S. Department of Justice, Record Group 60, National Archives, College Park, MD. Johnson was special counsel to the attorney general, working on a suit to be brought by the United States against the ACM. E. P. Mathewson, the smelter superintendent, had told Johnson about the large amount of copper that the ACM had been able to recover from the hills back of the Old Works. This was to convey to the attorney general "the enormous metallic waste from a smelter plant."

17. Marcosson, *Anaconda*, 76; Malone, *Battle for Butte*, 166.

18. *E&MJ* 73 (1 March 1902): 311–313; "Anaconda Smelter—1903," photo no. 940–409, Montana Historical Society Photo Archives, Helena.

19. Schwerin, "Method of Concentrating at Anaconda," 388–389; Austin, "Washoe Plant," 440–442; Richards, *Ore Dressing*, vol. IV, 1932. Richards reported that the Washoe concentrator required about 21 tons of water per ton of ore.

20. "Kills Horses and Cattle," *Silver State*, 17 September 1902.

21. Library of Congress, "About the New North-West (Deer Lodge, Mont.) 1869–1897," Chronicling America webpage, https://www.loc.gov/item/sn84038125/; Montana Historical Society, "Union List of Montana Newspapers in Montana Repositories," 1986; Sanders, *History of Montana*, vol. I, 770–772. The *Silver State* changed its name to the *Silver State Post* in 1925 and is still being published.

22. "Their Cattle Dying," *Silver State*, 15 October 1902; "Fumes Kill Cattle," *Silver State*, 15 October 1902.

23. "Their Cattle Dying," *Silver State*, 15 October 1902; "The Smelter Poison," *Silver State*, 26 November 1902.

24. "Investigations in Anaconda," *Helena Independent*, 23 November 1902.

25. "Meeting to Consider Ways and Means," *Powell County Record*, 26 December 1902; "Deer Lodge Valley Held Mass Meeting," *Butte Miner*, 30 December 1902; "Seeking Information," *Silver State*, 17 December 1902; "The People Aroused!" *Silver State*, 26 November 1902; the quote is from "A Big Meeting," *Silver State*, 31 December 1902.

26. "The Committee's Report," *Silver State*, 7 January 1903.

27. "Smoke Nuisance Is Investigated," *Butte Miner*, 19 December 1902; "Echoes from the Hills," *Silver State*, 28 January 1903; the quote is in "Board of Health Discusses Smoke," *Butte Miner*, 18 January 1903.

28. "Business Men Talked Smoke," *Butte Miner*, 20 January 1903.

29. "Big Smelter Stack Dynamited Yesterday," *Butte Miner*, 27 January 1903.

30. "Echoes from the Hills," *Silver State*, 28 January 1903.

31. "Senate Made Good Headway," *Butte Miner*, 30 January 1903; "Senator Kohrs' Bills," *Silver State*, 11 February 1903; "Ruin of the Deer Lodge Valley," *The Press*, 14 February 1903.

32. "The Conception of Kohrs," *Butte Miner*, 2 February 1903.

33. "Senator Kohrs' Four Good Bills," *The Press*, 7 February 1903; "Ruin of the Deer Lodge Valley," *The Press*, 14 February 1903. The quotes are small bullets that follow "Corporation Hands in Legislation," *The Press*, 14 February 1903.

34. "Smelter Smoke Nuisance," *Silver State*, 11 February 1903.

35. "Kohrs' Smoke Bills Made Them Be Good," *Reveille*, 13 February 1903.

36. "Smoke Problem Solved at the Washoe Smelter," *Butte Miner*, 11 February 1903. The *Miner* was one of the "kerosene newspapers" publishing this article, which was a press release from Amalgamated. For overall descriptions of the new flue-and-stack system built at the Washoe smelter, see *E&MJ* 76 (4 July 1903): 26; "The New Flues and Stack at Anaconda," *E&MJ* 76 (6 December 1903): 962–965; McDougall, "Flues and Flue Dust," 26–34.

37. "Should Fool No One," *Silver State*, 11 February 1903; "Kohrs' Smoke Bills Made Them Be Good," *Reveille*, 13 February 1903; "Ruin of the Deer Lodge Valley," *The Press*, 14 February 1903.

38. "Kohrs' Smoke Bills Made Them Be Good" and "Commendable Work," *Butte Miner*, 13 February 1903; "Three Smoke Bills Die in the Senate," *Reveille*, 13 February 1903; Editorial, *Silver State*, 18 February 1903; "Men Who Corrupt Montana Politics," *The Press*, 21 February 1903.

39. "Four Days More of the Session," *Butte Miner*, 2 March 1903; "Change of Venue Bill Becomes a Law," *Butte Miner*, 3 March 1903; "Substitute Bill Now a Law," *Western News*, 4 March 1903; Malone, *Battle for Butte*, 177.

40. "Smelter Smoke Nuisance," *Silver State*, 11 February 1903; "The 'Smoke' Meeting," *Silver State*, 15 April 1903.

41. "Fifty Thousand Damages," *Silver State*, 12 December 1902; "Papers Filed," *Silver State*, 18 March 1903; "Would Close Smelters," *Anaconda Standard*, 21 February 1903. The Bielenberg quote is from *Silver State*, 13 May 1903.

42. "Heinze Forces Are behind Bulkheads," *Butte Miner*, 2 March 1903; "Both Sides Are Now Enjoined," *Butte Miner*, 3 March 1903.

43. "'Twill Be a Mighty Giant Three Hundred Feet in Height," *Anaconda Standard*, 11 February 1903; "They Need Laborers at the New Washoe Plant," *Anaconda Standard*, 17 February 1903; "Building the Monster Smokestack," *Anaconda Standard*, 9 March 1903; "Big Chimney Is Growing," *Anaconda Standard*, 11 May 1903; "To Dine the Farmers," *Anaconda Standard*, 24 June 1903.

44. "To Dine the Farmers," *Anaconda Standard*, 24 June 1903; "Smelter Guest Who Saw the New Stack's Last Brick Laid," *Anaconda Standard*, 26 June 1903; "Jolly Deer Lodgers Feast in the Smelter's Big Flue," *Anaconda Standard*, 26 June 1903.

45. "To Suspend Work Temporarily at the Washoe Smelting Plant," *Anaconda Standard*, 28 June 1903; "Nine Mines Close," *Anaconda Standard*, 1 July 1903; "Stop Making of Copper," *Anaconda Standard*, 1 July 1903; "Gagnon Mine Shut Down," *Anaconda Standard*, 1 July 1903.

46. "Amalgamated Mines All Running," *Anaconda Standard*, 18 September 1903; "In Mine and Smelter Many New Sites Will Greet These Eastern Visitors Today," *Anaconda Standard*, 19 September 1903.

47. "Running Full Blast," *Anaconda Standard*, 19 September 1903.

48. "Thousands of Miners Enjoy Big Barbecue," *Butte Miner*, 14 October 1903.

49. "Thousands of Miners Enjoy Big Barbecue," *Butte Miner*, 14 October 1903.

50. Malone, *Battle for Butte*, 172–173.

51. "Thousands of Miners Enjoy Big Barbecue," *Butte Miner*, 14 October 1903.

52. "Injunction against B. & M. Reinforced by Judge Clancy," *Butte Miner*, 23 October 1903; Malone, *Battle for Butte*, 167–173.

53. "Montana Must Face an Industrial Crisis," "15,000 Men Suffer by Shutdown," and "Smelters Doomed to Idleness," *Butte Miner*, 23 October 1903; Malone, *Battle for Butte*, 174–177.

54. "Amalgamated Resumes," *Silver State*, 11 November 1903; Malone, *Battle for Butte*, 177–179.

55. "A New Party," *Silver State*, 2 December 1903; "The Case in a Nutshell," *Silver State*, 17 February 1904.

56. *E&MJ* 88 (24 July 1909): 168–169; Wiggin, "Great Falls System of Concentration," 210–212; Richards, *Ore Dressing*, vol. IV, 1841, 1847; Bakken and Bakken, "Goldfish Died," 38–51. On the complaints from downstream, see folders 4–13, box 193, Records of the ACM, M-169, MHS Archives.

57. Bakken and Bakken, "Goldfish Died," 44–51.

58. Hiram W. Hixon, "Hunting Metals for Their Hides," *E&MJ* 88 (24 July 1909): 168–169.

59. Peters, *Practice of Copper Smelting*, 311–343; W. A. Kidney, testimony in MVC, vol. VIII, 2917–2920, 2927, 2939–2940, 2945; David T. Gilbert, testimony in MvC, vol. VIII, 2964, 2971–2977.

60. Gage, testimony in MvC, vol. VIII, 2760–2767, 2770–2772; Arthur M. Day, testimony in MvC, vol. VIII, 2778, 2802–2812, 2816, 2851–2852, 2857–2858; Henry Webking, testimony in MvC, vol. VIII, 2889–2892, 2902–2903; *E&MJ* 68 (16 December 1899): 739; Freeman, *Brief History of Butte*, 84; "Butte & Boston Tailings Dump Looking NE," photo in exhibits accompanying "Magone transcripts," National Archives, Seattle Branch; U.S. Geological Survey, "Butte Special Map," *Butte Special Folio*, Folio 38 (1897); "Sanborn Map for Butte" (1900), sheet 81.

61. *E&MJ* 68 (22 July 1899): 108, (2 September 1899): 289; *E&MJ* 69 (19 May 1900): 599; *E&MJ* 82 (28 July 1906): 176, (29 September 1906): 611; John Gillie, testimony in *William E Wall and Louis Foss v. Parrot Silver and Copper Company and Anaconda Copper Mining Company* 8, 78–80; *Mineral Industry, 1900*, vol. IX, 194; "Sanborn Map for Butte" (1900), sheet 82; Bushnell, "Precipitating Copper from Butte Mine-Water," 1229–1230; Ed McCool, testimony in MvC, vol. VIII, 3065–3066, 3073–3074; George F. Bartlett, testimony in MvC, vol. VIII, 3103; Rombauer, testimony in MvC, vol. VII, 2626; Robert A. McArthur, testimony in MvC, vol. VIII.

62. *E&MJ* 80 (12 August 1905): 273; *E&MJ* 82 (28 July 1906): 176, (29 September 1906): 611; Febles, "Precipitation of Copper," 191–192; Stevens, *Copper Handbook*, 757.

63. *E&MJ* 53 (2 January 1892): 57; *E&MJ* 54 (5 November 1892): 447; *E&MJ* 57 (5 May 1894): 494–495; Goodale, "Concentration of Ores," 603, 619–629; C. W. Goodale, testimony in MvC, vol. VIII, 2824, 2831, 2834–2836, 2841–2844; Lou P. Smith, testimony in MvC, vol. VII, 2746; Richard R. Vail, testimony in MvC, vol. VIII, 3137–3140, 3143–3148; Michael Curnow, testimony in MvC, vol. VIII, 3156, 3159–3166, 3183–3185; James R. Anderson, testimony in MvC, vol. VIII, 3192–3196, 3199–3204; "Sanborn Map for Butte" (1900), sheets 90, 98.

64. A. H. Wethey, testimony in MvC, vol. IV, 37–38.

65. Doull and Pratt, "Report on the Deposition of Tailings along Silver Creek in Vicinity of Silver Bow, Montana," 1–5 (Quivik Collection); James Doull, testimony in MvC, vol. IV, 570–571.

66. A. H. Wethey, testimony in MvC, vol. IV, 53–57.

67. A. H. Wethey, testimony in MvC, vol. IV, 9–10, 15, 25, 50–54, 57–64, 67, 101; F. A. Pratt, testimony in MvC, vol. IV, 489–490; James Doull, testimony in MvC, vol.

IV, 549–552, 586–587, 594–597; Jerry Rourke, testimony in MvC, vol. IV, 188–190, 209, 219–220, 232–234; "Outlet of Colusa-Parrot (Butte Reduction Works) Tailings Dump," photo presented in MvC.

68. A. H. Wethey testimony, MvC, vol. IV, 8, 11; Jerry Rourke, testimony in MvC, vol. IV, 197–200, 209–212, 228–231; U.S. Geological Survey, *Mineral Resources of the United States, Calendar Year 1907*, Part I—Metallic Products, 336; Wethey, "Smelting Plant of the Butte," 1153–1155; "Slag-Wall and Culvert Construction," *E&MJ* 89 (5 March 1910): 501.

69. A. H. Wethey, testimony in MvC, vol. IV, 12, 25, 69–70; James Doull, testimony in MvC, vol. IV, 567; Hugh McLeod, testimony in MvC, vol. VI, 1469–1470.

70. "Farmers against Smelter Owners," *Anaconda Standard*, 7 January 1903; Curtis, *Gambling on Ore*, 173–176. Note that Curtis misidentifies the Colusa-Parrot as the owner of the Parrot smelter, not of Clark's Butte Reduction Works.

71. In the early twentieth century, the Deer Lodge River was the name of the stream that flowed the full length of the Deer Lodge Valley from where Silver Bow Creek emerges from Durant Canyon to the river's confluence with the Little Blackfoot River to form the Clark Fork. Today, the upper portion of that stream retains the name Silver Bow Creek past the emergence from Durant Canyon to the confluence with Warm Springs Creek, at which point it takes the name Clark Fork all the way downstream to the Clark Fork's confluence with the Columbia River.

72. *E&MJ* 76 (12 September 1903): 389; Austin, "Washoe Plant," 442, 483; Laist and Wiggin, "Flotation Concentration at Anaconda, Mont.," 486; S. J. Gormly, testimony in MvC, vol. VIII, 3288–3290; Frank Noble, testimony in MvC, vol. VIII, 3252–3257, 3265–3266, 3278–3282; ACM Co., Office of Civil Engineer, "Map Showing Slime and Tailings Fields and Dams" (Anaconda, MT, 29 September 1918), box 278, straight numerical files (Enclosures), Entry 112, D. J. Central Files, RG-60, General Records of the U.S. Department of Justice, National Archives, Washington, D.C.; "New Works, 27 May 1924," photo no. PAc 82–62.278, Montana Historical Society Photo Archives, Helena. The method of excavation is depicted in "Slum ponds—loading bucket, Bradley Plant—East Anaconda, May 1912," photo no. PAc 82–62.195, Montana Historical Society Photo Archives, Helena.

73. Austin, "Washoe Plant," 442, 460–462, 474–475, 483; *E&MJ* 83 (6 April 1907): 660.

74. John Gillie, testimony in MvC, vol. IX, 3462–3465, 3509–3520; S. J. Gormly, testimony in MvC, vol. VIII, 3284–3288, 3306; Frank Noble, testimony in MvC, vol. VIII, 3244–3251, 3256–3259, 3266–3271. Photographic depictions of the tailings dump include "Anaconda Smelter," photo no. PAc 82–62.MM14, and "Anaconda Copper Mining Company Reduction Works [no date]," photo no. 940–411, both at Montana Historical Society Photo Archives, Helena.

75. John Gillie, testimony in MvC, vol. IX, 3462–3465, 3509–3520; S. J. Gormly, testimony in MvC, vol. VIII, 3284–3288, 3306; Frank Noble, testimony in MvC, vol. VIII, 3244–3251, 3256–3259, 3266–3271; W. F. Flynn, "Report on Tailings Disposal at Anaconda Reduction Works," ACM Co. report dated 26 January 1937, 1, file 10, box 90, MC-169, ACM.

76. Charles H. Williams, testimony in MvC, vol. I, 62, 70–74; James F. Saville, testimony in MvC, vol. I, 390, 394–398; Peter Valiton, testimony in MvC, vol. I, 562, 568, 573–575, 579–581.

77. *William Norton, et al, v. Colorado Smelting and Mining Company, et al.*, complaint filed 22 June 1901 in the District Court of the Second Judicial District of the State of Montana in and for the County of Silver Bow. See also *Peter Valiton v. Anaconda Copper Mining Company*, docket no. 48, Third Judicial District Court, State of Montana, County of Powell; *Elizabeth Ann Staffanson v. Anaconda Copper Mining Company*, docket no. 2390, District Court of the Third Judicial District of the State of Montana, Deer Lodge County; *Henry Mason v. Anaconda Copper Mining Company*, docket no. 2391, District Court of the Third Judicial District of the State of Montana, Deer Lodge County.

78. "Important Suit Begun," *Silver State*, 9 September 1903.

79. Hugh Magone, testimony in MvC, vol. III, 590–598, 667.

80. Hugh Magone, testimony in MvC, vol. III, 598–607, 677; Traphagen, "Recent Litigation Involving Questions," 251.

81. "Ranchers, Attention!" *Silver State*, 6 April 1904; "The Ranchers' Meeting," *Silver State*, 13 March 1904; news note about Clowes Miles in *Silver State*, 13 March 1904; Special Correspondence from San Francisco, *E&MJ* 88 (11 December 1909): 1190; Kelley, "Mining Debris Controversy," 331–346.

82. Oliver T. Crane, "Report of the Master," in *Hugh Magone v. Colorado Smelting & Mining Company, et al.*, The Circuit Court of the United States, Ninth Circuit, District of Montana (26 March 1906), 1–2; "Don't Look Good for Deer Lodge Valley People," *Silver State*, 13 April 1904; "Smelter Tailings Suit," *Silver State*, 15 February 1905; "May Join Butte in Big Litigation," *Butte Evening News*, 27 March 1905.

83. Frederick, *Rugged Justice*, 123–124; Sanders, *History of Montana*, 1227.

84. Piatt, ed., *Story of Butte*, 60; Sanders, *History of Montana*, 864–865.

85. Crane, "Report of the Master," 1–2; "Transcript of Testimony," in MvC, vol. I, 4; Traphagen, "Recent Litigation," 251.

86. Robert W. Ellis, testimony in MvC, vol. I, 6–61; "Big Money Is Involved," *Anaconda Standard*, 28 March 1905.

87. C. H. Williams, testimony in MvC, vol. I, 62–236; "Many Farmers Attend Trial," *Butte Evening News*, 29 March 1905.

88. C. H. Williams, testimony in MvC, vol. I, 62–236.

89. Peter Valiton, testimony in MvC, vol. I, 562–563, 619–620.

90. James F. Saville, testimony in MvC, vol. I, 390, 417.

91. Crane, "Report of the Master," 3; Traphagen, "Recent Litigation," 251–252; "Make a Record in Magone Suit," *Butte Evening News*, 9 April 1905.

92. Traphagen, "Recent Litigation," 252; the numerous old-time miners and concentrator foremen and superintendents have been cited throughout the earlier chapters of this book. Their testimony provides a major source for this work.

93. "Make a Record in Magone Suit," *Butte Evening News*, 9 April 1905; "Master Will Not Visit Magone Land," *Butte Miner*, 27 May 1905; "Magone Hearing May Be Resumed Tomorrow," *Anaconda Standard*, 20 August 1905; "Magone Hearing Re-

sumed," *Anaconda Standard*, 10 October 1905; "Magone-Smelter Case Now up to the Master," *Anaconda Standard*, 13 November 1905.

94. John Jacobson, testimony in MvC, vol. X, 4087–4091, 4095–4096, 4116–4118; Ephraim Staffanson, testimony in MvC, vol. X, 4129–4132; Frank Threlkeld, testimony in MvC, vol. X, 4144–4145, 4150–4155. On the balancing doctrine, see Rosen, "Differing Perceptions of the Value," 303–381.

95. Crane, "Report of Master"; Traphagen, "Recent Litigation," 252.

96. *Mountain Copper Co. v. United States*, no. 1,203, United States Court of Appeals, Ninth Circuit, 26 February 1906; Frederick, *Rugged Justice*, 115–116.

97. On the historical propensity of farmers joining organizations, see Tontz, "Memberships of General Farmers' Organizations," 143–156; Grant, "Rocky Mountain Husbandman," 34–43; Remele, "'God Helps Those Who Help,'" 22–33; Rothstein, "Farmer Movements and Organization," 161–181.

98. Letter from a committee of the Deer Lodge Valley Farmers' Association to the Anaconda Copper Mining Company dated 4 March 1905, quoted in *The Federal Reporter*, vol. 186, 818.

99. News note, *Silver State*, 13 April 1904; "The Poisoned Sheep," *Silver State*, 20 April 1904.

100. News note, *Silver State* 18 May 1904; W. D. Harkins, testimony in BvA, vol. 13, 4851–4854; "The Poisoned Sheep," *Silver State*, 20 April 1904; Swain and Harkins, "Papers on Smelter Smoke," 970.

101. "Smoke Meeting Held," and "Citizens' Mass Meeting!" *Silver State*, 28 December 1904; "The Smoke Meeting," *Silver State*, 4 January 1905; "The Big Suit Begun," *Silver State*, 12 April 1905.

102. "Salt Lake's Smelter Smoke," *Silver State*, 28 April 1904; "The People Aroused," *Silver State*, 22 February 1905; Lamborn and Peterson, "Substance of the Land," 312–320.

103. Letter from a committee of the Deer Lodge Valley Farmers' Association to the Anaconda Copper Mining Company dated 4 March 1905, quoted in *The Federal Reporter*, vol. 186, 817–819 (see also 790); "The Big Suit Begun," *Silver State*, 12 April 1905.

104. "Companies File Answer," *Anaconda Standard*, 20 June 1905.

105. "Anaconda Smoke Case," *Silver State*, 21 June 1905; "Views the Valley," *Silver State*, 23 August 1905; "Seeing the Hay and the Cattle," *Anaconda Standard*, 17 August 1905; "Judge Hunt Views Fields and Cattle," *Butte Miner*, 17 August 1905.

106. MacMillan, "History of the Struggle," 129–131; Salmon, "Arsenical Poisoning from Smelter Smoke," *American Veterinary Review* 39 (April 1911): 21; McIntosh, "Million for a Smoke Eater," 162.

107. Angus D. Smith, testimony in BvA, vol. 2, 641–736; Kenneth D. Smith, testimony in BvA, vol. 2, 760–809, vol. 3, 810–897; Bryon Howells, testimony in BvA, vol. 3, 1007–1067; Conrad Kohrs, testimony in BvA, vol. 7, 2590–2622; Ephraim Staffanson, testimony in BvA, vol. 8, 2911–2978; George Parrott, testimony in BvA, vol. 8, 3152–3184, vol. 9, 3185–3215; William Evans, testimony in BvA, vol. 9, 3468–3529; Morgan Evans, testimony in BvA, vol. 11, 3980–4024; W. C. Staton, testimony in BvA, vol. 15, 5956–5968.

108. John W. Collins, testimony in BvA, vol. 7, 2503–2517. William Evans testified (p. 3487) that in 1906 there were only about 10 percent the number of cattle and horses there had been prior to the Washoe smelter starting up in 1902.

109. B. F. Notestine, testimony in BvA, vol. 8, 2831–2906; Ephraim Staffanson testimony, vol. 8, 2911–2915; William Evans, testimony in BvA, 3471; William M. Evans, testimony in BvA, vol. 10, 3918.

110. John Bohn, testimony in BvA, vol. 10, 3639–3709.

111. Charles T. Harrington, testimony in BvA, vol. 26, 10088–10142; Sebastian Rodony, testimony in BvA, vol. 29, 11365–11375; Manuel Strozzi, testimony in BvA, vol. 29, 11415–11427; Felix Tatti, testimony in BvA, vol. 29, 11422–11425; Gy Berti, testimony in BvA, vol. 29, 11442–11447; Peter Antonioli, testimony in BvA, vol. 31, 12228–12247.

112. John Wilson Hammer, testimony in BvA, vol. 19, 7337–7615; Ben T. Crosswhite, testimony in BvA, vol. 46, 18073–18203.

113. Peter Jessen, testimony in BvA, vol. 26, 10211–10264.

114. A. M. Day, testimony in BvA, vol. 23, 8947–9014.

115. The English judicial system began accepting expert scientific testimony (as distinct from eyewitness testimony) in the late eighteenth century, and U.S. courts, which had adopted the English legal system, adopted the practice in the mid-nineteenth century. A good overview of the wedding, in the U.S. and England, of two otherwise distinct worlds of human endeavor, the legal world and the scientific world, is Golan, *Laws of Men*. His history shows how the legal and scientific approaches to evidence and conclusions of fact were mutually beneficial, as experts testified at trials.

116. W. D. Harkins, testimony in BvA, vol. 13, 4851–4976; Robert E. Swain, testimony in BvA, vol. 4, 1206–1260; Daniel E. Salmon, testimony in BvA, vol. 50, 19867–20024; Bakken, "Was There Arsenic," 39–41; Weigle, "Report on Conditions of Plant"; Haywood, *Injury to Vegetation* (1908); Formad, "Effect of Smelter Fumes."

117. Leonard, *Who's Who in Engineering*, 1402; Marcosson, *Metal Magic*, 58–65 ; Fell, *Ores to Metals*, 222–224, 231–236; Rickard, *Interviews with Mining Engineers*, 335–350.

118. Leonard Pearson, testimony in BvA, vol. 48, 18891–19038; Theobald Smith, testimony in BvA, vol. 47, 18428–18575; V. A. Moore, testimony in BvA, vol. 44, 17319–17506; Duncan McEachran, testimony in BvA, vol. 45, 17716–17787; Bakken, "Was There Arsenic," 39–41; Hughes, "Cornell Veterinarians," 299–302; Leonard, *Cornell Heritage*, 81–85, 91–96, 96–100; F. W. Traphagen, testimony in BvA, vol. 35, 13731–13813; Harry Snyder, testimony in BvA, vol. 60, 23914–23998. See also D. McEachran to Veranus A. Moore, letter dated 2 January 1906; E. P. Mathewson to Veranus A. Moore, letter dated 2 January 1906; D. McEachran to D. E. Salmon, letter dated 2 January 1906; E. P. Mathewson to D. E. Salmon, letter dated 2 January 1906; Duncan McEachran to E. P. Mathewson, letter dated 27 January 1906; all in file 4, box 445, ACM.

119. Leonard, *Who's Who in Engineering*, 719–720, 1402; E. P. Mathewson to Prof. S. Fortier, letters dated 10 and 24 February 1906, file 4, box 445, ACM; E. P.

Mathewson to Prof. R. W. Smith, letter dated 24 February 1906, file 4, box 445, ACM; E. P. Mathewson to Prof. Ralph E. Smith, letter dated 6 March 1906, file 4, box 445, ACM.

120. Quivik, "Tragic Montana Career," 32–47, 92–94.

121. Bierer, *Short History of Veterinary Medicine*, 94–97; Jones, *Valuing Animals*, 40–43; Greene, *Horses at Work*, 269–270.

122. V. A. Moore Collection, no. 24–1–119, Rare and Manuscript Collections, Carl A. Kroch Library, Cornell University, Ithaca, NY, hereafter cited as VAM; Gossel, "Emergence of American Bacteriology," 114–120, 134–136.

123. Sellers, "Factory as Environment," 55–83; Golan, *Laws of Men*, 176–210.

124. Box 73 of the V. A. Moore Papers contains all of Moore's personal financial ledgers, showing both income and expenses for nearly every year of his life, VAM.

125. Sweet to Moore, letter dated 15 March 1907; Sweet to Moore, letter dated 2 April 1907; Hardcastle to Moore, letter dated 23 April 1907, VAM.

126. Bakken, "Was There Arsenic," 38; Mathewson to Moore, letter dated 26 December 1905; Moore to Kelley, letter dated 16 January 1907; Kelley to Moore, letter dated 31 January 1907, VAM.

127. Moore to Gardiner, letter dated 20 January 1906; Flocken to Moore, latter dated 14 April 1906, VAM.

128. McEachran to Moore, letter dated 3 January 1907, VAM.

129. William H. Hunt, decree in *Bliss v. Anaconda*, 25 January 1909, cited in *The Federal Reporter*, vol. 186, 810, 815–816.

130. It should be stated that "the science of environmental impact" is a concept from the late twentieth century, not from the time of the Bliss case. The concept at the time could be called the science of personal injury, or the science of injury to property.

131. Gardiner to Moore, letter dated 2 June 1910; Moore to Kelley, letter dated 8 June 1910; Moore to Gardiner, letter dated 23 July 1910; Moore to Gardiner, 1 February 1911, VAM.

132. Gossel, "Emergence of American Bacteriology," 118–129; Bierer, *Short History of Veterinary Medicine*, 66–75.

133. Young, *Pure Food*, 221–224.

134. *New York Times* (1 August 1905): 2, (2 August 105): 1, (6 August 1905): 1, (8 August 1905): 7, (10 August 1905): 2, (31 August 1905): 4, (7 September 1905): 4; Jones, *Valuing Animals*, 84.

135. Farrington to Moore, letter dated 2 January 1906, VAM; "Retirement of Dr. Salmon," *American Veterinary Review* 29 (October 1905): 674–675.

136. Bakken, "Was There Arsenic," 38; Flocken to Moore, letter dated 15 April 1906; Kelley to Moore, letter dated 22 November 1906; Kelley to Moore, letter dated 1 January 1907; Pearson to Moore, letter dated 15 January 1907, VAM.

137. W. Horace Hoskins, "Obituary, Daniel Elmer Salmon," *American Veterinary Review* 46 (October 1914): 93–95; *Butte Daily Post* (2 September 1914): 2; Smithcors, *Veterinarian in America*, 93. For more details on Salmon's work for the farmers and his last year in Butte, see Quivik, "Tragic Montana Career."

138. Bell to Moore, letter dated 4 October 1906; Bell to Moore, letter dated 26 November 1906, VAM.

139. "The Montana Smoke Case," *American Veterinary Review* 32 (November 1907): 249–253; D. Arthur Hughes to the Editors of the *American Veterinary Review*, letter dated 20 June 1908, *American Veterinary Review* 33 (July 1908): 493–497.

140. Salmon, "Arsenical Poisoning from Smelter Smoke," nine issues in volumes 39, 40, and 41.

141. Salmon, "Arsenical Poisoning from Smelter Smoke," *American Veterinary Review* 39 (April 1911): 14.

142. For parallel developments in industrial hygiene, especially in the beginnings of recognition of dangers from chronic exposure to toxins, see Donegan, "For the Good of Us All," 107–116; Bale, "Compensation Crisis," 399–403; Angela Nugent Young, "Interpreting the Dangerous Trades," 22–33.

143. Salmon, "Arsenical Poisoning from Smelter Smoke," *American Veterinary Review* 39 (April 1911): 15; Rosenberg, *Trial of the Assassin Guiteau*, 66–67; Golan, *Laws of Men*, 107–143.

144. Daniel E. Salmon, testimony in *Bliss v. ACM*, vol. 51, 20036–20038. Describing the encounter between the ACM attorneys and Salmon, Bakken, in "Was There Arsenic," 40, attributes the cross-examination to Con Kelley and employs the more sensational language used by the newspapers in reporting the incident. Bakken describes Kelley asking Salmon why President Roosevelt had fired him.

145. For an analysis of the Anaconda company's legal strategy, see Bakken, "Was There Arsenic," 30–41.

146. Hunt, decree in *Bliss v. Anaconda Copper Mining Company*, 25 January 1909, 167 F. 342, United States Circuit Court for the District of Montana.

147. Erskine Ross, order in *McCarthy v. Bunker Hill & Sullivan Mining & Concentrating Co*, 164 F. 927, 6 November 1908, United States Court of Appeals for the Ninth Circuit, no. 1, 397; Frederick, *Rugged Justice*, 117.

148. William H. Hunt, opinion in the case *Fred J. Bliss v. Anaconda Copper Mining Company and Washoe Copper Company*, dated 25 January 1909.

149. Frederick, *Rugged Justice*, 118; Rosen, "Differing Perceptions of the Value." Although the farmers appealed to the U.S. Supreme Court, that body refused to hear the appeal on a technicality.

150. Swain and Harkins, "Papers on Smelter Smoke," 970–998; Swain and Harkins, "Papers on Smelter Smoke," 915–946.

151. Salmon, "Arsenical Poisoning from Smelter Smoke," in *American Veterinary Review* 39 (April 1911): 14–22, (June 1911): 245–260, (August 1911): 517–536, 40 (November 1911): 164–178, (February 1912): 579–590, (March 1912): 739–747, and 41 (May 1912): 164–171, (June 1912): 300–308, (July 1912): 395–421.

152. Haywood, "Injury to Vegetation and Animal Life by Smelter Fumes," 998–1009; Haywood, *Injury to Vegetation and Animal Life by Smelter Wastes* (1908); Haywood, *Injury to Vegetation and Animal Life by Smelter Wastes* (Revised, 1910).

153. Formad, "Effect of Smelter Fumes."

154. Salmon, "Arsenical Poisoning from Smelter Smoke," *American Veterinary Review* 41 (July 1912): 421.

CHAPTER 7. INTEGRATION INTO A GLOBAL CORPORATION

1. Richter, "Copper-Mining Industry," 264.

2. Richter, "Copper-Mining Industry," 263–267, 270; Young, "American Copper Frontier," 7.

3. The developments of the ACM's complex technological system follow the trajectory generally analyzed in Hughes, *Networks of Power*; Hughes, *American Genesis*; and Chandler, *Visible Hand.*

4. Marcosson, *Anaconda*, 100–101; Malone, *Battle for Butte*, 166.

5. Marcosson, *Anaconda*, 67–78; Malone, *Battle for Butte*, 19.

6. "Thomas Cole's Career," *Butte Miner*, 14 February 1906; *E&MJ* 79 (29 June 1905): 1254; Richter, "Amalgamated Copper Company," 391–392.

7. "Olive Branch Now Reposes upon the Hill," *Anaconda Standard*, 14 February 1906; "Marks End of Mining War and Beginning of New Era for Butte," *Butte Miner*, 14 February 1906; *E&MJ* 81 (24 February 1906): 390–391; *E&MJ* 82 (28 July 1906): 176; Richter, "Amalgamated Copper Company," 391–392.

8. Fay, ed., *Mineral Industry*, 166; *Mineral Resources of the United States, Calendar Year 1910* (Washington, D.C.: Government Printing Office, 1911), 158, 201; Marcosson, *Anaconda*, 138; Richter, "Amalgamated Copper Company," 394–397.

9. Fay, ed., *Mineral Industry*, 166; *Mineral Resources of the United States, Calendar Year 1910*, 158, 201; Richter, "Amalgamated Copper Company," 394–397.

10. Schmitz, "World's Largest Industrial Companies," 85–96.

11. John Gillie, testimony in MvC, vol. IX, 3437–3438; John Gillie, testimony in *Wall and Foss v. Parrot Silver and Copper Company and Anaconda Copper Mining Company*, transcripts of the District Court of the United States, Ninth Circuit, District of Montana, dated 17 January 1913, 2–3, National Archives, Seattle Branch; "John Gillie," *Mining Journal* (15 June 1941): 14.

12. Daly, "Evolution of Mining Practice," 280.

13. Gillie, "Use of Electricity in Mining," 817.

14. Gillie, "Use of Electricity in Mining," 817; Bird, *Story of Montana Power*, 13; Kirk, *History of the Montana Power*, 51–68. For information on the development of long-distance transmission technology, see Hughes, *Networks of Power*, 131–139, 162–163, 262–284; Sharlin, "Electrical Generation and Transmission," 578–592; Rushmore and Lof, *Hydroelectric Power Stations*, 5–13.

15. Gillie, "Use of Electricity in Mining," 818; Shovers, "Miners, Managers, and Machines," 21, 34.

16. Reynolds, "Narrow Window of Opportunity," 9–10; Johnson, "Electrical Power, Copper," 24–37; Quivik, "Steel Transmission Towers," 67–69.

17. Johnson, "Electrical Power, Copper," 28–29; Hebgen, "Great Falls Hydroelectric Power Plant," 273–288.

18. Richards, "New Development in Compressed Air," 409–410; Nordberg, "Compressed Air System," 826–841; Read, "Compressed-Air Hoisting at Butte," 554–556.

19. Gillie, "Use of Electricity in Mining," 818–819; Gillie does not explicitly state whether his figures are for the ACM only or for all mining operations in Butte, but his narrative seems to imply that they are for all of Butte.

20. Johnson, "Electrical Power, Copper," 30–31.

21. "Butte, Anaconda & Pacific Electrification," *E&MJ* 97 (March 14, 1914): 559; Fiege, "General Electric, Montana Power."

22. Kirk, *History of the Montana Power*, 134–136, 227–230, 235–236, 241–247; "Construction of the Great Falls," 3–4; Bird, *Story of Montana Power*, 64, 68.

23. *Copper Handbook*, vol. XII, 147–149; Daly, "Evolution of Mining Practice," 280–284.

24. Young, *Elements of Mining*, 1, 13–15; Nystrom, *Seeing Underground*, 7–11.

25. Weed, *Geology and Ore Deposits*, 114; Daly, "Evolution of Mining Practice," 280.

26. Weed, *Geology and Ore Deposits*, 115, 127; Linforth, "Applied Geology in the Butte," 112.

27. Weed, *Geology and Ore Deposits*, 15–16; Sales, "Ore Discovery and Development," 277; "The Reminiscences of Reno Sales, January 1961," no. 437, vol. 3, p. 3, Mineral Engineering Project, Columbia Center for Oral History Center, Columbia University, New York City; Nystrom, *Seeing Underground*, 88–95.

28. Gow, "Mine Surveying Methods," 1209–1210; Sanders, *History of Montana*, vol. II, 1164; Sanders, *History of Montana*, vol. III, 1747; "Reminiscences of Reno Sales," 4–5, 7; Nystrom, *Seeing Underground*, 88–95.

29. Sales, "Ore Discovery and Development," 278–279; Daly, "Evolution of Mining Practice," 281–282; "Reminiscences of Reno Sales," 14–15, 20–21; Eric Nystrom, "Underground Mine Maps," 110.

30. *Mineral Resources of the United States, 1910*, Part I—Metals (Washington, D.C.: Government Printing Office, 1911), 173, 175; Fay, *Mineral Industry in 1910*, 152–153.

31. *Copper Handbook*, vol. IV, 610, 695; Marcosson, *Anaconda*, 80; Richter, "Amalgamated Copper Company," 392.

32. *Copper Handbook*, vol. X, 986; Marcosson, *Anaconda*, 258; Richter, "Amalgamated Copper Company," 392–393.

33. *Copper Handbook*, vol. III, 320–326; *Copper Handbook*, vol. VIII, 476–478, 746–757; Marcosson, *Anaconda*, 251–258; Richter, "Amalgamated Copper Company," 394.

34. *Copper Handbook*, vol. XI, 468–469; Parsons, *Porphyry Coppers*, 2–18, 70–77, 233–239; Richter, "Amalgamated Copper Company," 399; Hovis and Mouat, "Miners, Engineers," 443–445; LeCain, *Mass Destruction*, 118–131.

35. *Copper Handbook*, vol. XI, 473–476; Richter, "Amalgamated Copper Company," 399–401.

36. Richter, "Amalgamated Copper Company," 401–403; Marcosson, *Anaconda*, 143–144.

37. The labor turmoil of the 1910s is a chapter in Butte's history that has been described by many historians. See, for example, Toole, *Twentieth-Century Montana*, 123–138; Calvert, *Gibraltar*, 71–91; Emmons, *Butte Irish*, 255–291.

38. McGlynn, "Flying the Red Flag," 6–8; Calvert, *Gibraltar*, 71–91.

39. *E&MJ* 98 (1 August 1914): 227, (15 August 1914): 326, (22 August 1914): 371, (29 August 1914): 412, (5 September 1914): 446, (21 November 1914): 932–933.

40. *Report of the Anaconda Copper Mining Company for the Year Ending December 31, 1914*, 4–5; *Report of the Anaconda Copper Mining Company for the Year Ending December 31, 1915*, 3–4; *E&MJ* 98 (15 August 1914): 326; *E&MJ* 99 (10 April 1915): 673, (24 April 1915): 755–756.

41. Calvert, *Gibraltar*, 103–105; Emmons, *Butte Irish*, 340–364; Punke, *Fire and Brimstone*, 5–7.

42. Harrington, *Lessons from the Granite Mountain*. The best overall history of the Granite Mountain disaster is Punke's *Fire and Brimstone*. See also Ammons, *Darkness Lit by Heroes*, which is a detailed, almost minute-by-minute account of the disaster itself.

43. More miners have been killed in single coal-mining disasters because of the added danger of explosive methane gas in many coal mines. There are no explosive gases normally associated with hardrock mines, hence the distinction of the Granite Mountain disaster as the country's worst in a hardrock mine.

44. Gutfeld, *Montana's Agony*, 23–33; Calvert, *Gibraltar*, 108–110; Punke, *Fire and Brimstone*, 202–212; Botkin, *Frank Little and the IWW*, 269–311.

45. Calvert, *Gibraltar*, 109–113; Murphy, *Mining Cultures*, 24–26, 210; Phipps, "A. S. Embree," 35–39.

46. Sales, "Ore Deposits at Butte, Mont.," 58–61.

47. *Mineral Resources of the United States, 1907*, Part I—Metals (Washington, D.C.: Government Printing Office, 1908): 672; *Mineral Resources of the United States, 1915*, Part I—Metals (Washington, D.C.: Government Printing Office, 1917): 969; *Mineral Resources of the United States, 1917*, Part I—Metals (Washington, D.C.: Government Printing Office,1921): 856, 877.

48. *E&MJ* 78 (15 December 1904): 962; *E&MJ* 79 (12 January 1905): 111, (30 March 1905): 641, (11 May 1905): 930; *E&MJ* 80 (5 August 1905): 224; *E&MJ* 82 (18 August 1906): 325; *E&MJ* 84 (26 October 1907): 804; *E&MJ* 85 (23 May 1908): 1075; *E&MJ* 87 (9 January 1909): 108, (1 May 1909): 912, (8 May 1909): 971; *E&MJ* 88 (23 October 1909): 844, (30 October 1909): 889, (13 November 1909): 995, (11 December 1909): 1190–1191, (18 December 1909): 1244, (25 December 1909): 1293.

49. *E&MJ* 89 (21 May 1910): 1080; *E&MJ* 92 (4 November 1911): 910, (11 November 1911): 960; *E&MJ* 93 (23 June 1912): 1289; *E&MJ* 94 (19 October 1912): 757; *E&MJ* 97 (27 June 1914): 1314.

50. *E&MJ* 92 (29 July 1911): 230, (16 December 1911): 1172; *E&MJ* 96 (11 October 1913): 713, (22 November 1913): 999; Simons, "Concentrator of the Timber Butte," 911; Rickard, *History of American Mining*, 404–406; Hovis and Mouat, "Miners, Engineers," 449.

51. Electro-winning is similar to electrolytic refining. In the latter, atoms of metal, like copper, move from an anode, through an electrolytic fluid, to be deposited on a cathode; in the former, atoms of metal, like zinc, move from being dissolved in the fluid to be deposited on a cathode.

52. *E&MJ* 99 (13 March 1915): 511; *E&MJ* 102 (5 August 1916): 277; Ingalls, "Electrolytic Zinc," 425–428; Laist et al., "Electrolytic Zinc Plant of Anaconda," 699–753; ACM Co., *Anaconda Reduction Works*, 21; Ralston, *Electrolytic Deposition and Hydrometallurgy*, 7.

53. *M&SP* 117 (21 December 1918): 814; Parsons, *Porphyry Coppers*, 239, 247–250, 331–332; Marcosson, *Anaconda*, 212–213. Ricketts's views on Thompson and Thornton are recorded in Rickard, *Interviews with Mining Engineers*, 443–445.

54. Weed, *Mines Handbook*, vol. 12, 1202–1203; Weed, *Mines Handbook*, vol. 13, 608–609.

55. Gahl, "History of the Flotation Process," 577; "Reminiscences of Reno Sales," 17; Parsons, *Porphyry Coppers*, 249–250, 327–332; Marcosson, *Anaconda*, 211–213.

56. Weed, *Mines Handbook*, vol. 14, 88, 206, 959–963, 1267, 1770.

Chapter 8. The United States Files Suit

1. John D. Ryan, quoted in Ingalls, "Business of Mining," his commencement address to the University of Missouri's School of Mines, Rolla, 26 May 1916. The address was published in *Mining & Scientific Press* 113 (19 August 1916): 276–281. Ingalls was a leading metallurgist of the period. About engineers, Ingalls told the students: "We have several kinds of engineers. There are first of all the great body of men who operated our mines and metallurgical works. They are the subalterns, captains, and colonels of our army. Fewer in number are those who advise about the development of mines, build metallurgical works, devise new metallurgical processes. They are the staff-officers. Many of them are great scientists, whose work is often inadequately requited. Finally there are the engineers in whom the business instinct is highly developed—men like Jackling, Hoover, Hammond, Bradley—who are our generals. I admire the engineers of each of these classes. Each in his own way, humble or high, is doing necessary work and contributing to the wealth that the mining industry bestows upon the world." Ingalls named Marcus Daly and John D. Ryan as great generals who, incidentally, had no engineering training. Such generals fulfilled "one of the great functions of the business generally, namely the control of expenditures within the limits of what can be afforded, having in mind first of all the interests of stockholders, the owners of the business" (277).

2. Goodwin, *Bully Pulpit*, 342–351. On Roosevelt's early efforts to regulate corporations through a Bureau of Corporations, see Johnson, "Theodore Roosevelt and the Bureau," 571–590; Bittlingmayer, "Antitrust and Business Activity," 363–401. On Roosevelt's commitments to forest conservation, see Forbes, "Vision for Cultivating a Nation,'" 226–231; Schulman, "Governing Nature, Nurturing Government," 375–403.

3. *Silver State*, 8 April 1903; "Mayor's Banquet to Nation's Chief," *Reveille*, 29 May 1903; Marcosson, *Anaconda*, 106–107; Malone, *Battle for Butte*, 164–165; Brands, *T.R.: The Last Romantic*, 474–475.

4. Haywood, *Injury to Vegetation by Smelter Fumes*, U.S. Bureau of Chemistry Bulletin No. 89, 65, 72–86.

5. Haywood, *Injury to Vegetation and Animal Life by Smelter Wastes*, U.S. Bureau of Chemistry Bulletin No. 113, 13–19; Maysilles, *Ducktown Smoke*, 94–101, 154–155, 164–169.

6. Weigle, "Report on Conditions of Plant"; Haywood, *Injury to Vegetation and*

Animal Life by Smelter Wastes, U.S. Bureau of Chemistry Bulletin No. 113, 34; MacMillan, *Smoke Wars*, 150–152.

7. Weigle, "Report on Conditions of Plant," 21–22.

8. The Roosevelt administration's preparations of the case against the ACM are described in MacMillan, "History of the Struggle," 190–257; Bloom, *Murder of a Landscape*, 86, 92, 96–98. Copies of correspondence among the attorney general, President Roosevelt, Ligon Johnson, ACM officials, technical experts, and others involved in the several copper smelter smoke cases are held in the Department of Justice Central Files, Record Group 60, Records of the U.S. Dept. of Justice, National Archives, Washington, D.C. Correspondence pertaining to the Anaconda case is found in boxes 273–278, straight numerical file (Enclosures); box 513, straight numerical file (Folded); box 514, straight numerical file (Folded); and boxes 886 and 886A. These documents are hereafter cited as box no., DOJ. See also Ligon Johnson to Hon. Chas. J. Bonaparte, Atty Gen'l, letter dated 7 November 1907, box 274, and "Memorandum relative to the status of smelter fumes litigation and matters appertaining thereto," document dated 29 August 1915, box 886, both in DOJ.

9. *Who Was Who in America*, vol. 3, 452; Theodore Roosevelt to Ligon Johnson, letter dated 11 January 1908, reel 347, vol. 77, series 2, Theodore Roosevelt Papers, Library of Congress, Washington, D.C. (hereafter cited as TR); Maysilles, *Ducktown Smoke*, 106–107, 153–154. On Roosevelt's commitment to forest conservation, see Hays, *Conservation and the Gospel of Efficiency*, 27–48; Brands, *T.R.: The Last Romantic*, 623–625.

10. Johnson to Bonaparte, 7 November 1907; "Memorandum relative to the status of smelter fumes litigation," 29 August 1915, box 274, DOJ. Johnson described the work he did for the State of Georgia on the Ducktown case and for the U.S. on the various western smelter cases in a 1914 letter: Ligon Johnson to W. J. Hughes, letter dated 21 August 1914, box 886, DOJ.

11. MacMillan, *Smoke Wars*, 159–168; Toole, *Montana: An Uncommon Land*, 212–214.

12. All of these observations are from affidavits by Deer Lodge Valley residents recorded in 1908 and filed in box 274, DOJ: Lizz Newsome (affidavit recorded 29 October 1908), 124–127; Margret Quinlan (affidavit recorded 29 October 1908), 121–123; Henry H. Eccleston (affidavit recorded 31 October 1908), 143–145.

13. Margret Quinlan (affidavit recorded 29 October 1908), 121–123; Henry H. Eccleston (affidavit recorded 31 October 1908), 143–145; Walter C. Staton (affidavit recorded 27 October 1908), 29–34; John E. Jones (affidavit recorded 29 October 1908), 106–109; Lizz Newsome (affidavit recorded 29 October 1908), 124–127.

14. MacMillan, *Smoke Wars*, 168–171; Johnson to Bonaparte, 7 November 1907; Johnson to the Attorney General, letter dated 24 November 1908, box 274, DOJ.

15. Murphy, *Comical History of Montana*, 261–262. Jerre C. Murphy had been the personal secretary of Robert LaFollette in Wisconsin until 1903, when he ventured west to become editor of the *Butte Inter Mountain*, one of the daily newspapers in Butte sympathetic to Amalgamated; see Drake, *Education of an Anti-Imperialist*, 91; Swibold, *Copper Chorus*, 115, 136, 259.

16. Charles Bonaparte to the President, letter dated 8 December 1908, reel 86, series 1, TR; Theodore Roosevelt to Charles Bonaparte, letter dated 9 December 1908, reel 353, series 2, TR; MacMillan, *Smoke Wars*, 171–175.

17. Peirce et al., "Report on the Effects" (unpublished report prepared for the U.S. Department of Justice and dated 5 February 1913), 90, box 84b, General Files Prior to 1954, Records of the U.S. Bureau of Mines, RG-70, National Archives, College Park, MD (hereafter cited as BOM). This report is in two sections, one on botanical investigations (pp. 1–90) and one on chemical investigations (pagination starting over again, pp. 1–80). The conclusion quoted here is from the last page of the section on botanical investigations. The section on chemical investigations ends with ten conclusions describing evidence of widespread levels of sulfur high enough to be harmful to plant life and of arsenic high enough to be harmful to animal life.

18. Lovejoy, "Report on Damage to Deerlodge" (unpublished U.S. Forest Service report dated 3 July 1911), box 84b, General Files Prior to 1954, Records of the U.S. Bureau of Mines, RG-70, NARA College Park.

19. Charles A. Doremus to C. F. Kelley, letter dated 21 December 1908, and F. J. Falding to C. F. Kelley, letter dated 28 December 1908, box 277, straight numerical files (Enclosures), DOJ; MacMillan, *Smoke Wars*, 178–179. For the amounts of ore treated and sulfur discharged at Ducktown, see Quinn, "Industry and Environment," 592.

20. In this regard, both Johnson and ACM officials were hoping to implement the engineering ideal, described by Gorman, in *Redefining Efficiency*, to reduce waste and thereby both reduce pollution and generate profit.

21. F. J. Falding to C. F. Kelley, letter dated 28 December 1908 (p 2), box 277, straight numerical files (Enclosures), DOJ.

22. J. Parke Channing to Ligon Johnson, letter dated 24 December 1908 (p. 2), box 277, straight numerical files (Enclosures), DOJ.

23. Theodore Roosevelt to Attorney General, letters dated 4 and 18 February 1909, reel 354, series 2, TR; MacMillan, *Smoke Wars*, 186–188.

24. For a general discussion of the Taft administration's actions leading up to the filing of the suit against the ACM, see MacMillan, *Smoke Wars*, 189–204. See also Ligon Johnson to the Attorney General, letters dated 10 November 1909 and 31 January 1910, box 274, DOJ; Complaint filed by the U.S. in the Circuit Court of the United States, Ninth Circuit, District of Montana, *United States v. Anaconda Copper Mining Company and Washoe Copper Company*, no. 967 in equity, document dated 10 March 1910, both in box 514, straight numerical files (Folded), DOJ.

25. Complaint in *U.S. v. ACM Co.*, 10 March 1910.

26. Malone, *Battle for Butte*, 205–206; *E&MJ* 89 (9 April 1910): 788–789; "Indenture between the Boston and Montana Consolidated Copper and Silver Mining Company of Montana and the Anaconda Copper Mining Company," document dated 31 May 1910 (Quivik Collection); "Minutes of a Special Meeting of the Stockholders of the Anaconda Copper Mining Company," document dated 23 March 1910, 12–14 (Quivik Collection); "Indenture between the Red Metal Mining Company and the Anaconda Copper Mining Company," document dated 31 May 1910 (Quivik

Collection); "Agreement between the Amalgamated Copper Company and the Anaconda Copper Mining Company," document dated 26 May 1910 (Quivik Collection). A copy of the contract between Clark and Ryan, dated 14 May 1910, is appended to the "Agreement between Amalgamated and the ACM" as Exhibit A.

27. See, for example, the correspondence between Attorney General Wickersham and the ACM's Kelley during the closing days of negotiations: C. F. Kelley to the Attorney General, letter dated 9 March 1911, and Attorney General to C. F. Kelley, letter dated 30 March 1911, box 886, DOJ.

28. For a general discussion of the negotiations between the government and the ACM following filing of the suit, culminating in the 1911 agreement, see MacMillan, *Smoke Wars*, 211–214.

29. "Agreement between United States Department of Justice and Anaconda Copper Mining Company" (agreement dated 13 April 1911), box 84b, BOM.

30. "Agreement between United States Department of Justice and Anaconda Copper Mining Company," 13 April 1911; *E&MJ* 91 (1 April 1911): 683, (6 May 1911): 892.

31. The classic historical treatment of this aspect of the Progressive Era, especially regarding the regulation of exploitation of natural resources, is Hays, *Conservation and the Gospel of Efficiency*. See also Layton, *Revolt of the Engineers*, 59–68; Noble, *America by Design*, 61–63; Hughes, *American Genesis*, 246–248; Nye, *Electrifying America*, 166–167, 180–181; Jordan, *Machine-Age*, 13–67; among others.

32. The quote is from John Hays Hammond to H. Foster Bain, letter dated 30 January 1924 (Quivik Collection). On the overall composition of the commission, see MacMillan, "History of the Struggle," 302–306. On Hammond, see Spence, *Mining Engineers*, 224–225; Hammond, *Autobiography of John Hays Hammond*, 562–563; Marcosson, *Anaconda*, 40.

33. MacMillan, "History of the Struggle," 303–305. On Ricketts, see *Sixth Annual Report*, 19; Marcosson, *Anaconda*, 40, 256–259. On Holmes, see *E&MJ* 90 (10 September 1910): 487; *E&MJ* 100 (17 July 1915): 117.

34. Ligon Johnson to the Attorney General, letter dated 7 November 1911, box 886, DOJ; Caldwell, memorandum dated 29 June 1914, box 886A, DOJ.

35. In a 1914 letter to the attorney general reviewing his service to the government, Johnson paraphrased the 1911 agreement thus: "By the stipulations, the company bound itself to install under the direction of the Commission, every device commercially feasible that might remedy or lessen the fume damage." Ligon Johnson to the Attorney General, letter dated 11 February 1914, box 886, DOJ.

36. Johnson to Attorney General, 7 November 1911, p. 9, box 886, DOJ.

37. Johnson to Attorney General, letter dated 12 December 1911, box 886, DOJ.

38. Joseph William Blankenship, testimony in BvA, vol. 38, 15082–15083; Arthur E. Wells, "Report of the Anaconda Smelter Smoke Commission" (unpublished report dated 1920), 17–19, box 278, straight numerical files, DOJ. Another version of this report exists in box 84b, BOM. It is a carbon copy of a different original, and the pagination of the version is different although the text is the same. For example, the passages cited here are on pp. 11–12. See also John Hays Hammond to the Attorney General, letter dated 13 December 1912, box 886, DOJ.

39. Wells, "Report of the Anaconda Smelter" (1920), 19–27. Dunn's earliest report in the Smoke Commission records, "Report of Dust Determination at Stack" (box 84b, BOM), is dated 3 April 1911, eight days before the ACM and the government signed the agreement creating the Smoke Commission. The report describes tests he and his staff conducted on Washoe smelter emissions in January, February, and March 1911.

40. Dunn, "Determination of Gases in Smelter," 648–689.

41. Leonard, *Who's Who in Engineering*, 460; Wells, "Report of the Anaconda Smelter" (1920), 19–20, 31–32; Dunn, "Determination of Gases in Smelter," 687; Dunn, "Report of Dust Determination" (cover page); "Memorandum of Services of Messrs. Elton, Kellogg and Welch (H.V.), with the Anaconda Commission," undated memo, box 84b, BOM; F. G. Cottrell to The Anaconda Smelter Commission, letter dated 17 June 1913, box 84b, BOM.

42. Wells, "Report of the Anaconda Smelter" (1920), 24–26, 29–30.

43. Wells, "Report of the Anaconda Smelter" (1920), 74–85.

44. John Hays Hammond, Louis D. Ricketts, and J. A. Holmes to the Attorney General, letter dated 24 February 1913, box 886, DOJ.

45. Wells, "Report of the Anaconda Smelter" (1920), 74–75, 87–95.

46. Wells, "Report of the Anaconda Smelter" (1920), 38, 45–47; Bender and Goe, "Production of Arsenic Trioxide," 325. For a comparison between the initial arsenic plant and the one installed to Smoke Commissions specifications, see ACM Co. Staff, *Brief Description of the Washoe Smelter* (1913), 30, and ACM Co. Staff, *Brief Description of the Washoe Smelter* (1916), 40. For a description the tests leading to the change in arsenic plant design, see Elton, "Arsenic Trioxide from Flue Dust," 692–702.

47. Herrick, "World's Largest Chimney," 257.

48. Goodale and Klepinger, "Great Falls Flue System," 643.

49. Higgins, "New 506-Ft. Chimney," 156; Herrick, "World's Largest Chimney," 257; Goodale and Klepinger, "Great Falls Flue System," 569, 572, 643; Frank Klepetko to C. W. Goodale, letter dated 9 May 1898, folder 4, box 161, ACM.

50. Herrick, "World's Largest Chimney," 258–259; Goodale and Klepinger, "Great Falls Flue System," 573, 597–615, 642; Hofman, *Metallurgy of Lead*, 392; Peters, *Practice of Copper Smelting*, 587–595.

51. Higgins, "New 506-Ft. Chimney," 156, 159; Herrick, "World's Largest Chimney," 257; Goodale and Klepinger, "Great Falls Flue System," 616–624.

52. Wells, "Report of the Anaconda Smelter" (1920), 39, 42–43.

53. "Cottrell Process for Condensing Smelter"; Cottrell, "Electrical Precipitation of Suspended Particles," 542–550; Cottrell, "Electrical Fume-Precipitation," 512–520; "Research Corporation and the Cottrell"; Cottrell, "Smoke Problem in the Mineral," 867–899.

54. Wells, "Report of the Anaconda Smelter" (1920), 52–57.

55. Wells, "Report of the Anaconda Smelter" (1920), 47. Wells drew this language from "Progress Report, Anaconda Smelter Commission," undated (ca. 1916) unpublished report, 30, box 84b, BOM.

56. There is ample correspondence between the smelter management and the

ACM's top corporate officials to demonstrate that the company was proceeding with the mid-decade improvements with no concern for the effects the alterations to the plant would have on emanations from the smelter stack, and that the company was proceeding without thought of the Smoke Commission. See, e.g., Frederick Laist to C. F. Kelley, letters dated 6 July 1915, 28 July 1915; 12 October 1915, 16 October 1915, 17 November 1915, 3 December 1915 (Quivik Collection).

57. Leonard, *Who's Who in Engineering*, 1206; Marcosson, *Anaconda*, 110; *E&MJ* 99 (16 January 1915): 163.

58. Frederick Laist, forward dated 2 November 1915 in Anaconda Copper Mining Company Reduction Department, *Report of the Research Department* (1915), n.p.

59. Anaconda Copper Mining Company Reduction Department, *Report of the Research Department* (1915), 2–105; ACM Co. Reduction Department, *Report of the Research Department for the Year 1915* (Anaconda, MT: Anaconda Copper Mining Company, 1916), 9–239; ACM Co. Reduction Department, *Report of the Research Department* (1917), 9–107; ACM Co. Reduction Department, *Report of the Research Department* (1918), 9–95.

60. Anaconda Copper Mining Company Reduction Department, "Report of the Concentration Department" (1915), 115–407; ACM Co. Reduction Department, "Report of the Concentration Department" (1916), 249–386; ACM Co. Reduction Department, "Report of the Concentration Department" (1917), 117–208; ACM Co. Reduction Department, "Report of the Concentration Department" (1918), 101–136.

61. Anaconda Copper Mining Company Reduction Department, "Report of the Testing Department" (1915), 415–496; ACM Co. Reduction Department, "Report of the Testing Department" (1916), 399–421; ACM Co. Reduction Department, "Report of the Testing Department" (1917), 215–235; ACM Co. Reduction Department, "Report of the Testing Department" (1918), 155–160.

62. *E&MJ* 93 (18 May 1912): 977; *E&MJ* 94 (21 December 1912): 1189; Wiggin, "Great Falls System of Concentration," 209–238 (see especially tables on 237).

63. Wiggin, "Great Falls System of Concentration," 236; Crowfoot, "Development of the Round Table," 417–426, 433–434, 440–441; Ingalls, "Recent Observations in Butte," 937–943; Hayden, "Concentration of Slimes at Anaconda," 239–265; Laist and Wiggin, "Slime-Concentrating Plant at Anaconda," 470–484; *E&MJ* 96 (25 October 1913): 805; *E&MJ* 97 (10 January 1914): 100, (11 April 1914): 783.

64. Simons, "Evolution of the Round Table," 351–352; Crowfoot, "Development of the Round Table," 420–441; Hayden, "Concentration of Slimes at Anaconda," 248, 257–258; Ingalls, "Recent Observations in Butte," 937–943.

65. *E&MJ* 83 (23 May 1907): 1014; *E&MJ* 93 (24 February 1912): 403, (13 April 1912): 759; *E&MJ* 95 (10 May 1913): 969; *E&MJ* 96 (20 September 1913): 552, (27 September 1913): 617, 96 (18 October 1913): 758, (22 November 1913): 962–963; Walker, "Metallurgy of Copper in 1911," 47–49; Walker, "Metallurgy of Copper in 1912," 103–104; Laist, "Roasting and Leaching Tailings," 362–377; Laist and Aldrich, "Experimental Leaching at Anaconda," 671–690.

66. As with the MacDougall roasting furnaces used to roast concentrates, this roasting furnace in the leaching plant discharged some copper and other metals with

the flue dust. The ACM added an experimental Cottrell treater in 1913 to the flue leaving the leaching plant in an effort to recover values in the flue dust and as part of its experimentation, under the supervision of the Anaconda Smelter Smoke Commission, with using the Cottrell electrostatic precipitation system to treat smoke at the smelter generally. See *E&MJ* 96 (13 December 1913): 1146.

67. *E&MJ* 96 (18 October 1913): 758, (15 November 1913): 952, (22 November 1913): 962–963; *E&MJ* 97 (3 January 1914): 42; Peterson, "Copper Leaching at Butte, Montana," 423; Peterson, "Bullwhacker Leaching Plant," 585–586; Ingalls, "Recent Observations in Butte," 937–943.

68. *E&MJ* 97 (3 January 1914): 42, (10 January 1914): 101, (7 February 1914): 344, (11 April 1914): 783, (9 May 1914): 937–943; *E&MJ* 99 (2 January 1915): 31; (12 June 1915): 1047; Laist and Aldrich, "2,000-Ton Leaching Plant," 866–878; ACM Co., *Brief Description of the Washoe Smelter*, 35, 38.

69. Laist and Wiggin, "Slime-Concentrating Plant at Anaconda," 470–471.

70. Laist, "Improvements at the Reduction Works," 418–419.

71. *M&SP* 104 (2 March 1912): 350; *M&SP* 106 (15 February 1913): 291, (19 April 1913): 574–575, (28 June 1913): 1001; *M&SP* 107 (29 November 1913): 866; *M&SP* 108 (17 January 1914): 154–155; *M&SP* 111 (3 July 1915): 7–10; *M&SP* 112 (26 February 1916): 320, (6 May 1916): 676; Rickard, "Miami, Arizona: The Milling of the Ore—V," 680; Gahl, "Development of Copper Ore Leaching," 636–638; Rickard, *History of American Mining*, 409; Parsons, *Porphyry Coppers*, 244.

72. "Anaconda" (editorial), *M&SP* 112 (26 February 1916): 296; "Agreement between Minerals Separation and the Inspiration-Anaconda Companies," *M&SP* 113 (16 September 1916): 296; Laist and Wiggin, "Flotation Concentration at Anaconda," 486–487; Rickard, *History of American Mining*, 408.

73. Laist, "Improvements at the Reduction Works," 418–419; U.S. Geological Survey, *Mineral Resources of the United States, Calendar Year 1915*, Part I—Metals (Washington, D.C.: Government Printing Office, 1917), 608; Rickard, *History of American Mining*, 408; ACM Co., *Anaconda Reduction Works* (1920), 19–21.

74. Laist, "Improvements at the Reduction Works," 418–419; U.S. Geological Survey, *Mineral Resources of the United States* (1917), 607–608.

75. Wells, "Report of the Anaconda Smelter" (1920), 39, 42–43.

76. ACM Co., "Dry Tons Tailings and Overflow to Waste from the Following Materials from the Time of Installation of Flotation at Anaconda Reduction Works," unpublished report dated ca. 1945 (Quivik Collection); "Tailings. Loading tailings for leaching plant, 2 July 1915," photo no. PAc 82–62.2438, Montana Historical Society Photo Archives, Helena.

77. Rickard, *History of American Mining*, 408.

78. *E&MJ* 99 (13 March 1915): 506.

79. *E&MJ* 101 (19 February 1916): 361–362; Frederick Laist to C. F. Kelley, letter dated 3 December 1915 (Quivik Collection); E. P. Mathewson to M. H. Gerry Jr., letter dated 10 February 1916 (Quivik Collection); ACM Co., *Report of the Anaconda Copper Mining Company for the Year Ending December 31st, 1915*, 3; *Mineral Industry*, vol. I (1893), 111; Rickard, *History of American Mining*, 408.

80. Austin, "Washoe Plant," 462–468.

81. Hofman, "Notes on the Metallurgy," 289–302; Mathewson, "Development of the Reverberatory Furnace," 781–805; *E&MJ* 99 (17 April 1915): 698; Bender, "Coal-Dust Fired Reverberatories," 750–751; Laist, "Changes in Smelting Practice," 635; Bender, "Development of Copper Smelting," 302; ACM Co. Staff, *Brief Description of the Washoe Smelter* (1916), 21; ACM Co., *Anaconda Reduction Works* (1920), 28; ACM Co., *Brief Description of the Washoe Smelter* (1928), 43; ACM Co., *Brief Description of the Washoe Smelter* (1938), 20.

82. Austin, "Washoe Plant," 442–449, 455–460; *E&MJ* 82 (17 November 1906): 928; *E&MJ* 83 (6 April 1907): 660; Offerhaus, "Copper Blast Furnace Smelting," 243–250; Peters, "Notes on Copper Smelting," 735.

83. Bender, "Development of Copper Smelting," 302; Laist, "History of Reverberatory Smelting," 53, 56.

84. Bender, "Development of Copper Smelting," 302–303.

85. Bender, "Development of Copper Smelting," 302–303; Kelly and Laist, "Development of Copper Converting," 122–131.

86. "New Works from cliffs across the valley, 24 June 1925," photo no. PAc 82–62.525, Montana Historical Society Photo Archives, Helena.

87. "Flues: Construction, Service and Efficiency of the Flue System at the Washoe Reduction Works," in ACM Co. Reduction Department, *Report of the Testing Department for the Years 1912, 1913 and 1914*, 455.

88. Bakken reviews the ACM's assertions in "Was There Arsenic," 40.

89. T. T. Black to H. C. Gardiner, letter dated 9 February 1911 (Quivik Collection).

90. *E&MJ* 96 (5 July 1913): 39, (18 October 1913): 729; *E&MJ* 97 (16 May 1914): 1024, (30 May 1914): 1125.

91. Fred Whiteside to H. C. Gardiner, letter dated 3 February 1911 (Quivik Collection); T. T. Black to Dr. H. C. Gardiner, letter dated 9 February 1911 (Quivik Collection); R. W. Thatcher to H. C. Gardiner, letter dated 17 March 1911 (Quivik Collection); Alfred Atkinson, Professor of Agronomy, to Dr. H. C. Gardiner, letter dated 20 April 1911 (Quivik Collection); Paris Gibson to Dr. H. C. Gibson, letter dated 20 April 1911 (Quivik Collection); J. E. Morse to H. C. Gardiner, letter dated 2 June 1911 (Quivik Collection); Blaskovich, "History of Opportunity," 24–27; *E&MJ* 96 (18 October 1913): 729; Swett, *Montana's Trolleys II*, 91. Students of Montana history will recognize Fred Whiteside as the senator who exposed W. A. Clark's bribery of the Montana Senate in 1899 to gain election to the United States Senate; see Malone, *Battle for Butte*, 114–130.

92. *Butte Daily Post*, advertisement (11 June 1914): 13.

93. Blaskovich, "History of Opportunity," 25–26; *Butte Daily Post*, advertisement (11 June 1914): 13; E. P. Mathewson to H. C. Gardiner, letter dated 7 March 1913 (Quivik Collection); Samuel Fortier to Frank Adams, letter dated 1 July 1914 (Quivik Collection); Frank Adams to H. C. Gardiner, letter dated 18 July 1914 (Quivik Collection); H. C. Gardiner to Frank Adams, letter dated 23 July 1914 (Quivik Collection); H. C. Gardiner to Frank Adams, letter dated 27 July 1914 (Quivik Collection); Leonard, *Who's Who in Engineering*, 719–720.

94. Tyer, *Opportunity, Montana.* Tyer describes the ways that the EPA has overlooked Opportunity's environmental conditions as the agency has overseen the larger Superfund remediation of the Deer Lodge Valley as part of the Clark Fork Superfund site. A group of Opportunity residents sued ARCO (successor to the ACM) in state court for damages caused by decades of smelter pollution depositing on their property. Before the case, *Christian et al. v. ARCO*, could be tried, however, ARCO appealed to the U.S. Supreme Court, arguing that such a trial could not be held in state court because any remedy ordered by the state court might be in conflict with the remedy chosen by the EPA for the Clark Fork Superfund site, including Opportunity. The Supreme Court ruled in ARCO's favor in 2020. As of this writing, no further action has been taken in *Christian v. ARCO*.

95. Wells, "Report of the Anaconda Smelter" (1920), 57–65; Anaconda Smelter Commission, "Report of Progress for the Year July 1, 1916, to June 30, 1917," unpublished report, 5–6, in box 886A, DOJ.

96. Anaconda Smelter Commission, "Progress Report," 35–38, unpublished, undated (1916) report, box 84b, BOM.

97. Anaconda Smelter Commission, "Report of Progress for the Year July 1, 1916, to June 30, 1917," 1–12; Frederick Laist to C. F. Kelley, letter dated 23 January 1917 (Quivik Collection).

98. Gardiner to C. F. Kelley, letter dated 5 April 1916 (Quivik Collection). People have believed they could discern patterns in the climate since ancient times. Trying to understand those patterns was not simply a preoccupation of pre-Enlightenment thinkers; it occupies the minds of many of the world's top climatologists today, as they try to understand what cyclical, natural phenomena, such as sunspot activity or oscillations in equatorial wind directions, might affect weather patterns, and as they try to apply chaos theory to the understanding of why patterns of climate do not always hold. See Burroughs, *Weather*, 1–10, 43–45.

But the significance of Gardiner's letter lies not in his belief that he could predict an imminent drought but in his understanding of the relationship between rainfall and damage due to smelter smoke. Anyone familiar with Montana weather knows that no matter how many recent years of relatively high precipitation there have been, dry years will return. Gardiner wanted the ACM to make modifications in its smoke handling apparatus so that when the dry years did return, damage from smelter smoke to the surrounding environment would be minimized. It is merely an interesting footnote to the history of this episode that Gardiner's prediction proved correct that drier than average climate was about to return. Average rainfall in Anaconda in the early twentieth century was about fourteen inches per year, and rainfall was near or exceeded that amount every year from 1906 until 1917. Average rainfall, however, dropped to ten and eleven inches in 1919 and 1920, respectively. At East Anaconda, less than eight inches of precipitation fell in 1919; see U.S. Department of Agriculture, Weather Bureau, *Summary of the Climatological Data*, Section 28: "Western Montana," 4–5.

99. For an example of Gardiner's awareness of damage spreading into the Big Hole drainage because of operational changes in the mid-1910s, see Gardiner to Kelley,

letter dated 15 November 1917 (Quivik Collection). For Gardiner's concerns over the continued deposition of solids even after the installation of the new stack the Cottrell treaters, see A. E. Wells to Dr. Manning, letter dated 28 October 1919, and Arthur E. Wells to F. C. Cottrell, letter dated 13 November 1919 (Quivik Collection).

100. Thatcher to Gardiner, 17 March 1911; Elizabeth M. Dedekind to Prof. R. W. Thatcher, letter dated 8 November 1917 and attached abstracts; Elizabeth M. Dedekind to R. D. Cole, letter dated 10 November 1917 (all in Quivik Collection).

101. William Moore, "Report of Trip to Investigate Insect Injury to Forest Trees in Vicinity of Anaconda, July 30–August 11, 1917," unpublished report attached to William Moore to Dr. H. C. Gardiner, letter dated 3 December 1917 (Quivik Collection).

102. E. P. Taylor to Dr. H. C. Gardiner, letter dated 1 October 1917 (Quivik Collection); E. P. Taylor, "Report of Smelter Smoke Inspection, Anaconda, July 20–21, 1917," unpublished report (Quivik Collection). Gardiner attributed the limited tree damage to the fact that the smelter had been closed for a portion of 1917 and when it did operate it was at 50 percent of capacity; see H. C. Gardiner to C. F. Kelley, letter dated 10 November 1917 (Quivik Collection). By the end of the year, however, the amount of ore treated in 1917 reached about 82 percent of what had been treated in 1916; see Quivik, "Expert Report" (15 August 1997), F-61.

103. C. F. Kelley to Dr. H. C. Gardiner, letter dated 3 November 1917 (Quivik Collection).

104. Gardiner to Kelley, 10 November 1917 (Quivik Collection).

105. Gardiner to Kelley, 10 November 1917 (Quivik Collection).

106. Blankenship, "Smoke Conditions in the Deer."

107. Blankenship, "Smoke Conditions in the Deer."

108. Wells, "Report of the Anaconda Smelter" (1920), 65–70.

109. Wells, "Report of the Anaconda Smelter" (1920), 72–74.

110. John Hays Hammond et al., "Report of the Anaconda Smelter Commission," unpublished report dated 1 May 1924, 1–7, box 886A, DOJ. It is worth noting that this 17 percent return on investment is almost exactly that predicted by Frederick Laist in 1917 when he recommended to ACM vice president Con Kelley that the company build the stack and treaters; Laist to Kelley, letter dated 23 January 1917 (Quivik Collection).

111. Wells, "Report of the Anaconda Smelter" (1920), 71, 74.

112. Hammond et. al, "Report of the Anaconda Smelter," 9.

113. Hammond et al., "Report of the Anaconda Smelter," 10–13.

114. Wells, "Report of the Anaconda Smelter" (1920), 95–96.

115. *Brief Description of the Washoe Smelter*, 34–35, describes it as a 100-ton plant, which could be operated at somewhat higher capacity when necessary. On sales to outside customers, see Frederick Laist to C. F. Kelley, letters dated 6 July 1915 and 17 November 1915 (Quivik Collection).

116. Anaconda Smelter Commission, "Report of Progress for the Year July 1, 1916, to June 30, 1917," 12; Wells, "Report of the Anaconda Smelter" (1920), 96–97.

117. Wells, "Report of the Anaconda Smelter" (1920), 96–99.

118. F. F. Frick, "Phosphate Fertilizer Work," unpublished ACM report dated 1916, box 84b, BOM; Wells, "Report of the Anaconda Smelter" (1920), 101–110.

119. Wells, "Report of the Anaconda Smelter" (1920), 112–114.

120. L. D. Ricketts to Van H. Manning, letter dated 6 January 1920 (Quivik Collection). See also Ricketts to Manning, letter dated 24 September 1918 (Quivik Collection). Ricketts was invoking the engineering ideal that Gorman analyzes in *Redefining Efficiency*, but he wanted to ensure that the ACM was not forced to implement technology that was not also profitable.

121. Arthur E. Wells, "The Phosphate Fertilizer Industry in the United States and Other Countries with Special Reference to the Prospective Markets for the 'Anaconda Treble Superphosphate,'" unpublished Smoke Commission report dated July 1920, box 278, DOJ; Wells, "Report of the Anaconda Smelter" (1920), 114–117.

122. H. C. Gardiner to Professor E. P. Taylor, letter dated 8 June 1920 (Quivik Collection); E. P. Taylor to Dr. H. C. Gardiner, letter dated 21 August 1920 (Quivik Collection).

123. Hammond et al., "Report of the Anaconda Smelter," 1, 14–16; ACM Co., *Brief Description of the Anaconda Reduction Works* (1928), 74; Morrow, "Both Copper and Zinc Ores," 295.

124. H. C. Gardiner to E. G. Leipheimer, letter dated 28 April 1945 (Quivik Collection).

125. On the vertical and horizontal integration of the Anaconda company, other copper-producing corporations, the steel industry, and other sectors of the U.S. economy, see Schmitz, *Growth of Big Business*, 54–68; Chandler, *Visible Hand*, 345–376.

126. Fisher, "Montana: Land of the Copper," 290–292; Toole, *Twentieth-Century Montana*, 252–271.

127. Wm. F. Flynn, "Tailings and Slag Disposal," no date (attached to cover letter, Frederick Laist to C. F. Kelley dated 30 September 1922), 2–5, file 10, box 90, ACM; W. C. Capron, "Report on Tailings Disposal Plant at Opportunity and Clarifying Plant at Warm Springs," unpublished ACM Co. report dated 1922, 1–6, file 10, box 90, ACM.

128. Flynn, "Tailings and Slag Disposal," 2–5; Capron, "Report on Tailings Disposal Plant," 1–6; Flynn, "Report on Tailings Disposal" (1937), 1–2; E. P. Dimock, "A Survey of Tailings Disposal from Anaconda Reduction Works, Anaconda, Montana," unpublished ACM report dated 11 September 1944, 1, file 10, box 90, ACM; "Tailings launder and dump, May 1938," photo in file 4, box 91, ACM.

129. Capron, "Report on Tailings Disposal Plant."

130. Frederick Laist to C. F. Kelley, letter dated 6 August 1917 (Quivik Collection); Samuel Fortier, "Preliminary Report: Regarding the Disposal of Tailings at the Plant of the Anaconda Copper Mining Company at Anaconda, Montana," unpublished, incomplete report (Quivik Collection). Unfortunately, the Fortier report is incomplete: page 4 is missing; it ends with page 11 in the middle of a sentence; it does not bear Fortier's name. It is, however, attached to Laist's letter, which in turn says that Fortier's report is enclosed. Moreover, the report reads as if written by some-

one like Fortier (outside the ACM, familiar with similar conditions throughout the West). On the Kennedy Mining and Milling Company's dam, see Jackson, *Building the Ultimate Dam*, 143–146.

131. Kelley to Laist, 11 December 1917 (Quivik Collection); Capron, "Report on Tailings Disposal Plant," 1–3; Laist to Kelley, 14 January 1918 (Quivik Collection); H. C. Gardiner to Samuel Fortier, letter dated 30 May 1918 (Quivik Collection); Johnson, "Electrical Power, Copper," 32 (photograph); *Montana Standard* (14 September 1933).

132. Johnson, "History and Legal Phases," 203.

133. Johnson, "History and Legal Phases," 202–205.

Chapter 9. Land Exchanges—The Environmental Controversies Draw to a Close

1. Fred Morrell to L. O. Evans, letter dated 28 July 1922, Land Exchange Records, Regional Office, U.S. Forest Service, Missoula, MT (hereafter cited as FSRO).

2. *Mineral Resources of the United States, 1918*, Part I—Metals, 894; *Mineral Resources of the United States, 1919*, Part I—Metals (Washington, D.C.: Government Printing Office, 1922), 553; *Mineral Resources of the United States, 1920*, Part I—Metals (Washington, D.C.: Government Printing Office, 1922), 466; *Mineral Resources of the United States, 1921*, Part I—Metals (Washington, D.C.: Government Printing Office, 1924), 243; *Mineral Resources of the United States, 1922*, Part I—Metals (Washington, D.C.: Government Printing Office, 1925), 264; *Mineral Resources of the United States, 1923*, Part I—Metals (Washington, D.C.: Government Printing Office, 1927), 199, 202, 214; *Mineral Resources of the United States, 1924*, Part I—Metals (Washington, D.C.: Government Printing Office, 1927), 376; *Mineral Resources of the United States, 1925*, Part I —Metals (Washington, D.C.: Government Printing Office, 1928), 143–144; *Mineral Resources of the United States, 1926*, Part I—Metals (Washington, D.C.: Government Printing Office, 1929), 572; *Mineral Resources of the United States, 1927*, Part I—Metals (Washington, D.C.: Government Printing Office, 1930), 687.

3. Murphy, *Mining Cultures*, 33–35. Subsequent chapters in Murphy's book describe the ways men and women attended to their leisure time in Butte prior to the U.S. entry into World War II.

4. District Forester to Mr. Wolff, memorandum dated 2 March 1922, FSRO; "An Act to Consolidate National Forest Lands," *Sixty-Seventh Congress, Session II*, Chapter 105 (20 March 1922), 465.

5. M. H. Wolff, memorandum dated 7 July 1922; John R. Preston, Acting Assistant Forester, to District Forester, letter dated 11 March 1922, both FSRO.

6. Wells, "Report of the Anaconda Smelter" (1920), 71, 74.

7. U.S. Department of Commerce, Bureau of Corporations, *Lumber Industry*, Part I, Standing Timber, 18; Rader, "Montana Lumber Strike of 1917," 190.

8. F. G. Clark, Deerlodge Forest Supervisor, to District Forester, letter dated 11 March 1922; Acting Forester to District Forester, letter dated 17 March 1922, all in FSRO. Quotation is from the last-cited letter.

9. Fred Morrell, District Forester, to The Forester, letter dated 2 October 1922; Morrell to L. O. Evans, letter dated 3 May 1922; M. H. Wolff, office memorandum dated 12 July 1922, all FSRO.

10. M. H. Wolff to Morrell, memorandum dated 21 April 1922.; M. H. Wolff, office memorandum dated 22 July 1922; L. F. Kneipp to District Forester, letter dated 18 August 1922, all FSRO.

11. L. F. Kneipp, Acting Forester, to District Forester, letter dated 18 August 1922, FSRO.

12. Fred Morrell, District Forester, to The Forester, letter dated 29 September 1922; Morrell to The Forester, letter dated 2 October 1922, both in FSRO.

13. Fred Morrell, District Forester, to L. O. Evans, letter dated 2 October 1922, FSRO. Gardiner's advocacy to the ACM's top officials in favor of constructing the stack and treaters can best be seen in his 5 April 1916 letter to C. F. Kelley (Quivik Collection).

14. Quote is from "Interview with Mr. Fred Morrell," memorandum dated 21 September 1922, FSRO. Gardiner's name does not appear on this document, but its contents strongly suggest that he, as an ACM official, was the author.

15. Morrell to The Forester, 29 September 1922; Morrell to The Forester, 2 October 1922, both in FSRO.

16. Morrell to The Forester, 2 October 1922, FSRO.

17. L. O. Evans to Fred Morrell, letter dated 6 October 1922, FSRO.

18. Morrell to Gardiner, 8 June 1923; Fred Morrell, District Forester, to the Forester, letter dated 8 June 1923; Fred Morrell, District Forester, to Dr. H. C. Gardiner, letter with attachment dated 14 June 1923, all FSRO.

19. Morrell to Gardiner, 8 June 1923; Fred Morrell, District Forester, to The Forester, letter dated 28 August 1923; E. A. Sherman, Acting Forester, to District Forester, letter dated 7 September 1923, all FSRO.

20. L. O. Evans to F. H. Morrell, letter dated 14 September 1923; Fred Morrell, District Forester, to L. O. Evans, letter dated 26 September 1923; Fred Morrell, District Forester, to The Forester, letter dated 27 September 1923; L. O. Evans to F. H. Morrell, letter dated 9 October 1923, all FSRO. It is not clear precisely what the ACM and Forest Service understood by the term "cut-over" land. Because the two parties were working toward an exchange of timber for timber, it is almost certain that they did not mean "clear-cut" land. It may be that the ACM had selectively cut certain mature trees from the "cut-over" land, leaving immature trees standing which nevertheless represented value in board feet of timber.

21. Marcosson, *Anaconda*, 163; J. R. Hobbins, vice president, to Fred Morrell, District Forester, letter dated 9 October 1923, FSRO.

22. E. A. Sherman, Acting Forester, to the Solicitor, letter dated 7 September 1923, FSRO; H. Foster Bain, Director, to E. A. Sherman, Associate Forester, letter dated 8 September 1923 (Quivik Collection); R. W. Williams, Solicitor, to E. A. Sherman, Acting Forester, letter dated 18 October 1923, FSRO; L. F. Kneipp, Assistant Forester, to District Forester, letter dated 20 October 1923, FSRO; John B. Taylor, Forest Supervisor, to District Forester, letter dated 19 November 1923, FSRO.

23. Fred Morrell to The Forester, letter dated 14 January 1924; Acting Forester to Ligon Johnson, letter dated 11 February 1924, both FSRO; Ligon Johnson to The Forester, letter dated 4 March 1924 (Quivik Collection); L. F. Kneipp, Acting Forester, to The Solicitor, letter dated 5 March 1924, FSRO; C. F. Marvin, Acting Secretary of Agriculture, to the Attorney General, letter dated 20 March 1924, FSRO; Harlan Stone, Attorney General, to Hon. H. C. Wallace, Secretary of Agriculture, letter dated 28 May 1924, FSRO.

24. Arthur E. Wells to Dr. H. Foster Bain, letter dated 12 January 1924 (Quivik Collection).

25. Wells to Bain, 12 January 1924.

26. H. Foster Bain, Director, to John Hays Hammond, letter dated 16 January 1924 (Quivik Collection); John Hays Hammond to Dr. H. Foster Bain, letter dated 30 January 1924 (Quivik Collection); Arthur E. Wells to H. Foster Bain, Director, letter dated 14 February 1924 (Quivik Collection); Arthur E. Wells to Dorsey A. Lyon, Acting Director, 14 February 1924 (Quivik Collection); Hammond et al., "Report of the Anaconda Smelter."

27. Hammond et al., "Report of the Anaconda Smelter."

28. Arthur E. Wells to H. Foster Bain, Director, letter dated 23 April 1924 (Quivik Collection); Arthur E. Wells to Dr. H. Foster Bain, Director, letter dated 2 May 1924 (Quivik Collection); Bain to D. A. Lyons, letter dated 28 May [1924] (Quivik Collection); B. M. Parmenter, Assistant Attorney General, to the Attorney General, memorandum dated 11 November 1925 (Quivik Collection); Attorney General to Scott Turner, Director, letter dated 11 November 1925 (Quivik Collection).

29. John B. Taylor, Forest Supervisor, to District Forester, letter dated 23 January 1926; John B. Taylor, Forest Supervisor, to District Forester, letter dated 6 February 1926, both FSRO.

30. Roscoe Haines, "Memorandum of Conference at Office of A.C.M. Co. Lumber Department, Bonner, Wednesday 29," memorandum to Dr. H. C. Gardiner dated 30 July 1925 (Quivik Collection); E. A. Sherman, Associate Forester, to The Forester, letter dated 30 July 1925 (Quivik Collection). Incidentally, the secretary of the interior had to approve any land exchange because the 1922 act required it. This is because the Department of the Interior was responsible for transacting federal land deals, especially for issuing patents to parties acquiring federal lands.

31. John B. Taylor, Forest Supervisor, to District Forester, letter dated 9 April 1926; John B. Taylor, Forest Supervisor, to J. R. Hobbins, letter dated 9 April 1926, both FSRO.

32. "Record of Land Exchange," card for case no. GF-071748; "Record of Land Exchange," card for case no. GF-072525; "Record of Land Exchange," card for case no. GF-072740, all in FSRO.

33. Morrow, "Both Copper and Zinc Ores," 295–296.

34. Morrow, "Both Copper and Zinc Ores," 296–298.

35. Bender, "Development of Copper Smelting," 301–302.

36. Morrow, "Both Copper and Zinc Ores," 296–298; Laist, "Modern Metallurgical Miracles," 288.

37. V. A. Moore to Dr. H. C. Gardiner, letter dated 7 September 1928 (Quivik Collection); V. A. Moore, "Report of the Inspection of the Ranches in Deer Lodge Valley and about the Smelter at Tooele, Utah," unpublished report dated 7 September 1928 (Quivik Collection).

38. E. A. Sherman, Acting Forester to the Solicitor, letter dated 29 June 1927 (Quivik Collection).

39. B. M. Parmenter, Assistant Attorney General, to William M. Jardine, Secretary of Agriculture, letter dated 22 July 1927 (Quivik Collection).

40. Forest Supervisor to Regional Forester, memorandum dated 9 August 1935, FSRO.

41. The land exchanges are summarized in Forest Supervisor to Regional Forester, 9 August 1935, FSRO. See also the "Record of Land Exchange," cards for each exchange, FSRO.

42. "Civil Docket—United States District Court," *United States of America v. Anaconda Copper Mining Company* (Quivik Collection).

43. Supreme Court of the State of Montana, Opinion, 1 September 2015, p. 5, on the appeal in *Gregory A. Christian et al. v. Atlantic Richfield Company*, Cause No. DV-08173 BN.

44. Forest Supervisor to Regional Forester, 9 August 1935; R. T. Ferguson to M. H. Wolff, memorandum dated 24 October 1935, FSRO.

45. Wolff, acting for Kelley, to the Chief, 16 April 1936; John E. Burch, Acting Chief, to Regional Forester, letter dated 23 April 1936; Evan W. Kelley, Regional Forester, to the Chief, letter dated 5 May 1936; Evan W. Kelley, Regional Forester, to the Chief, letter dated 3 March 1937; "Record of Land Exchange" card, case no. GF-080093, all in FSRO.

46. R. U. Harmon to Forest Supervisor, memorandum dated 20 February 1946; F. K. Stewart, office memorandum dated 8 February 1951, both FSRO.

Chapter 10. The Decades of Anaconda Company Hegemony

1. "Anaconda: I," *Fortune*, 82–94, 210–220; "Anaconda: II," *Fortune*, 70–77, 134–149.

2. Lewis Mumford's *Technics and Civilization* in 1934; the *Fortune* articles appeared in 1936 and 1937. The author of the *Fortune* articles may have been aware of Mumford's work.

3. "Anaconda: I," 85.

4. "Anaconda: I," 86.

5. "Anaconda: I," 88–89; Stone, "Giesche Company," 682–684, 695–696.

6. The most thorough scholarly work on "the War of the Copper Kings" is Malone, *Battle for Butte*. Other important treatments include Glasscock, *War of the Copper*; Howard, *Montana: High, Wide, and Handsome*; Toole, "History of the Anaconda Copper"; McNelis, *Copper King at War*; Toole, *Twentieth Century*.

7. Recent scholarly treatments of Butte's labor history and of the often violent conflict between labor and capital include: Calvert, *Gibraltar*; Frisch, "'Gibraltar

of Unionism'"; Finn, *Tracing the Vein*; Basso, *Meet Joe Copper*. Other important treatments include Jensen, *Heritage of Conflict*; Melvyn Dubofsky, *We Shall Be All*; Gutfeld, *Montana's Agony*.

8. Dobb, "Pennies from Hell," 39–55.

9. Sklar, *Corporate Reconstruction of American Capitalism*.

10. This concept of incorporating potential sources of conflict into an enterprise's control combines ideas of vertical integration in corporate development, as described in Hovenkamp, *Enterprise and American Law*, 331–332, and ideas of system-building in Hughes, *Networks of Power*, 404–408.

11. Ore, "Labor and the New Deal," 47–95; Finn, *Tracing the Vein*.

12. "Anaconda—city looking east from west of athletic field, 1923," photo no. PAc 82–62.283 (2 of 2) shows an extraordinary amount of dust blowing off the Anaconda tailings fields.

13. Chris Jergensen to Frederick Laist, letter dated 23 March 1917 (Quivik Collection); H. C. Gardiner to Hyrum Rodgers, letter dated 23 March 1917 (Quivik Collection). The ACM paid Jergensen for his forty-two cattle, which he could no longer support on his land. In exchange for the payment, the ACM also obtained a release from Jergensen for liability for other livestock losses; see H. C. Gardiner to R. D. Cole, letter dated 19 September 1917 (Quivik Collection); H. C. Gardiner to L. O. Evans, letter dated 19 September 1917 (Quivik Collection).

14. Flynn, "Report on Tailings Disposal" (1937), 3–4; Dimock, "Survey of Tailings Disposal" (1944), 11–12; Dimock, "Report on Tailings Disposal at the Anaconda Reduction Works," unpublished ACM report dated 20 February 1947, 1–2, file 1, box 91, ACM; Research Department, Anaconda Reduction Department, ACM Co., "Tailings Disposal—Review and Recommendation," unpublished report dated 10 January 1947, file 1, box 91, ACM; "Appropriation for Tailings Disposal Project," draft ACM Co. document dated 25 March 1947, file 1, box 91, ACM; Day, "Disposal of Metallurgical Wastes," 56; Rossillon, "Yellow Ditch Flume."

15. Quivik, "Expert Report," pp. C-45, C-84, C-124, C-141, C-170.

16. Flynn, "Tailings and Slag Disposal," 2–5; Capron, "Report on Tailings Disposal Plant," 6–7; Flynn, "Report on Tailings Disposal" (1937), 2–3; Robert L. Dent, deposition testimony (1995), 25–29, 37–49, 57–60, 69–71, 77–78; Frederick Laist to C. F. Kelley, letter dated 30 September 1922 (exhibit 5 in the Dent deposition), file 10, box 90, ACM; W. C. Capron to Frederick Laist, letter dated 30 September 1922, file 10, box 90, ACM; U.S. Environmental Protection Agency, "Statutory Five-Year Review Report" (23 March 2000), 4.

17. Dent deposition, 51–57, 73–74; Day, "Disposal of Metallurgical Wastes," 10–12; U.S. Environmental Protection Agency, "Statutory Five-Year Review Report" (2000), 4–5.

18. U.S. Environmental Protection Agency, "Statutory Five-Year Review Report" (2000), 5.

19. *State of Montana v. ARCO*, "Deposition of Emil Kramlick" (Butte, MT: Nordhagen Court Reporting, December 1992), 109.

20. Dimock, "Report on Tailings Disposal"; Ralph Costantini, "Report on Tailings

Disposal Operations at the Anaconda Reduction Department, Anaconda, Montana," unpublished ACM Co. report dated 19 September 1957, 1, file 9, box 90, ACM; Day, "Disposal of Metallurgical Wastes," 7–8.

21. Corbett, "Changing Face of Mining," 20–25; Marcosson, *Anaconda*, 310.

22. Leech, *City That Ate Itself*.

23. Averett, "Macro-Invertebrates of the Clark," 1–3; Gurnham, "Pollution Control at Anaconda's Copper," 125–127.

24. Gurnham, "Pollution Control at Anaconda's Copper," 127–130.

25. Gurnham, "Pollution Control at Anaconda's Copper," 126.

26. Gurnham, "Pollution Control at Anaconda's Copper," 129.

27. "Pollution—Not for Cities Alone," 686; Dent deposition, 64–65; "River of Poison," 70; Farling, "Montana's Clark Fork River," 11.

28. John A. Grant and William F. Flynn, "Resume of the Present Tailings Disposal," unpublished ACM Co. report dated 1952, 1, file 2, box 91, ACM; W. E. Mitchell, "Tailings," unpublished ACM Co. report dated approximately April 1953, 1, file 2, box 91, ACM; W. F. Flynn to W. E. Mitchell, ACM Co. memorandum dated 10 April 1956, 1, file 2, box 91, ACM; Marcosson, *Anaconda*, 259–260; Fulmor, "Anaconda's Butte Concentrator," 54–73.

29. Juan de Onis, "Chilean Senate Gives Allende Power to Nationalize U.S. Copper Interests," *New York Times*, 11 February 1971; "Anaconda Reaches Accord with Chile," *New York Times*, 25 July 1974; Finn, "Intimate Strangers," 22–39.

30. Gene Smith, "ARCO Plans to Buy Rest of Anaconda," *New York Times*, 3 July 1976; "Anaconda Holders Back ARCO Merger," *New York Times*, 2 November 1976; "ARCO Acquires Anaconda after Court Refuses to Bar Tie," *New York Times*, 13 January 1977; *Minerals Yearbook 1980*, vol. I, Metals (Washington, D.C.: Government Printing Office, 1982), 263.

31. "Anaconda Closing Open-Pit Copper Mine at Butte," *New York Times*, 25 April 1982; Lydia Chavez, "When ARCO Left Town," *New York Times*, 25 July 1982; *Minerals Yearbook 1980*, vol. II, Area Reports (Washington, D.C.: Government Printing Office, 1982), 322; *Minerals Yearbook 1983*, vol. I, Metals (Washington, D.C.: Government Printing Office, 1984), 281; Czehura, "Butte: A World Class," 14; Susan Dunlap, "Earth Day Significant for Butte in 1982," *Montana Standard*, 22 April 2015; Duaime and McGrath, *Butte, Montana*, 1, 3.

Epilogue: Butte and Anaconda as Cultural Landmarks

1. Dobb "Pennies from Hell," 41.

2. Dobb "Pennies from Hell," 39.

3. Dobb, "Viewpoint," 1–12.

4. Duaime and McGrath, *Butte, Montana*, 4; Frank Gardner, interview by Rayelynn Connole, 5 April 2014, Clark Fork Watershed Education Program, https://cfwep.org/frank-gardner-is-the-former-president-of-montana-resources-and-was-the-general-manager-of-the-berkeley-pit/; Michael Cast, "Montana Resources Reaping the Benefit of Sky-High Copper, Molybdenum Prices," *Montana Standard*, 24 June 2021.

5. Duaime and McGrath, *Butte, Montana*, 4.

6. Gorman, *Redefining Efficiency*, 288, 301–302.

7. Spindler, "Clean-Up of Silver Bow Creek," 58–63.

8. Beck, "Love Canal Tragedy."

9. Barnett, *Toxic Debts and the Superfund*, 67–68; Hird, *Superfund*, 9–10.

10. Barnett, *Toxic Debts and the Superfund*, 11, 59–60; Hird, *Superfund*, 16–17.

11. Barnett, *Toxic Debts and the Superfund*, 72–73, 228–229; Hird, *Superfund*, 11–17, 169–170, 186–190, 206–207.

12. U.S. EPA, National Priorities List Sites in Montana, https://www.epa.gov/superfund/national-priorities-list-npl-sites-state#MT.

13. The author worked as an expert witness under contract to the DOJ during this litigation in the 1990s, conducting extensive historical research and writing two expert reports. That historical research provided some of the basis for this book. The most recent consent decrees, as of this writing were filed in 2020 and 2022; Consent Decree for the Butte Priority Soils Operable Unit, *U.S. v. Atlantic Richfield Company et al.*, United States District Court for the District of Montana, Civil Action No. CV 89–039-BU-SEH, 16 September 2020; Consent Decree for the Anaconda Smelter NPL Site, *U.S. v. Atlantic Richfield Company*, United States District Court for the District of Montana, Civil Action No. CV89-039-BU-SHE, 16 December 2022.

14. EPA, Silver Bow Creek / Butte Area, Butte, Montana, Operable Units, https://cumulis.epa.gov/supercpad/SiteProfiles/index.cfm?fuseaction=second.ous&id=0800416.

15. Zick et al., "Butte Mine Flooding Water Treatment."

16. Zick et al., "Butte Mine Flooding Water Treatment."

17 Zick et al., "Butte Mine Flooding Water Treatment"; U.S. District Court for the District of Montana, *U.S. v. Atlantic Richfield Company et al.*, Consent Decree for the Butte Mining Flooding Site, August 2002.

18. EPA, "Fourth Five-Year Review Report," 41–46; EPA, "Fifth Five-Year Review Report," 32, 36–38; Susan Dunlap, "Water Leaves the Berkeley Pit for First Time in 37 Years," *Montana Standard*, 13 May 2019; Dunlap, "Polishing Plant to Control Berkeley Pit Water Almost Ready," *Montana Standard*, 6 August 2019; Matt Vincent, "Superfund during the Covid-19 Pandemic," Pit Watch, 22 April 2020, https://pitwatch.org/category/blog/.

19. EPA, "Fourth Five-Year Review Report," 41; Berkeley Pit Water Levels, https://pitwatch.org/current-water-level/.

20. Susan Dunlap, "Parrot Tailings: EPA Resists Calls to Remove Mine Waste in Center of Butte," *Montana Standard*, 18 June 2015; Dunlap, "Deep in the Heart of Butte: A Special Report," *Montana Standard*, 19 July 2016; U S. District Court for the District of Montana, *U.S. v. Atlantic Richfield Company et al.*, Consent Decree for the Butte Priority Soils Operating Unit, May 2020.

21. Shovers, "Old Works Golf Course," 64–70; EPA, "Sixth Five-Year Review Report," 1–3.

22. Brooks, "Remove the Dam, Restore," 3–24; EPA, "Third Five-Year Review Report," 1–11.

23. The remediation of the three Clark Fork Superfund sites has produced voluminous documentation, but the amount ARCO has spent on remediation is not public information. I am grateful to Joe Griffin, a retired employee of the State of Montana, who worked for many years for the state involvement in the Superfund remediation and continues to work on those issues as a concerned citizen of Butte, who keeps his ear to the ground, and he has given me his best estimate of the amount ARCO has spent.

24. Derek Strahn, Butte-Anaconda Historic District, National Historic Landmark Nomination Form, National Park Service, January 2006.

25. Shovers, "Remaking the Wide-Open Town," 46–47; Leech, *City That Ate Itself*, 277–290.

26. Shovers, "Remaking the Wide-Open Town," 44–49; National Park Service, *Butte, Montana: A Project Report*; "Regional Historic Preservation Plan" (RHPP).

27. Butte–Silver Bow Public Archives, "The Building," https://buttearchives.org/visit-the-archives/the-building/.

28. Renewable Technologies Inc., *Butte-Anaconda Historical Park System*. The author, Fred Quivik, was one of the three coauthors of the RTI plan, along with Dennis Glick and Mark Fiege.

29. "Regional Historic Preservation Plan," acknowledgments and pp. 1-1 to 1-3; EPA, "Building on Mining History," 3, https://semspub.epa.gov/work/08/1570747.pdf.

30. EPA, "Building on Mining History," 11; The Lakota Group, "Anaconda-Deer Lodge County Historic Preservation."

31. Dobb, "Dirty Old Town," 256.

32. LaFever, "Rethinking Industrial Heritage," 3–4, 113–115.

33. Strahn, Butte-Anaconda Historic District, National Historic Landmark Nomination; Roberta Forsell Stauffer, "Interior Secretary Approves District Expansion," *Montana Standard*, 26 April 2006.

34. Pelts, "Preface," in *Extraction: Art on the Edge of the Abyss*, 11–13; Obituary for Edwin Charles Dobb Jr., *Montana Standard*, 20 August 2019.

35. Fahlman, *Landscapes of Extraction*.

36. Nye, *American Technological Sublime*, xi–xx, 1–16, 32–43.

37. Neiman, *Learning from the Germans*.

38. Goodell, *Heat Will Kill You First*, 20, 309–317.

39. Two scholars have critically addressed the intersections of tourism and environmental remediation in the damaged landscapes of Butte and Anaconda. Barry, in "Toxic Tourism," analyzes the ideas of certain preservation advocates in Butte for interpreting landscape features of Butte's toxic legacy, but she frames the analysis in terms of "either/or," either preserve contaminated features or remediate them. She does not consider the "both/and" alternative, of seeing the problem as a design challenge to remediate sites in a manner that addresses the profound environmental issues while also preserving certain significant historical features so that visitors to Butte can learn something of its industrial past, which created such a vast landscape of environmental damage. Moreover, she frames tourism largely as a potential tool

for economic development, without addressing the possibility that tourism can serve serious educational objectives.

Mary Curran, in "Dialogues of Difference," explores the ways the discourses of tourism and environmental remediation construct images of Butte by means of maps. She is particularly interested in the extent to which two maps identify (or do not identify) the low-income neighborhoods of Butte. The tourism lens she presents is that of the Chamber of Commerce and not of a program with serious educational objectives.

Bibliography

Manuscript Collections

Baker Library, Harvard University, Cambridge, Mass.
Historical Collections, R. G. Dun & Co. Collection.
Bureau of Land Management, Butte, MT.
General Land Office Survey Records.
Butte-Silver Bow Courthouse, Butte, MT.
Recorder's Office, Deed Records.
Mining Claim Records.
Butte-Silver Bow Public Archives, Butte, MT.
Anaconda Copper Mining Company Records.
Butte, Anaconda & Pacific Railroad Records.
Butte Public Library Collection.
Carl A. Kroch Library, Rare and Manuscript Collections, Cornell University, Ithaca, NY.
V. A. Moore Collection (VAM).
Columbia Center for Oral History Center, Columbia University, New York City.
Mineral Engineering Project.
Library of Congress, Washington, D.C.
Sanborn Fire Insurance Maps, Digital Collections.
Theodore Roosevelt Papers (TR).
Montana Historical Society, Helena, MT.
Library Collections.
Photo Archives.
Records of the Anaconda Copper Mining Company, MC-169 (ACM).
National Archives and Records Administration, College Park, MD.
RG-60, General Records of U.S. Dept. of Justice, Entry 112, D. J. Central Files (DOJ).
RG-70, Records of the U.S. Bureau of Mines, General Files Prior to 1954 (BOM).
RG-95, Records of the U.S. Forest Service.
National Archives, Seattle Branch, Seattle, WA.
Records of the Circuit Court of the United States, Ninth Circuit, Record Group 21, *Hugh Magone v. Colorado Mining & Smelting Company, et al.* (MvC).
William E. Wall and Louis Foss v. Parrot Silver and Copper Company and Anaconda Copper Mining Company, no. 464, 1913.

Northwest Energy, Butte, MT.
Montana Power Company Records.
Quivik Collection, Saint Paul, MN.
Documents produced in *U.S. v. ARCO*, the original provenance is unknown.
Thomas Couch Papers.
Regional Office, U.S. Forest Service, Missoula, MT.
Land Exchange Records (FSRO).

Newspapers and Magazines

Anaconda Standard (Anaconda, MT).
Anaconda Weekly Review (Anaconda, MT).
The Boston Globe (Boston, MA).
Butte Daily Miner (Butte, MT).
The Butte Daily Post (Butte, MT).
Butte Evening News (Butte, MT).
Butte Miner (Butte, MT).
Daily Inter Mountain (Butte, MT).
Engineering & Mining Journal (*E&MJ*).
Helena Independent (Helena, MT).
Mines & Minerals.
Mining & Scientific Press (*M&SP*).
Mining & Engineering World.
Mining Journal.
Montana Post (Virginia City, MT).
The New Northwest (Deer Lodge, MT).
New York Times (New York City).
Powell County Record (Deer Lodge, MT).
The Press (Helena, MT).
Reveille (Butte, MT).
Silver State (Deer Lodge, MT).
The Weekly Missoulian (Missoula, MT).
Western News (Hamilton, MT).

Court Cases

Circuit Court of the United States, Ninth Circuit, District of Montana. *Hugh Magone v. Colorado Smelting & Mining Company, et al.*, no. 222 in equity, 1903.
Circuit Court of the United States, Ninth Circuit, District of Montana. *United States v. Anaconda Copper Mining Company and Washoe Copper Company*, no. 967 in equity, March 1910.
District Court of the Second Judicial District of the State of Montana in and for the County of Silver Bow. *William Norton, et al., v. Colorado Smelting and Mining Company, et al.*, 1901.
District Court of the Third Judicial District of the State of Montana, Deer Lodge

County. *Elizabeth Ann Staffanson v. Anaconda Copper Mining Company*, no. 2390, 1905.

District Court of the Third Judicial District of the State of Montana, Deer Lodge County. *Henry Mason v. Anaconda Copper Mining Company*, no. 2391, 1905.

District Court of the Third Judicial District of the State of Montana, Deer Lodge County. *Peter Valiton v. Anaconda Copper Mining Company*, docket no. 48, 1908.

District Court of the United States, Ninth Circuit, District of Montana. *William E. Wall and Louis Foss v. Parrot Silver and Copper Company and Anaconda Copper Mining Company*, no. 464, 1913.

Montana Supreme Court. *MacGinnis, Respondent, v. Boston & Montana Consolidated Copper & Silver Mining Company, et al., Appellants*, case no. 2001, 1904.

Supreme Court of the State of Montana, on the appeal in *Gregory A. Christian, et al., v. Atlantic Richfield Company*, Cause No. DV-08173 BN, September 1, 2015.

United States Court of Appeals, Ninth Circuit. *Mountain Copper Co. v. United States.* No. 1, February 26, 1906.

United States Circuit Court for the District of Montana. *Fred J. Bliss v. The Washoe Copper Company and the Anaconda Copper Mining Company*, no. 167 Fed. 342, 1905.

United States Circuit Court of Appeals for the Ninth Circuit. *Fred J. Bliss v. The Washoe Copper Company and the Anaconda Copper Mining Company*, no. 1738, 1910.

United States Court of Appeals, Ninth Circuit. *McCarthy v. Bunker Hill & Sullivan Mining & Concentrating Co.*, 164 F. 927, November 6, 1908.

United States District Court for the District of Montana, Butte Division. *State of Montana v. Atlantic Richfield Company*, No. CV-83–317-H-PGH.

United States District Court for the District of Montana, Butte Division. *United States of America v. Atlantic Richfield Company and Cleveland Wrecking Company, Inc.*, No. CV-89–39-BU-PGH.

Depositions

Dent, Robert L. Deposition testimony given in *U.S. v. ARCO*. Denver, CO: Agren Reporting & Video, Inc., 1995.

Kramlick, Emil. Deposition testimony given in *State of Montana v. ARCO*. Butte, MT: Nordhagen Court Reporting, December 1992.

Online Documents about the Clark Fork Superfund Complex

U.S. Environmental Protection Agency, National Priorities List Sites in Montana. https://www.epa.gov/superfund/national-priorities-list-npl-sites-state#MT.

U.S. Environmental Protection Agency. "Building on Mining History: Cleanup, Reuse, and Community Resilience at the Silver Creek / Butte Area Superfund Site in Butte, Montana." May 2014. https://semspub.epa.gov/work/08/1570747.pdf.

U.S. Environmental Protection Agency. "Statutory Five-Year Review Report: Silver Bow Creek / Butte Area Superfund Site, Clark Fork Basin, Montana, with Emphasis on the Warm Springs Ponds Operable Units." March 23, 2000.

U.S. Environmental Protection Agency. "Third Five-Year Review Report for

Milltown Reservoir Sediments / Clark Fork River Superfund Site, Missoula, Granite, Powell, and Deer Lodge Counties, Montana." August 2021. https://semspub.epa.gov/work/08/100010818.pdf.

U.S. Environmental Protection Agency. "Fourth Five-Year Review Report for Silver Bow Creek / Butte Area Superfund Site." January 2016. https://semspub.epa.gov/work/08/1551629.pdf.

U.S. Environmental Protection Agency. "Fifth Five-Year Review Report: Silver Bow Creek / Butte Area Superfund Site, Silver Bow and Deer Lodge Counties, Montana." August 30, 2021. https://semspub.epa.gov/work/08/100010834.pdf.

U.S. Environmental Protection Agency. "Sixth Five-Year Review Report for Anaconda Co. Smelter Superfund Site." September 2020. https://semspub.epa.gov/work/08/100008701.pdf.

U.S. Department of Justice, webpage describing *U.S. v. Atlantic Richfield Co.* https://www.justice.gov/enrd/us-v-atlantic-richfield-co.

Pit Watch website. https://pitwatch.org.

Berkeley Pit Water Levels.

Unpublished Sources

Averett, Robert C. "Macro-Invertebrates of the Clark Fork River, Montana." Montana State Board of Health and Montana Fish & Game Department (1961), Montana State Library, Helena, MT.

Bale, Anthony. "Compensation Crisis: The Value and Meaning of Work-Related Injuries and Illnesses in the United States, 1842–1932." PhD diss., Brandeis University, 1986.

Barry, Bridget. "Toxic Tourism: Promoting the Berkeley Pit and Industrial Heritage in Butte, Montana." Master's thesis, University of Nebraska, Lincoln, 2012.

Blankenship, J. W. "Smoke Conditions in the Deer Lodge Valley, Montana, and Parts Adjacent, for the Year 1918." Undated report to Dr. H. C. Gardiner (Quivik Collection).

"The Construction of the Great Falls Hydroelectric Development of the Great Falls Power Company." Report on file at the office of William O'Keefe, Rainbow Shop, The Montana Power Company, Great Falls, n.d.

Donegan, Craig. "For the Good of Us All: Early Attitudes toward Occupational Health with Emphasis on the Northern United States from 1787–1870." PhD diss., University of Maryland, 1984.

Doull, James, and F. A. Pratt. "Deposition of Tailings along Silver Bow Creek in Vicinity of Silver Bow, Montana." Report submitted to A. H. Wethey dated November 16, 1899 (Quivik Collection).

Fiege, Mark T. "General Electric, Montana Power, the Anaconda Copper Mining Company, and Steam Railroad Electrification in Montana, 1912–1917." Paper presented at the Pacific Northwest History Conference, Spokane, WA, April 1987.

Fogarty, Kate Hammond. "A History of Butte, Montana," (n.d., n.p.), 1, 6, Butte Public Library Collection, #AW F014, Butte-Silver Bow Public Archives, Butte, MT.

Finn, Janet Leigh. "Mining Community: The Cultural Politics of Copper, Class and Gender in Butte, Montana, Unites States of America, and Chuquicamata, Chile." PhD diss., University of Michigan, 1995.

Frisch, Paul Andrew. "The 'Gibraltar of Unionism': The Working Class at Butte, Montana, 1878–1906." PhD diss., University of California at Los Angeles, 1992.

Gossel, Patricia Peck. "The Emergence of American Bacteriology, 1875–1900." PhD diss., Johns Hopkins University, 1988.

Hutchinson, G. A. "Sampling Spitzlutten Products: Testing Old Wilfley Tables Fed with Same." ACM Co. report dated 1901 (Quivik Collection).

The Lakota Group. "Anaconda-Deer Lodge County Historic Preservation Plan." Draft plan, June 2021.

LaFever, Alison M. "Rethinking Industrial Heritage: A Discussion of the Preservation of Compromised and Contested Landscapes in Butte, Montana." Master's thesis, Graduate School of Architecture, Planning, and Preservation, Columbia University, 2012.

Lovejoy, P. S. "Report on Damage to Deerlodge National Forest Caused by Fumes from Washoe Smelter." Prepared for the U.S. Forest Service, report dated July 3, 1911.

MacMillan, Donald. "A History of the Struggle to Abate Air Pollution from Copper Smelters in the Far West 1885–1933." Ph.D. diss., University of Montana, 1973.

Ore, Janet. "Labor and the New Deal in Butte, Montana: The International Union of Mine, Mill and Smelter Workers' Strike of 1934." Master's thesis, Washington State University, 1987.

Peirce, G. J., R. E. Swain, and J. P. Mitchell. "Report on the Effects of Smelter Smoke on Vegetation and the Conditions of the National Forests in the Vicinity of Anaconda, Montana." Prepared for the U.S. Department of Justice and dated February 5, 1913 (DOJ).

Quivik, Fredric L. "Expert Report." Unpublished report prepared for the Department of Justice, dated August 15, 1997, and submitted in *U.S. v. ARCO*, the Clark Fork Superfund case in U.S. District Court for the District of Montana; a copy of the report is in the Montana Historical Society Library.

Quivik, Fredric L. "Smoke and Tailings: An Environmental History of Copper Smelting Technologies in Montana, 1880–1930." PhD diss., University of Pennsylvania, 1998.

"Regional Historic Preservation Plan" (RHPP) for the Anaconda-Butte Heritage Corridor. Report prepared for the RHPP Joint Committee, 1993.

Renewable Technologies Inc. *Butte-Anaconda Historical Park System Master Plan.* Butte, MT: Butte Historical Society, 1985.

Rossillon, Mitzi. "Yellow Ditch Flume." Unpublished report prepared for Anaconda-Deer Lodge, Montana, dated October 3, 2012.

Shovers, Brian Lee. "Miners, Managers, and Machines: Industrial Accidents and Occupational Disease in the Butte Underground, 1880–1920." Master's thesis, Department of History, Montana State University, 1987.

Strahn, Derek. Butte-Anaconda Historic District, National Historic Landmark Nomination Form. National Park Service, January 2006.

Toole, K. Ross. "The History of the Anaconda Copper Mining Company: A Study of the Relationship between a State, Its People, and a Corporation." PhD diss., University of California at Los Angeles, 1954.

Weigle, W. G. "Report on Conditions of Plant and Animal Life in and around Deerlodge Valley, Montana." October 6, 1906, box 239, Research Compilation Files, 1897–1935, entry 115, Records of the U.S. Forest Service, RG-95, National Archives, College Park, MD.

Wells, Arthur E. "Report of the Anaconda Smelter Smoke Commission." Report dated 1920, 17–19, box 278, straight numerical files, DOJ (see also box 84b, BOM).

Young, Angela Nugent. "Interpreting the Dangerous Trades: Workers' Health in America and the Career of Alice Hamilton, 1910–1935." PhD diss., Brown University, 1982.

Zick, Robert L., David A. Velego Jr., Mark W. Hess, and Martin Foote. "Butte Mine Flooding Water Treatment Facility: Implementation of Major Component of Selected Remedy for Historic Contamination at Berkeley Pit Site." Paper presented at the American Society for Mining and Reclamation, Morgantown, West Virginia, April 2004. https://projects.itrcweb.org/miningwaste-guidance/References/2079-ZickPA.pdf.

Government and Corporate Publications

Annual Government Publications

Mineral Resources of the United States (Washington, D.C.).

Minerals Yearbook (Washington, D.C.).

Other Government Publications

The Federal Reporter.

Browne, J. Ross. *Report on the Mineral Resources of the States and Territories West of the Rocky Mountains*. Executive Document No. 202, U.S. House of Representatives, Second Session, 40th Congress, 1867–1868. Washington, D.C.: Government Printing Office, 1868.

Douglas, James, Jr. "The Metallurgy of Copper." In *Mineral Resources of the United States*, 257–280. Washington, D.C.: Government Printing Office, 1883.

Duaime, Terence E., and Steven F. McGrath. *Butte, Montana: The Berkeley Pit: Changes in Water Quality and Water Sampling Methods, 1982–2017*. Montana Bureau of Mines and Geology, Bulletin 138. Butte: Montana Bureau of Mines and Geology, 2017.

Emmons, S. F., and G. F. Becker. "Amalgamating Mills." In *Statistics and Technology of the Precious Metals*, U.S. Dept. of the Interior, Census Office, 242–285. Washington, D.C.: Government Printing Office, 1885.

Formad, Robert J. "The Effect of Smelter Fumes upon the Livestock Industry in the Northwest." Reprinted from *The Twenty-Fifth Annual Report of the Bureau of Animal Industry, 1908*. Washington, D.C.: Government Printing Office, 1910.

Harrington, Daniel. *Lessons from the Granite Mountain Shaft Fire, Butte*. U.S. Bureau of Mines Bulletin No. 188. Washington, D.C.: Government Printing Office, 1922.

Haywood, J. K. *Injury to Vegetation by Smelter Fumes*. USDA Bureau of Chemistry Bulletin No. 89. Washington, D.C.: Government Printing Office, 1905.

Haywood, J. K. *Injury to Vegetation and Animal Life by Smelter Wastes*. USDA Bureau of Chemistry Bulletin No. 113. Washington, D.C.: Government Printing Office, 1908.

Haywood, J. K. *Injury to Vegetation and Animal Life by Smelter Wastes*. USDA Bureau of Chemistry Bulletin No. 113 (Revised). Washington, D.C.: Government Printing Office, 1910.

Howe, Henry M. *Copper Smelting*, U.S. Geological Survey Bulletin No. 26. Washington, D.C.: Government Printing Office, 1885.

Kirchhoff, C., Jr. "Copper: The Copper Industry of the United States." In *Mineral Resources of the United States*, 213–305. Washington, D.C.: Government Printing Office, 1883.

Kirchhoff, C., Jr. "Copper: The Copper Industry in the United States." *Mineral Resources of the United States, 1883 and 1884*, 322–410. Washington, D.C.: Government Printing Office, 1885.

Kirchhoff, C., Jr. "Copper: The Copper Industry in the United States." *Mineral Resources of the United States, 1885*, 206–243.Washington, D.C.: Government Printing Office, 1886.

Kirchhoff, C., Jr. "Copper." *Mineral Resources of the United States, 1886*, 109–139. Washington, D.C.: Government Printing Office, 1887.

Kirchhoff, C., Jr. "Copper." *Mineral Resources of the United States, 1888*, 43–77. Washington, D.C.: Government Printing Office.

Kirchhoff, Charles, Jr. "Copper." In *Mineral Resources of the United States, 1900*, 141–190. Washington, D.C.: Government Printing Office, 1901.

National Park Service. *Butte, Montana: A Project Report*. Washington, D.C.: U.S. Department of the Interior, 1981.

Parker, Edward W. "Coal." In *Mineral Resources of the United States, 1910*, Part II, Non-Metals, 5–242. Washington, D.C.: Government Printing Office, 1911.

Parker, Edward W. "Coal." In *Mineral Resources of the United States, 1913*, Part II, Non-Metals, 709–928. Washington, D.C.: Government Printing Office, 1914.

Peters, Edward D. "The Mines and Reduction Works of Butte City, Montana." In *Mineral Resources of the United States, 1883 and 1884*, 374–396. Washington, D.C.: Government Printing Office, 1885.

Peters, Edward D. "The Roasting of Copper Ores and Furnace Products." In *Mineral Resources of the United States*, 280–297. Washington, D.C.: Government Printing Office, 1883

Raymond, Rossiter W. *Statistics of Mines and Mining in the States and Territories West of the Rocky Mountains*. Vol. 10, No. 207, Executive Documents of the House of Representatives, 2nd Session, 40th Congress, 1869–1870. Washington, D.C.: Government Printing Office, 1870.

Sixth Annual Report by the Director of the Bureau of Mines to the Secretary of the Interior for the Fiscal Year Ended June 1916. Washington, D.C.: Government Printing Office, 1916.

U.S. Department of Agriculture, Weather Bureau. *Summary of the Climatological Data for the United States, by Sections*. Washington, D.C.: Government Printing Office, ca. 1922.

U.S. Department of Commerce, Bureau of Corporations. *The Lumber Industry*, Part I, Standing Timber. Washington, D.C.: Government Printing Office, 1913.

U.S. Department of the Interior, Census Office. *Statistics and Technology of the Precious Metals*. Washington, D.C.: Government Printing Office, 1885.

U.S. Geological Survey. *Butte Special Folio, Montana*, Geologic Folio 38. Washington, D.C.: Government Printing Office, 1897.

U.S. Judiciary. *Federal Reporter: Cases Argued and Determined in the Circuit Courts of Appeals and Circuit and District Courts of the United States*. Saint Paul, MN: West Publishing Company, 1895-1917.

Weed, Walter Harvey. *Geology and Ore Deposits of the Butte District*, USGS Professional Paper 74. Washington, D.C.: Government Printing Office, 1912.

Corporate Publications

Alice Gold and Silver Mining Co. *Annual Report from January 1, 1894, to January 1, 1895*. Salt Lake City: Salt Lake Lithographing Co., Printers, 1895.

Alice Gold and Silver Mining Co. *Annual Report from January 1, 1898, to January 1, 1899*. Salt
Lake City: T. C. Crawford, Printer, 1899.

Anaconda Copper Mining Company. *Report of the Anaconda Copper Mining Company, Covering a Period of about 15 Years, Ending June 30, 1895*. Anaconda, MT: Standard Publishing Company, 1895.

Anaconda Copper Mining Company. *Report of the Anaconda Copper Mining Company for the Year Ending December 31, 1914*.

Anaconda Copper Mining Company. *Report of the Anaconda Copper Mining Company for the Year Ending December 31, 1915*.

Anaconda Copper Mining Company Reduction Department. *Report of the Research, Concentration and Testing Departments for the Years 1912, 1913 and 1914*. Anaconda, MT: Anaconda Copper Mining Company, 1915.

Anaconda Copper Mining Company Reduction Department. *Report of the Research, Concentration, and Testing Departments for the Year 1915*. Anaconda, MT: Anaconda Copper Mining Company, 1916.

Anaconda Copper Mining Company Reduction Department. *Report of the Research, Concentration, and Testing Departments for the Year 1916*. Anaconda, MT: Anaconda Copper Mining Company, 1917.

Anaconda Copper Mining Company Reduction Department. *Report of the Research, Concentration, and Testing Departments for the Year 1917*. Anaconda, MT: Anaconda Copper Mining Company, 1918.

Anaconda Copper Mining Company Staff. *The Anaconda Reduction Works*. Anaconda, MT: Anaconda Copper Mining Company, 1920.

Anaconda Copper Mining Company Staff. *A Brief Description of the Washoe Smelter*. Anaconda, MT: Anaconda Copper Mining Company, 1913.

Anaconda Copper Mining Company Staff. *A Brief Description of the Washoe Smelter*. Anaconda, MT: Anaconda Copper Mining Company, 1916.

Anaconda Copper Mining Company Staff. *A Brief Description of the Washoe Smelter*. Anaconda, MT: Anaconda Copper Mining Company, 1928.

Anaconda Copper Mining Company Staff. *A Brief Description of the Washoe Smelter*. Anaconda, MT: Anaconda Copper Mining Company, 1938.

Boston and Montana Consolidated Copper and Silver Mining Company. *Report of the Boston and Montana Consolidated Copper and Silver Mining Company for the Year Ending December 31, 1897*.

Boston and Montana Consolidated Copper and Silver Mining Company. *Report of the Boston and Montana Consolidated Copper and Silver Mining Company for the Year Ending December 31, 1898*.

Articles, Books, and Other Published Sources

Agricola, Georgius. *De Re Metallica*. Translated from the 1556 Latin edition by Herbert Clark Hoover and Lou Henry Hoover. New York: Dover Publications. 1950 reprint of the original 1912 translation that was published by The Mining Magazine, London.

Aiken, Katherine G. "'Not Long Ago a Smoking Chimney Was a Sign of Prosperity': Corporate and Community Response to Pollution at the Bunker Hill Smelter in Kellogg, Idaho." *Environmental History Review* 18 (Summer 1994): 67–86.

Aiken, Katherine G. "'The Environmental Consequences . . . Were Calamitous': Smelter Smoke Controversies in Progressive Era America, 1899–1918." *Technology and Culture* 60 (January 2019): 132–164.

Ammons, Doug. *A Darkness Lit by Heroes: The Granite Mountain-Speculator Mine Disaster of 1917*. Missoula, MT: Water Nymph Press, 2017.

"Anaconda: I." *Fortune* 14 (December 1936): 82–94, 210–220.

"Anaconda: II." *Fortune* 15 (January 1937): 70–77, 134–149.

Andrews, Thomas G. *Killing for Coal: America's Deadliest Labor War*. Cambridge, MA: Harvard University Press, 2008.

Austin, L. S. "The Washoe Plant of the Anaconda Company in 1905." *Transactions of the American Institute of Mining Engineers* 37 (1907): 431–485.

Austin, Leonard S. *The Metallurgy of the Common Metals: Gold, Silver, Iron (and Steel), Copper, Lead, and Zinc*. New York: John Wiley & Sons, 1921.

Bakken, Gordon Morris. "Was There Arsenic in the Air? Anaconda Versus the Farmers of Deer Lodge Valley." *Montana: The Magazine of Western History* 41 (Summer 1991): 30–41.

Bakken, Gordon Morris, and J. Elwood Bakken. "The Goldfish Died: Great Falls, Fort Benton, and the Great Flood of 1908." *Montana: The Magazine of Western History* 51 (Winter 2001): 38–51.

Baldwin, William Lee. *Antitrust and the Changing Corporation*. Durham, NC: Duke University Press, 1961.

Barnett, Harold C. *Toxic Debts and the Superfund Dilemma*. Chapel Hill: University of North Carolina Press, 1994.

Basso, Matthew L. *Meet Joe Copper: Masculinity and Race on Montana's World War II Home Front*. Chicago: University of Chicago Press, 2013.

Beck, Eckardt C. "The Love Canal Tragedy." *EPA Journal*, January 1979. https://archive.epa.gov/epa/aboutepa/love-canal-tragedy.html.

Bender, Louis V. "Coal-Dust Fired Reverberatories at Washoe Reduction Works." *Transactions of the American Institute of Mining Engineers* 51 (1916): 743–751.

Bender, Louis V. "Development of Copper Smelting at Anaconda." *E&MJ* 128 (August 24, 1929): 301–305.

Bender, Louis V., and Harold H. Goe. "Production of Arsenic Trioxide at Anaconda." *Transactions of the American Institute of Mining Engineers* 106 (1933): 324–328.

Bierer, B. W. *A Short History of Veterinary Medicine in America*. East Lansing: Michigan State University Press, 1955.

Bining, Arthur Cecil. *Pennsylvania Iron Manufacturing in the Eighteenth Century*. Harrisburg: Pennsylvania Historical Commission, 1938.

Bird, Frank. *Story of Montana Power*. Butte: The Montana Power Company, 1941.

Bittlingmayer, George. "Antitrust and Business Activity: The First Quarter Century." *The Business History Review* 70 (Autumn 1996): 363–401.

Blake, William P. "The Rainbow Lode, Butte City, Montana." *Transactions of the American Institute of Mining Engineers* 16 (1888): 65–82.

Blake-Coleman, B. C. *Copper Wire and Electrical Conductors—The Shaping of a Technology*. Philadelphia: Harwood Academic Publishers, 1992.

Blaskovich, Mary Popovich. "History of Opportunity." In *In the Shadow of a Mount Haggin: The Story of Anaconda and Deer Lodge County from 1863 to 1976*. Anaconda, MT: Deer Lodge County History Group, 1975.

Bloom, Khaled J. *Murder of a Landscape: The California Farmer-Smelter War, 1897–1916*. Norman, OK: The Arthur H. Clark Company, 2010.

Botkin, Jane Little. *Frank Little and the IWW: The Blood That Stained an American Family*. Norman: University of Oklahoma Press, 2017.

Brands, H. W. *T.R.: The Last Romantic*. New York: Basic Books, 1997.

Brechin, Gray. *Imperial San Francisco: Urban Power, Earthly Ruin*. Berkeley: University of California Press, 1999.

Brooks, David. "Remove the Dam, Restore the River: How Public Participation Redefined Superfund Law at Milltown, Montana." *Montana: The Magazine of Western History* 65 (Autumn 2015): 3–24.

Brown, John K. *The Baldwin Locomotive Works, 1831–1915: A Study in American Industrial Practice*. Baltimore: Johns Hopkins University Press, 1995.

Bryant, Keith L. "Entering the Global Economy." In *The Oxford History of the American West*, edited by Clyde A. Milner II, Carol A. O'Connor, and Martha A. Sandweiss. New York: Oxford University Press, 1994.

Burroughs, William James. *Weather Cycles: Real or Imaginary?* New York: Cambridge University Press, 1992.

Bushnell, A. F. "Precipitating Copper from Butte Mine-Water." *E&MJ* 83 (June 29, 1907): 1229–1230.

Butler, B. S. "Copper." In *Mineral Resources of the United States, 1910*. Washington, D.C.: Government Printing Office, 1911.
Calvert, Gerry W. *The Gibraltar: Socialism and Labor in Butte, Montana, 1895–1920*. Helena: Montana Historical Society Press, 1988.
Chandler, Alfred D., Jr. *The Visible Hand: The Managerial Revolution in American Business*. Cambridge, MA: Harvard University Press, 1977.
Chicago Association of Commerce, Committee of Investigation on Smoke Abatement and Electrification of Railway Terminals. *Smoke Abatement and Electrification of Railway Terminals in Chicago*. Chicago: Rand McNally & Company, 1915.
Colten, Craig, and Peter N. Skinner. *The Road to Love Canal: Managing Industrial Waste before EPA*. Austin: University of Texas Press, 1996.
Corbett, Robert. "The Changing Face of Mining in Butte." *The Speculator: A Journal of Butte and Southwest Montana History* 1 (Summer 1984): 20–25.
Cottrell, F. G. "Electrical Fume-Precipitation." *Transactions of the American Institute of Mining Engineers* 43 (1913): 512–520.
Cottrell, F. G. "The Electrical Precipitation of Suspended Particles." *The Journal of Industrial and Engineering Chemistry* 3 (August 1911): 542–550.
Cottrell, F. G. "The Smoke Problem in the Mineral Industries." In *The Mineral Industry: Its Statistics, Technology and Trade during 1914*, edited by G. A. Roush. New York: McGraw-Hill, 1915.
"Cottrell Process for Condensing Smelter Fumes." *E&MJ* 86 (August 22, 1908): 375–377.
Cowan, Ruth Schwartz. *A Social History of American Technology*. New York: Oxford University Press, 1997.
Coxe, Eckley B. "Presidential Address." *Transactions of the American Institute of Mining Engineers* 8 (1879): 126–127.
Cronon, William. "Kennecott Journey: The Paths out of Town." In *Under an Open Sky: Rethinking America's Western Past*, edited by William Cronon, George Miles, and Jay Gitlin, 28–51. New York: W. W. Norton & Company, 1992.
Cronon, William. *Nature's Metropolis: Chicago and the Great West*. New York: W. W. Norton & Company, 1991.
Crowfoot, Arthur. "Development of the Round Table at Great Falls." *Transactions of the American Institute of Mining Engineers* 49 (1915): 417–469.
Curran, Mary. "Dialogues of Difference: Contested Mappings of Tourism and Environmental Protection in Butte, Montana." In *Mapping Tourism*, edited by Stephen P. Hanna and Vincent J. Del Casino Jr. Minneapolis: University of Minnesota Press, 2003.
Curtis, Kent A. *Gambling on Ore: The Nature of Metal Mining in the United States, 1860–1910*. Boulder: University Press of Colorado, 2013.
Czehura, Steve J. "Butte: A World Class Ore Deposit." *Mining Engineering* 58 (September 2006): 16–17.
Daly, William B. "Evolution of Mining Practice at Butte." *Engineering & Mining Journal* 128 (August 24, 1929): 280–285.
Day, Frank H. "Disposal of Metallurgical Wastes." *Mining Congress Journal* 47 (November 1961): 52–56.

Day, Joan, and R. F. Tylecote, eds. *The Industrial Revolution in Metals*. London: The Institute of Materials, 1991.

Deane, Phyllis, and Jessica Kuper, eds. *A Lexicon of Economics*. New York: Routledge, 1988.

Diamond, Jared. *Collapse: How Societies Choose to Fail or Succeed*. New York: Viking, 2005.

Dobb, Edwin. "Dirty Old Town: Addiction and Betrayal in the Mining City." In *Coming Home: A Special Issue Devoted to the Historic Built Environment and Landscapes of Butte and Anaconda, Montana*, ed. Patty Dean, published as vol. 3, no. 1 of *Drummon Views* (Spring 2009): 249–258.

Dobb, Edwin. "Introduction: The Age of Sacrifice Zone." In *Extraction: Art on the Edge of the Abyss*, edited by Samuel Pelts, 14–18. Berkeley, CA: The CODEX Foundation, 2020.

Dobb, Edwin. "Pennies from Hell: In Montana, the Bill for America's Copper Comes Due." *Harper's Magazine* 293 (October 1996): 39–54.

Dobb, Edwin. "Viewpoint: Location, Occupation, Juxtaposition, Interpretation: Notes on an Erotics of the Mining City." *Buildings and Landscapes: Journal of the Vernacular Forum* 17 (Spring 2010): 1–12.

Douglas, James, Jr. "Historical Sketch of Copper Smelting in the United States." In *The Mineral Industry, 1895*, vol. IV, edited by Richard P. Rothwell. New York: The Scientific Publishing Company, 1896.

Douglas, James, Jr. "Recent American Methods and Appliances Employed in the Metallurgy of Copper, Lead, Gold, and Silver." *Journal of the Society of Arts* 43 (August 23, 1895): 834–844.

Drake, Richard. *The Education of an Anti-Imperialist: Robert La Follette and U.S. Expansion*. Madison: University of Wisconsin Press, 2013.

Dubofsky, Melvyn. *We Shall Be All: A History of the Industrial Workers of the World*. Chicago: Quadrangle Books, 1969.

Dunn, Edgar M. "Determination of Gases in Smelter Flues, and Notes on the Determination of Dust Losses at the Washoe Reduction Works, Anaconda, Mont." *Transactions of the American Institute of Mining Engineers* 46 (1914): 648–689.

"Edward Dyer Peters." *Transactions of the American Institute of Mining Engineers* 60 (1919): 735–740.

Egleston, T. "Boston and Colorado Smelting Works." *Transactions of the American Institute of Mining Engineers* 4 (1876): 276–298.

Elton, James O. "Arsenic Trioxide from Flue Dust." *Transactions of the American Institute of Mining Engineers* 46 (1914): 690–702.

Emmons, David M. *The Butte Irish: Class and Ethnicity in an American Mining Town, 1875–1925*. Urbana: University of Illinois Press, 1989.

Fahlman, Betsy, ed. *Landscapes of Extraction: The Art of in the American West*. Phoenix: Phoenix Art Museum, 2021.

Farling, Bruce. "Montana's Clark Fork River: An Industrial Drain." *High County News* 20 (March 14, 1988): 1, 8–14.

Fay, Albert Hill, ed. *The Mineral Industry: Its Statistics, Technology and Trade during*

1910. New York: McGraw-Hill, 1911. This annual volume was first published for the year 1892 by the Scientific Publishing Co. It continued well into the twentieth century, with McGraw-Hill taking over as publisher.

Febles, J. C. "The Precipitation of Copper from the Miner Waters of the Butte District." *Transactions of the American Institute of Mining Engineers* 46 (1914): 177–195.

Fell, James E., Jr. *Ores to Metals: The Rocky Mountain Smelting Industry*. Lincoln: University of Nebraska Press, 1979.

Fiege, Mark. *The Republic of Nature: An Environmental History of the United States Lucifers*. Seattle: University of Washington Press, 2012.

Finn, Janet L. "Intimate Strangers: The Interlocking Histories of Butte, Montana, and Chuquicamata, Chile." *Montana: The Magazine of Western History* 48 (Autumn 1998): 22–39.

Finn, Janet L. *Mining Childhood: Growing Up in Butte, Montana: 1900–1960*. Helena: Montana Historical Society Press, 2012.

Finn, Janet L. *Tracing the Veins: Of Copper, Culture, and Community from Butte to Chuquicamata*. Berkeley: University of California Press, 1998.

Fisher, Arthur. "Montana: Land of the Copper Collar." *The Nation* 117 (September 19, 1923): 290–292.

Forbes, Linda C. "A Vision for Cultivating a Nation: Gifford Pinchot's 'The Fight for Conservation.'" *Organization & Environment* (June 17, 2004): 226–231.

Frederick, David C. *Rugged Justice: The Ninth Circuit Court of Appeals and the American West, 1891–1941*. Berkeley: University of California Press, 1994.

Freeman, Henry O. *A Brief History of Butte*. Chicago: Henry O. Shepherd Co., 1900.

Freese, Barbara. *Industrial-Strength Denial: Eight Stories of Corporations Defending the Indefensible, from the Slave Trade to Climate Change*. Oakland: University of California Press, 2021.

Freyer, Tony. "Legal Restraints on Economic Coordination: Antitrust in Great Britain and America, 1880–1920." In *Coordination and Information: Historical Perspectives on the Organization of Enterprise*, edited by Naomi R. Lamoreaux and Daniel M. G. Raff. Chicago: University of Chicago Press, 1995.

Freyfogle, Eric T. *The Land We Share: Private Property and the Common Good*. Washington, D.C.: Island Press, 2003.

Friedel, Robert. *A Culture of Improvement: Technology and the Western Millennium*. Cambridge, MA: MIT Press, 2007.

Fulmor, T. G. "Anaconda's Butte Concentrator." *Mining Engineering* 16 (May 1964): 54–73.

Gahl, Rudolph. "Development of Copper Ore Leaching in Arizona." *E&MJ* 127 (April 20, 1929): 636–638.

Gahl, Rudolph. "History of the Flotation Process at Inspiration." *Transactions of the American Institute of Mining Engineers* 55 (1917): 576–645.

Gates, William B., Jr. *Michigan Copper and Boston Dollars: An Economic History of the Michigan Copper Mining Industry*. Cambridge, MA: Harvard University Press, 1951.

Gillie, John. "Use of Electricity in Mining in the Butte District." *Transactions of the American Institute of Mining Engineers* 46 (1914): 817–819.

Glasscock, Carl B. *The War of the Copper Kings: Builders of Butte and Wolves of Wall Street*. Indianapolis: Bobbs-Merrill, 1935.

Goetzmann, William H. *Exploration and Empire: The Explorer and the Scientist in the Winning of the American West*. New York: W. W. Norton & Co., 1978.

Gohman, Sean. "It's Not Time to Be Wasted: Identifying, Evaluating, and Appreciating Mine Wastes in Michigan's Copper Country." *IA: The Journal of the Society for Industrial Archeology* 39, nos. 1 and 2 (2013): 5–22.

Golan, Tal. *Laws of Men and Laws of Nature: The History of Scientific Expert Testimony in England and America*. Cambridge, MA: Harvard University Press, 2004.

Goodale, Charles W. "The Concentration of Ores in the Butte District, Montana." *Transactions of the American Institute of Mining Engineers* 26 (1896): 599–639.

Goodale, C. W., and J. H. Klepinger. "The Great Falls Flue System and Chimney." *Transactions of the American Institute of Mining Engineers* 46 (1914): 567–647.

Goodell, Jeff. *The Heat Will Kill You First: Life and Death on a Scorched Planet*. New York: Little, Brown & Company, 2023.

Goodwin, Doris Kearns. *The Bully Pulpit: Theodore Roosevelt, William Howard Taft, and the Golden Age of Journalism*. New York: Simon & Schuster, 2013.

Gordon, Robert J. *The Rise and Fall of American Growth: The U.S. Standard of Living since the Civil War*. Princeton, NJ: Princeton University Press, 2016.

Gorman, Hugh. *Redefining Efficiency: Pollution Concerns, Regulatory Mechanisms, and Technological Change in the U.S. Petroleum Industry*. Akron, OH: University of Akron Press, 2001.

Gow, Paul A. "Mine Surveying Methods Employed at Butte, Mont." *E&MJ* 90 (December 17, 1910): 1209–1211.

Grant, Frank. "Rocky Mountain Husbandman: Embattled Voice of the Montana Farmer." *Montana: The Magazine of Western History* 24 (Spring 1974): 34–43.

Great Falls and Cascade County: Historic Scenic and Biographical. Great Falls, MT: The Tribune, 1899.

Greene, Ann Norton. *Horses at Work: Harnessing Power in Industrial America*. Cambridge, MA: Harvard University Press, 2008.

Gurnham, C. Fred. "Pollution Control at Anaconda's Copper Mines and Smelter." *Industrial Water & Wastes* 6 (September–October 1961): 125–130.

Gutfeld, Arnon. *Montana's Agony: Years of War and Hysteria, 1917–1921*. Gainesville: University of Florida Press, 1979.

Hammond, John Hays. *The Autobiography of John Hays Hammond*. New York: Farrar & Rinehart, 1935.

Hardesty, Donald. "Evaluating Site Significance in Mining Districts." *Historical Archaeology* 24, no. 2 (1990): 42–51.

Harkins, W. D., and R. E. Swain, "Papers on Smelter Smoke." *The Journal of the American Chemical Society* 29 (July 1907): 970–998.

Harvey, Charles E. *The Rio Tinto Company: An Economic History of a Leading International Mining Concern, 1873–1954*. Penzance, Cornwall, UK: Alison Hodge, 1981.

Hayden, Ralph. "Concentration of Slimes at Anaconda, Mont." *Transactions of the American Institute of Mining Engineers* 46 (1914): 239–265.

Hays, Samuel P. *Conservation and the Gospel of Efficiency: The Progressive Conservation Movement, 1890–1920*. Cambridge, MA: Harvard University Press, 1959.

Haywood, J. K. "Injury to Vegetation and Animal Life by Smelter Fumes." *The Journal of the American Chemical Society* 29 (July 1907): 998–1009.

Hebgen, Max. "The Great Falls Hydroelectric Power Plant." *Cassier's Magazine* 40 (August 1911): 273–288.

Herfindahl, Orris C. *Copper Costs and Prices: 1870–1957*. Baltimore: Johns Hopkins University Press, 1959.

Hewitt, Abram S. "A Century of Mining and Metallurgy in the United States." *Transactions of the American Institute of Mining Engineers* 5 (1876–1877): 164–191.

Herrick, R. L. "The World's Largest Chimney." *Mines and Minerals* 30 (December 1909): 257–260.

Higgins, Edwin. "New 506-Ft. Chimney at Great Falls Smelter." *E&MJ* 87 (January 16, 1909): 156–159.

Hildebrand, George H., and Garth L. Mangum. *Capital and Labor in American Copper, 1845–1990: Linkages between Product and Labor Markets*. Cambridge, MA: Harvard University Press, 1992.

Hills, Richard L. *Power in the Industrial Revolution*. Manchester, UK: Manchester University Press, 1970.

Hird, John A. *Superfund: The Political Economy of Environmental Risk*. Baltimore: Johns Hopkins University Press, 1994.

Hofman, H. O. *Metallurgy of Copper*. New York: McGraw-Hill, 1914.

Hofman, H. O. *The Metallurgy of Lead and the Desilverization of Base Bullion*. New York: The Scientific Publishing Company, 1899.

Hofman, H. O. "Notes on the Metallurgy of Copper in Montana." *Transactions of the American Institute of Mining Engineers* 34 (1904): 258–316.

Hoskins, W. Horace "Obituary, Daniel Elmer Salmon." *American Veterinary Review* 46 (October 1914): 93–95.

Hounshell, David. *From the American Systems to Mass Production, 1800–1932: The Development of Manufacturing Technology in the United States*. Baltimore: Johns Hopkins University Press, 1985.

Hovenkamp, Herbert. *Enterprise and American Law, 1836–1937*. Cambridge, MA: Harvard University Press, 1991.

Hovis, Logan, and Jeremy Mouat. "Miners, Engineers, and the Transformation of Work in the Western Mining Industry, 1880–1930." *Technology and Culture* 37 (July 1996): 429–456.

Howard, Joseph Kinsey. *Montana: High, Wide, and Handsome*. New Haven, CT: Yale University Press, 1943.

Hughes, D. Arthur. "Cornell Veterinarians." *Cornell Alumni News* 7 (February 15, 1905): 299–302.

Hughes, Thomas P. *American Genesis: A Century of Invention and Technological Enthusiasm*. New York: Viking, 1989.

Hughes, Thomas P. *Networks of Power: Electrification in Western Society, 1880–1930*. Baltimore: Johns Hopkins University Press, 1983.

Hyde, Charles K. *Copper for America: The United States Copper Industry from Colonial Times to the 1990s*. Tucson: University of Arizona Press, 1998.

Ingalls, W. R. "The Business of Mining." *Mining and Scientific Press* 113 (August 19, 1916): 276–281.

Ingalls, W. R. "Electrolytic Zinc." *E&MJ* 101 (March 4, 1916): 425–428.

Ingalls, W. R. "Recent Observations in Butte and Anaconda." *E&MJ* 97 (May 9, 1914): 937–943.

Jackson, Donald C. *Building the Ultimate Dam: John S. Eastwood and the Control of Water in the West*. Lawrence: University Press of Kansas, 1995.

Jensen, Vernon H. *Heritage of Conflict: Labor Relations in the Nonferrous Metals Industry up to 1930*. Ithaca, NY: Cornell University Press, 1950.

The Jewish Encyclopedia, vol. VIII. New York: Funk & Wagnalls Company, 1901–1906.

Johnson, Arthur M. "Theodore Roosevelt and the Bureau of Corporations." *The Mississippi Valley Historical Review* 45 (March 1959): 571–590.

Johnson, Bob. *Carbon Nation: Fossil Fuels in the Making of American Culture*. Lawrence: University Press of Kansas, 2014.

Johnson, Carrie. "Electrical Power, Copper, and John D. Ryan." *Montana: The Magazine of Western History* 38 (Autumn 1988): 24–37.

Johnson, Ligon. "The History and Legal Phases of the Smoke Problem." *Transactions of the American Institute of Mining Engineers* 58 (1918): 198–214.

Jones, Christopher F. *Routes of Power: Energy in Modern America*. Cambridge, MA: Harvard University Press, 2014.

Jones, Susan D. *Valuing Animals: Veterinarians and Their Patients in Modern America*. Baltimore: Johns Hopkins University Press, 2003.

Jordan, John M. *Machine-Age Ideology: Social Engineering and American Liberalism, 1911–1939*. Chapel Hill: University of North Carolina Press, 1994.

Kelley, Robert. *Gold vs. Grain: The Hydraulic Mining Controversy in California's Sacramento Valley*. Glendale, CA: Arthur H. Clarke, Co., 1959.

Kelley, Robert. "The Mining Debris Controversy in the Sacramento Valley." *Pacific Historical Review* 25 (November 1956): 331–346.

Kelly, William, and Frederick Laist. "Development of Copper Converting at Butte and Anaconda." *Transactions of the American Institute of Mining Engineers* 106 (1933): 122–131.

Kirk, Cecil H. *A History of the Montana Power Company*. Pleasant Hill, OR: Donn B. Kirk, 2008.

Kirsch, Stuart. *Mining Capitalism: The Relationship between Corporations and Their Critics*. Oakland: University of California Press, 2014.

Knowles, L. C. A. *Economic Development in the Nineteenth Century: France, Germany, Russia, and the United States*. London: Routledge and Kegan Paul, 1932.

Laist, Frederick. "Changes in Smelting Practice of Anaconda Copper Mining Co." *E&MJ* 102 (October 7, 1916): 635–638.

Laist, Frederick. "History of Reverberatory Smelting in Montana, 1879 to 1933." *Transactions of the American Institute of Mining Engineers* 106 (1933): 23–87.

Laist, Frederick. "Improvements at the Reduction Works of the Anaconda Company," *E&MJ* 99 (February 27, 1915): 418–419.

Laist, Frederick. "Modern Metallurgical Miracles." *E&MJ* 128 (August 24, 1929): 287–290.

Laist, Frederick. "Roasting and Leaching Tailings at Anaconda, Mont." *Transactions of the American Institute of Mining Engineers* 46 (1914): 362–382.

Laist, Frederick, and Harold W. Aldrich. "Experimental Leaching at Anaconda." *Transactions of the American Institute of Mining Engineers* 49 (1915): 671–690.

Laist, Frederick, and Harold W. Aldrich. "The 2,000-Ton Leaching Plant at Anaconda." *Transactions of the American Institute of Mining Engineers* 55 (1917): 866–882.

Laist, Frederick, and Albert E. Wiggin. "Flotation Concentration at Anaconda, Mont." *Transactions of the American Institute of Mining Engineers* 55 (1917): 485–526.

Laist, Frederick, and Albert E. Wiggin. "The Slime-Concentrating Plant at Anaconda." *Transactions of the American Institute of Mining Engineers* 49 (1915): 470–484.

Laist, Frederick, F. F. Frick, J. O. Elton, and R. B. Caples. "Electrolytic Zinc Plant of Anaconda Copper Mining Co., at Great Falls, Mont." *Transactions of the American Institute of Mining Engineers* 64 (1921): 699–753.

Lamborn, John E., and Charles S. Peterson. "The Substance of the Land: Agriculture v. Industry in the Smelter Cases of 1904 and 1906." *Utah Historical Quarterly* 53 (Fall 1985): 312–320.

Lamoreaux, Naomi R. *The Great Merger Movement in American Business, 1895–1904*. New York: Cambridge University Press, 1985.

Lamoreaux, Naomi R. "Sklar's 'The Corporate Reconstruction of American Capitalism': An Economic Historian's Perspective." *Business and Economic History* 21 (1992): 36–39.

Lankton, Larry. *Cradle to Grave: Life, Work, and Death at the Lake Superior Copper Mines*. New York: Oxford University Press, 1991.

Lankton, Larry. *Hollowed Ground: Copper Mining and Community Building on Lake Superior, 1840s–1990s*. Detroit: Wayne State University Press, 2010.

Lawson, Thomas W. *Frenzied Finance: The Crime of Amalgamated*. New York: The Ridgeway-Thayer Company, 1905.

Layton, Edwin T., Jr. *The Revolt of the Engineers: Social Responsibility and the American Engineering Profession*. Baltimore: Johns Hopkins University Press, 1986.

LeCain, Timothy. *Mass Destruction: The Mean and Giant Mines That Wired America and Scarred the Planet*. New Brunswick, NJ: Rutgers University Press, 2009.

Leeson, M. A. *History of Montana, 1739–1885*. Chicago: Warner, Beers & Company, 1885.

Leech, Brian James. *The City That Ate Itself: Butte, Montana, and Its Expanding Berkeley Pit*. Reno: University of Nevada Press, 2018.

Leonard, Ellis Pierson. *A Cornell Heritage: Veterinary Medicine, 1868–1908*. Ithaca, NY: New York State College of Veterinary Medicine, 1979.

Leonard, John William, ed. *Who's Who in Engineering*. New York: Who's Who Publications Inc., 1925.

Lincoln, David Francis. *School and Industrial Hygiene*. Philadelphia: Presley Blakiston, 1880.

Linforth, Frank A. "Applied Geology in the Butte Mines." *Transactions of the American Institute of Mining Engineers* 46 (1914): 110–122.

Lingenfelter, Richard E. *The Hardrock Miners: A History of the Mining Labor Movement in the American West, 1863–1893*. Berkeley: University of California Press, 1974.

MacMillan, Donald. "The Butte 'Smoke Messiahs' and Their War against Air Pollution." *The Speculator: A Journal of Butte and Southwest Montana History* 1 (Summer 1984): 48–55.

MacMillan, Donald. *Smoke Wars: Anaconda Copper, Montana Air Pollution, and the Courts, 1890–1920*. Helena: Montana Historical Society Press, 2000.

Malm, Andreas. *Fossil Capital: The Rise of Steam Power and the Roots of Global Warming*. New York: Verso, 2016.

Malone, Michael P. *The Battle for Butte: Mining and Politics on the Northern Frontier, 1864–1906*. Seattle: University of Washington Press, 1981.

Malone, Michael P., and F. Ross Peterson, "Politics and Protests." In *The Oxford History of the American West*, edited by Clyde A. Milner. New York: Oxford University Press, 1994.

Malone, Michael P., and Richard B. Roeder. *Montana: A History of Two Centuries*. Seattle: University of Washington Press, 1976.

Marcosson, Isaac F. *Anaconda*. New York: Dodd, Mead & Company, 1957.

Marcosson, Isaac F. *Metal Magic: The Story of the American Smelting & Refining Company*. New York: Farrar, Straus & Company, 1949.

Mathewson, E. P. "The Development of the Reverberatory Furnace for Smelting Copper-Ores." *Transactions of the American Institute of Mining Engineers* 44 (1913): 781–805.

Maxwell, James R. "Mountain Railroads of the Mining Regions." *The Engineering Magazine* 10 (October 1895): 61–75.

Mayer, Carl J., and George A. Riley. *Public Domain, Private Domain: A History of Public Mineral Policy in America*. San Francisco: Sierra Club Books, 1985.

Maysilles, Duncan. *Ducktown Smoke: The Fight over One of the South's Greatest Environmental Disasters*. Chapel Hill: University of North Carolina Press, 2011.

McClernan, Henry. "Sixty Million Years of History: The Formation of Butte Copper." *The Speculator: A Journal of Butte and Southwest Montana History* 1 (Summer 1984): 16–19.

McDougall, C. H. "Flues and Flue Dust at Anaconda." *The Canadian Mining Review* 24 (February 1905): 26–34.

McGlynn, Terrance D. "Flying the Red Flag in Butte: The Life and Times of Butte's Socialist Mayor, Lewis Duncan." *The Speculator: A Journal of Butte and Southwest Montana History* 2 (Summer 1985): 3–8.

McIntosh, John Houston. "A Million for a Smoke Eater." *The Technical World Magazine* 13 (April 1910): 156–165.

McNeill, J. R. "The First Hundred Thousand Years." In *The Turning Points of Environmental History*, edited by Frank Uekoetter. Pittsburgh: University of Pittsburgh Press, 2010.

McNeill, J. R., and Peter Engelke. *The Great Acceleration: An Environmental History of the Anthropocene since 1945*. Cambridge, MA: Harvard University Press, 2014.

McNeill, J. R., and George Vrtis, eds. *Mining North America: An Environmental History since 1522*. Oakland: University of California Press, 2017.

McNelis, Sarah. *Copper King at War: The Biography of F. Augustus Heinze*. Missoula: University of Montana Press, 1968.

McPherson, James M. *Battle Cry of Freedom: The Civil War Era*. New York: Ballantine Books, 1989.

Mercier, Laurie. *Anaconda: Labor, Community, and Culture in Montana's Smelter City*. Urbana: University of Illinois Press, 2001.

Miller, Char. *Gifford Pinchott and the Making of Modern Environmentalism*. Washington, D.C.: Island Press, 2001.

Mix, Michael C. *Leaded: The Poisoning of Idaho's Silver Valley*. Corvallis: Oregon State University Press, 2013.

"The Montana Smoke Case." *American Veterinary Review* 32 (November 1907): 249–253.

Morin, Bode J. *The Legacy of American Copper Smelting: Industrial Heritage versus Environmental Policy*. Knoxville: University of Tennessee Press, 2013.

Morrow, Bayard S. "Both Copper and Zinc Ores Treated by Selective Flotation in Concentrators at Anaconda, Montana." *E&MJ* 128 (August 24, 1929): 295–300.

Mumford, Lewis. *Technics and Civilization*. New York: Harcourt, Brace & Company, 1934.

Murphy, Jerre C. *The Comical History of Montana: A Serious Story for Free People*. San Diego, CA: E. L. Schofield, 1912.

Murphy, Mary. "Bootlegging Mothers and Drinking Daughters: Gender and Prohibition in Butte, Montana." *American Quarterly* 46 (June 1994): 174–194.

Murphy, Mary. *Mining Cultures: Men, Women, and Leisure in Butte, 1914–41*. Urbana: University of Illinois Press, 1997.

Myers, Rex. "The Butte Rail Connection: Mining and Transportation, 1880–1980." *The Speculator: A Journal of Butte and Southwest Montana History* 1 (Summer 1984): 30–37.

Neiman, Susan. *Learning from the Germans: Race and the Memory of Evil*. New York: Farrar, Straus and Giroux, 2019.

Newell, Edmund. "Atmospheric Pollution and the British Copper Industry, 1690–1920." *Technology and Culture* 38 (July 1997): 655–689.

Noble, David F. *America by Design: Science, Technology, and the Rise of Corporate Capitalism*. New York: Alfred A. Knopf, 1977.

Nordberg, Bruno V. "The Compressed Air System of the Anaconda Copper Mining Co., Butte, Montana." *Transactions of the American Institute of Mining Engineers* 46 (1914): 826–841.

Novak, William J. *The People's Welfare: Law and Regulation in Nineteenth-Century America*. Chapel Hill: University of North Caroline Press, 1996.

Nye, David E. *American Technological Sublime*. Cambridge, MA: MIT Press, 1994.

Nye, David E. *Electrifying America: Social Meanings of a New Technology*. Cambridge, MA: MIT Press, 1991.

Nystrom, Eric C. *Seeing Underground: Maps, Models, and Mining Engineering in America*. Reno: University of Nevada Press, 2014.

Nystrom, Eric C. "Underground Mine Maps and the Development of the Butte System at the Turn of the 20th Century." *IA: The Journal of the Society for Industrial Archeology* 37, nos. 1 and 2 (2011): 97–113.

Offerhaus, C. "Copper Blast Furnace Smelting at Anaconda." *E&MJ* 88 (August 7, 1909): 243–250.

Parsons, A. B. *The Porphyry Coppers*. New York: American Institute of Mining Engineers, 1933.

Paul, Rodman W. *The Far West and the Great Plains in Transition, 1859–1900*. New York: Harper & Row, Publishers, 1988.

Peele, Robert. *Mining Engineers' Handbook*. New York: John Wiley & Sons Inc., 1918.

Pelts, Samuel, ed. *Extraction: Art on the Edge of the Abyss*. Berkeley, CA: The CODEX Foundation, 2020.

Peters, Edward D. *Modern American Methods of Copper Smelting*. New York: Scientific Publishing Company, 1887.

Peters, Edward D. "The Mount Lincoln Smelting Works, at Dudley, Colorado." *Transactions of*

the American Institute of Mining Engineers 2 (1872): 310–314.

Peters, Edward D. "Notes on Copper Smelting in the West." *E&MJ* 88 (October 9, 1909): 735–736.

Peters, Edward D. *The Practice of Copper Smelting*. New York: McGraw-Hill Book Company, 1911.

Peterson, Peter E. "Bullwhacker Leaching Plant, Butte, Mont." *Mining and Engineering World* 39 (October 4, 1913): 585–586.

Phipps, Stanley S. "A. S. Embree: Labor Leader and Prisoner of Conscience." *The Speculator: A Journal of Butte and Southwest Montana History* 2 (Summer 1985): 35–41.

Piatt, Guy X., ed. *The Story of Butte*. Butte, MT: The Butte Bystander, 1897.

Piggot, A. Snowden. *The Chemistry and Metallurgy of Copper*. Philadelphia: Lindsay and Blakiston, 1858.

"Pollution—Not for Cities Alone." *Environmental Science & Technology* 1 (September 1967): 686.

Pomeroy, Earl. *The Pacific Slope: A History of California, Oregon, Washington, Idaho, Utah, and Nevada*. Lincoln: University of Nebraska Press, 1991.

Posner, Richard A. "A Theory of Negligence." *The Journal of Legal Studies* 1 (January 1972): 29–96.

Punke, Michael. *Fire and Brimstone: The North Mining Disaster of 1917*. New York: Hyperion, 2006.

Quinn, M.-L. "Industry and Environment in the Appalachian Copper Basin, 1890–1930." *Technology and Culture* 34 (July 1993): 575–612.

Quivik, Fredric L. "The Historical Significance of Tailings and Slag: Industrial Waste as Cultural Resource." *IA: The Journal of the Society for Industrial Archeology* 33, no. 2 (2007): 35–54.

Quivik, Fredric L. "Nuisance, Source of Wealth, or Potentially Practical Material: Visions of Tailings in Idaho's Coeur d'Alene Mining District, 1888–2001." *IA: The*

Journal of the Society for Industrial Archeology 39, nos. 1 and 2 (2013): 41–64.

Quivik, Fredric L. "Renaturalization and Industrial Heritage in America's Largest Superfund Site: The Case of the Warm Springs Ponds in Montana's Clark Fork Superfund Site." In *Restoring Layered Landscapes: History, Ecology, and Culture*, edited by Marion Hourdequin and David G. Havlick, 202–221. New York: Oxford University Press, 2016.

Quivik, Fredric L. "Steel Transmission Towers and Energy for Montana's Copper Industry." *Montana: The Magazine of Western History* 38 (Autumn 1988): 67–69.

Quivik, Fredric L. "The Tragic Montana Career of D. E. Salmon." *Montana: The Magazine of Western History* 57 (Spring 2007): 32–47, 92–94.

Rader, Benjamin G. "The Montana Lumber Strike of 1917." *Pacific Historical Review* 36 (May 1967): 189–207.

Radkau, Joachim. *Nature and Power: A Global History of the Environment*. New York: Cambridge University Press, 2008.

Ralston, Oliver. *Electrolytic Deposition and Hydrometallurgy of Zinc and Cadmium*. New York: McGraw-Hill, 1921.

Read, Thomas T. "Compressed-Air Hoisting At Butte." *Mining & Scientific Press* 105 (November 2, 1912): 554–556.

Remele, Larry. "'God Helps Those Who Help Themselves': The Farmers Alliance and Dakota Statehood." *Montana: The Magazine of Western History* 37 (Autumn 1987): 22–33.

Repath, C. H. "The Mechanical Engineer." *Mining and Scientific Press* 85 (September 13, 1902): 143–145. Also published in *Mines and Minerals* 23 (November 1902): 173–177.

Repath, C. H. "Reminiscences of Early Years at the Great Falls Smelter." *Chemical and Metallurgical Engineering* 22 (January 21, 1920): 111–117.

"The Research Corporation and the Cottrell Process." *E&MJ* 97 (May 30, 1914): 1107–1109.

Reynolds, Terry S. "A Narrow Window of Opportunity: The Rise and Fall of the Fixed Steel Dam." *IA: The Journal of the Society for Industrial Archeology* 15, no. 1 (1989): 1–20.

Richards, Frank. "New Development in Compressed Air Storage." *American Machinist* 35 (August 31, 1911): 409–410.

Richards, Robert H. *Ore Dressing*, vols. I and II. New York: The Engineering and Mining Journal, 1906.

Richards, Robert H. *Ore Dressing*, vols. III and IV. New York: McGraw-Hill, 1909.

Richards, Robert H. *A Text Book of Ore Dressing*. New York: McGraw-Hill, 1909.

Richter, F. Ernest. "The Amalgamated Copper Company: A Closed Chapter in Corporation Finance." *The Quarterly Journal of Economics* 30 (1916): 387–407.

Richter, F. Ernest. "The Copper-Mining Industry in the United States, 1845–1925." *The Quarterly Journal of Economics* 41 (February 1927): 236–291.

Rickard, T. A. *A History of American Mining*. New York: McGraw-Hill, 1932.

Rickard, T. A. *Interviews with Mining Engineers*. San Francisco: Mining and Scientific Press, 1922.

Rickard, T. A. "Miami, Arizona: The Milling of the Ore—V." *Mining and Scientific Press* 115 (November 10, 1917): 679–884.
"River of Poison." *Newsweek* 70 (August 28, 1967): 70.
Robbins, William G. *Colony and Empire: The Capitalist Transformation of the American West*. Lawrence: University Press of Kansas, 1994.
Rohe, Randall. "Chinese River Mining in the West." *Montana, the Magazine of Western History* 46 (Autumn 1996): 14–29.
Rohe, Randall. "Man and the Land: Mining's Impact in the Far West." *Arizona and the West* 28 (Winter 1986): 299–338.
Rohrbough, Malcolm J. *Aspen: The History of a Silver Mining Town, 1879–1893*. New York: Oxford University Press, 1986.
Rohrbough, Malcolm J. *Days of Gold: The California Gold Rush and the American Nation*. Berkeley: University of California Press, 1997.
Rome, Adam W. "Coming to Terms with Pollution: The Language of Environmental Reform, 1865–1915." *Environmental History* 1 (July 1996): 6–28.
Rosen, Christine. "Differing Perceptions of the Value of Pollution Abatement across Time and Place: Balancing Doctrine in Pollution Nuisance Law, 1840–1906." *Law and History Review* 11 (Fall 1993): 307–381.
Rosenberg, Charles E. *The Trial of the Assassin Guiteau: Psychiatry and Law in the Gilded Age*. Chicago: University of Chicago Press, 1968.
Rosenberg, Nathan. *Inside the Black Box: Technology and Economics*. New York: Cambridge University Press, 1982.
Rothstein, Morton. "Farmer Movements and Organization: Numbers, Gains, Losses." *Agricultural History* 62 (Summer 1988): 161–181.
Rothwell, Richard P., ed. *The Mineral Industry, Its Statistics, Technology and Trade, in the United States and Other Countries, 1892*, vol. I. New York: The Scientific Publishing Company, 1893. This annual volume was published well into the twentieth century, with others assuming the role of editor, and with McGraw-Hill taking over as publisher.
Rushmore, David B., and Eric A. Lof. *Hydroelectric Power Stations*. 2nd ed. New York: John Wiley and Sons, 1923.
Safford, Jeffrey J. *The Mechanics of Optimism: Mining Companies, Technology, and the Hot Springs Gold Rush, Montana Territory, 1864–1868*. Boulder: University Press of Colorado, 2004.
Sales, Reno H. "Ore Deposits at Butte, Mont." *Transactions of the American Institute of Mining Engineers* 46 (1914): 3–109.
Sales, Reno. "Ore Discovery and Development of Fundamental Importance: How Anaconda's Geological Staff Is Employed." *E&MJ* 128 (August 24, 1929): 277–279.
Salmon, D. E. "Arsenical Poisoning from Smelter Smoke in Deer Lodge Valley, Montana." *American Veterinary Review* 39 (April 1911): 14–22, (June 1911): 245–260, (August 1911): 517–536; 40 (November 1911): 164–178, (February 1912): 579–590, (March 1912): 739–747; 41 (May 1912): 164–171, (June 1912): 300–308, (July 1912): 395–421.

Sanders, Helen Fitzgerald. *A History of Montana*, vols. I–III. Chicago: Lewis Publishing Company, 1913.
Sandlos, John, and Arn Keeling. "Claiming the New North: Development and Colonialism at the Pine Point Mine, Northwest Territories, Canada." *Environment and History* 18 (February 2012): 5–34.
Scheidler, Natalie F. "Failure to Protect: Legal Interpretations of Rape and Wife Assault in Butte, Montana, 1900–1920." *Montana: The Magazine of Western History* 65 (Winter 2015): 29–40.
Schlebecker, John T. *Whereby We Thrive: A History of American Farming, 1607–1972*. Ames: Iowa State University Press, 1975.
Schmitz, C. J. *The Growth of Big Business in the United States and Western Europe, 1850–1939*. New York: Cambridge University Press, 1995.
Schmitz, Christopher. "The World's Largest Industrial Companies of 1912." *Business History* 37 (1995): 85–96.
Schramm, E. "Experts in the Smelter Smoke Debate." In *The Silent Countdown: Essays in European Environmental History*, edited by Peter Brimblecombe and Christian Pfister. Berlin: Springer-Verlag, 1990.
Schulman, Bruce J. "Governing Nature, Nurturing Government: Resource Management and the Development of the American State, 1900–1912." *Journal of Policy History* 17, no. 4 (2005): 375–403.
Schwerin, M. "Method of Concentrating at Anaconda." *E&MJ* 76 (September 12, 1903): 388–389.
Sellers, Christopher. "Factory as Environment: Industrial Hygiene, Professional Collaboration and the Modern Sciences of Pollution." *Environmental History Review* 18 (Spring 1994): 55–83.
Sharlin, Harold I. "Electrical Generation and Transmission." In *The Emergence of Modern Industrial Society*, vol. 1 of *Technology in Western Civilization*, edited by Melvin Kranzberg and Carroll W. Pursell Jr. New York: Oxford University Press, 1967.
Shovers, Brian. "The Old Works Golf Course, Anaconda, Montana." *Montana: The Magazine of Western History* 54 (Autumn 2004): 64–70.
Shovers, Brian. "Remaking the Wide-Open Town: Butte at the End of the Twentieth Century." *Montana: The Magazine of Western History* 48 (Autumn 1998): 40–53.
Sieferle, Rolf Peter. *The Subterranean Forest: Energy Systems and the Industrial Revolution*. Cambridge, UK: The White Horse Press, 2001.
Simons, Theodore. "The Concentrator of the Timber Butte Milling Co., Butte, Mont." *Transactions of the American Institute of Mining Engineers* 52 (1916): 910–931.
Simons, Theodore. "The Evolution of the Round Table for the Treatment of Metalliferous Slimes." *Transactions of the American Institute of Mining Engineers* 46 (1914): 338–361.
Singer, Charles, E. J. Holmyard, A. R. Hall, and Trevor Williams. *A History of Technology*. New York: Oxford University Press, 1958.
Sklar, Martin J. *The Corporate Reconstruction of American Capitalism, 1890–1916: The Market, the Law, and Politics*. New York: Cambridge University Press, 1988.

Smith, Duane A. *Mining America: The Industry and the Environment, 1800–1980*. Lawrence: University Press of Kansas, 1987.

Smith, Michael L. *Pacific Visions: California Scientists and the Environment, 1850–1915*. New Haven, CT: Yale University Press, 1987.

Smith, Ralph I. *History of the Early Reduction Plants of Butte, Montana*. Butte: Montana School of Mines, 1953.

Smithcors, J. F. *The Veterinarian in America, 1625–1975*. Goleta, CA: American Veterinary Publications Inc., 1975.

Snow, Bradley D. *Living with Lead: An Environmental History of Idaho's Coeur D'Alenes, 1885–2011*. Pittsburgh: University of Pittsburgh Press, 2017.

Snyder, S. A. *Scenic Routes and Byways Montana*. Lanham, MD: Rowman & Littlefield Publishing Group, 2021.

Spence, Clark C. *Mining Engineers and the American West: The Lace-Boot Brigade, 1849–1933*. Moscow: University of Idaho Press, 1993.

Spindler, John C. "The Clean-Up of Silver Bow Creek." *Mining Congress Journal* 63 (June 1977): 58–63.

Stearns, Peter N. *The Industrial Revolution in World History*. Boulder, CO: Westview Press, 2013.

Steinberg, Theodore. *Nature Incorporated: Industrialization and the Waters of New England*. New York: Oxford University Press, 1991.

Stevens, Horace J. *The Copper Handbook*. Vol. X. Houghton, MI: Horace J. Stevens, 1911. This handbook was published regularly, eventually changing its name to *The Mines Handbook*.

Stone, Daniel. "The Giesche Company: Anaconda Copper's Subsidiary in Interwar Poland." *Slavic Review* 56 (Winter 1997): 679–697.

Stradling, David. *Smokestacks and Progressives: Environmentalists, Engineers, and Air Quality in America, 1881–1951*. Baltimore: Johns Hopkins University Press, 2003.

Struthers, Joseph. *The Mineral Industry, Its Statistics, Technology and Trade, in the United States and Other Countries, 1901*, vol. X. New York: The Engineering and Mining Journal, 1902.

Sullivan, Marianne. *Tainted Earth: Smelters, Public Health, and the Environment*. New Brunswick, NJ: Rutgers University Press, 2014.

Swain, R. E., and W. D. Harkins. "Papers on Smelter Smoke: Arsenic in Vegetation Exposed to Smelter Smoke." *Journal of the American Chemical Society* 30 (June 1908): 915–928.

Swett, Ira L. *Montana's Trolleys II: Butte, Anaconda, BA&P*. Special issue of *Interurbans Magazine* 26 (Winter 1969).

Swibold, Dennis L. *Copper Chorus: Mining, Politics, and the Montana Press, 1889–1958*. Helena: Montana Historical Society Press, 2006.

Tarr, Joel A. *The Search for the Ultimate Sink: Urban Pollution in Historical Perspective*. Akron, OH: University of Akron Press, 1996.

Tarr, Joel A. "Searching for a 'Sink' for an Industrial Waste: Iron-Making Fuels and the Environment." *Environmental History Review* 18 (Spring 1994): 9–34.

Tarr, Joel A. "Urban Environmental History." In *The Turning Points of Environmental*

History, edited by Frank Uekoetter. Pittsburgh: University of Pittsburgh Press, 2010.

Thorseim, Peter. *Inventing Pollution: Coal, Smoke, and Culture in Britain since 1800*. Athens: Ohio University Press, 2006.

Tontz, Robert L. "Memberships of General Farmers' Organizations, United States, 1874–1960." *Agricultural History* 38 (July 1964): 143–156.

Toole, K. Ross. "The Anaconda Copper Mining Company: A Price War and a Copper Corner." *Pacific Northwest Quarterly* 41 (October 1950): 312–329.

Toole, K. Ross. *Montana: An Uncommon Land*. Norman: University of Oklahoma Press, 1959.

Toole, K. Ross. *Twentieth-Century Montana: A State of Extremes*. Norman: University of Oklahoma Press, 1972.

Traphagen, F. W. "Recent Litigation Involving Questions of Alleged Damage from Tailings, Tailings Waters, and Smelter Fumes." *Report of Proceedings of the American Mining Congress, Ninth Annual Session* (1907): 251–255.

Tyer, Brad. *Opportunity, Montana: Big Copper, Bad Water, and the Burial of an American Landscape*. Boston: Beacon Press, 2013.

Tylecote, R. F. *A History of Metallurgy*. London: Institute of Materials, 1992.

Uekoetter, Frank. *The Age of Smoke: Environmental Policy in Germany and the United States, 1880–1970*. Pittsburgh: University of Pittsburgh Press, 2009.

Ulke, Titus. "Characteristic American Metal Mines." *The Engineering Magazine* 13 (July 1897): 512–534.

Ulke, Titus. *Modern Electrolytic Copper Refining*. New York: John Wiley & Sons, 1911.

Walker, Arthur L. "The Metallurgy of Copper in 1911." *E&MJ* 93 (January 6, 1912): 47–49.

Walker, Arthur L. "The Metallurgy of Copper in 1912." *E&MJ* 95 (January 11, 1913): 103–104.

Walker, George L. *The Copper Mines of Butte and the Amalgamated Copper Company*. Boston: Boston Financial News, 1900.

West, Elliot. *The Contested Plains: Indians, Goldseekers, and the Rush to Colorado*. Lawrence: University Press of Kansas, 1998.

Wethey, A. H. "Smelting Plant of the Butte Reduction Works." *E&MJ* 88 (December 11, 1909): 1153–1155.

White, Paul. "Heads, Tails, and Decisions In-Between: The Archaeology of Mining Wastes." *IA: The Journal of the Society for Industrial Archeology* 29, no. 2 (2003): 47–66.

White, Richard. *"It's Your Misfortune and None of My Own": A History of the American West*. Norman: University of Oklahoma Press, 1991.

Who Was Who in America, vol. 3. Chicago: A. N. Marquis Company, 1960.

Wiggin, Albert E. "The Great Falls System of Concentration." *Transactions of the American Institute of Mining Engineers* 46 (1914): 209–238.

Wirth, John D. *Smelter Smoke in North America: The Politics of Transborder Pollution*. Lawrence: University Press of Kansas, 2000.

Woodward, W. E. *A New American History*. New York: The Literary Guild, 1937.

Wyman, Mark. *Hard Rock Epic: Western Miners and the Industrial Revolution, 1860–1910*. Berkeley: University of California Press, 1979.

Young, George J. *Elements of Mining*. New York: McGraw-Hill, 1916.

Young, James Harvey. *Pure Food: Securing the Federal Food and Drugs Act of 1906*. Princeton, NJ: Princeton University Press, 1989.

Young, Otis E., Jr. "The American Copper Frontier, 1640–1893." *The Speculator: A Journal of Butte and Southwest Montana History* 1 (Summer 1984): 4–15.

Zallen, Jeremy. *American Lucifers: The Dark History of Artificial Light, 1750–1865*. Chapel Hill, NC: University of North Carolina Press, 2019.

Index

Page numbers in italics indicate illustrations.

About the Author

FREDRIC L. QUIVIK serves as an expert industrial historian in environmental litigation, including Superfund litigation involving the extractive industries, especially mining. He worked for the US Department of Justice on the Clark Fork Superfund case in Montana, which lies at the heart of *Smoke and Tailings*. Other cases he has been involved with include the Bunker Hill Superfund case in the Coeur d'Alene mining district, the Libby Superfund case involving vermiculite and asbestos in Montana, the Midnite Mine Superfund case involving uranium in Washington, and the *Deepwater Horizon* case in the Gulf of Mexico. A long-time member of the Society for Industrial Archeology, Quivik has served as its president and as editor of its journal, *IA*. He retired in 2015 as professor of history from Michigan Technological University where he taught history of technology, environmental history, and industrial archeology. He lives in Saint Paul, Minnesota.

